An Introduction to the
MICROBIOME IN HEALTH AND DISEASES

The Microbiome in Health and Diseases

An Introduction to the
MICROBIOME IN HEALTH AND DISEASES

Edited by

CHARLES OLUWASEUN ADETUNJI

Applied Microbiology, Biotechnology and Nanotechnology Laboratory, Department of Microbiology, and Directorate of Research and Innovation, Edo State University Uzairue, Iyamho, Auchi, Edo State, Nigeria

OLUGBENGA SAMUEL MICHAEL

Department of Medical Pharmacology and Physiology, University of Missouri, Columbia, MO, United States

NWADIUTO ESIOBU

Microbiology and Biotechnology Laboratory, Biological Sciences Department, Microbiome Innovation Center, Florida Atlantic University, Boca Raton, FL, United States

ROTIMI E. ALUKO

Department of Food and Human Nutritional Sciences, University of Manitoba, Winnipeg, MB, Canada

ELSEVIER

ACADEMIC PRESS

An imprint of Elsevier

Academic Press is an imprint of Elsevier
125 London Wall, London EC2Y 5AS, United Kingdom
525 B Street, Suite 1650, San Diego, CA 92101, United States
50 Hampshire Street, 5th Floor, Cambridge, MA 02139, United States

Notices

Knowledge and best practice in this field are constantly changing. As new research and experience broaden our understanding, changes in research methods, professional practices, or medical treatment may become necessary.

Practitioners and researchers must always rely on their own experience and knowledge in evaluating and using any information, methods, compounds, or experiments described herein. In using such information or methods they should be mindful of their own safety and the safety of others, including parties for whom they have a professional responsibility.

To the fullest extent of the law, neither the Publisher nor the authors, contributors, or editors, assume any liability for any injury and/or damage to persons or property as a matter of products liability, negligence or otherwise, or from any use or operation of any methods, products, instructions, or ideas contained in the material herein.

ISBN: 978-0-323-91190-0

For information on all Academic Press publications visit our website at https://www.elsevier.com/books-and-journals

Publisher: Jonathan Simpson
Acquisitions Editor: Kattie Washington
Editorial Project Manager: Manisha Rana
Production Project Manager: Kumar Anbazhagan
Cover Designer: Matthew Limbert

Typeset by TNQ Technologies

Working together to grow libraries in developing countries

www.elsevier.com • www.bookaid.org

Contents

4. Microbiome characterization and identification: key emphasis on molecular approaches   49

Frank Abimbola Ogundolie, Charles Oluwaseun Adetunji, Olulope Olufemi Ajayi, Michael O. Okpara, Olugbenga Samuel Michael, Juliana Bunmi Adetunji, Ohunayo Adeniyi Success and Oluwafemi Adebayo Oyewole

5. COVID-19 and microbiome   71

Olugbenga Samuel Michael, Juliana Bunmi Adetunji, Olufunto Olayinka Badmus, Emmanuel Damilare Areola, Ayomide Michael Oshinjo, Charles Oluwaseun Adetunji and Oluwafemi Adebayo Oyewole

6. Introduction to plant microbiome   **97**

Olulope Olufemi Ajayi, Charles Oluwaseun Adetunji, Olugbenga Samuel Michael,
Frank Abimbola Ogundolie, Juliana Bunmi Adetunji and Oluwafemi Adebayo Oyewole

7. Introduction to animal microbiome   **109**

Olulope Olufemi Ajayi, Charles Oluwaseun Adetunji, Olugbenga Samuel Michael
and Juliana Bunmi Adetunji

8. Patents, bioproducts, commercialization, social, ethical, and economic policies on microbiome   **117**

Olugbenga Samuel Michael, Juliana Bunmi Adetunji, Ebenezer Olusola Akinwale,
Olufemi Idowu Oluranti, Olulope Olufemi Ajayi, Charles Oluwaseun Adetunji,
Ayodele Olufemi Soladoye and Oluwafemi Adebayo Oyewole

9. Introduction to gut microbiome and epigenetics: Their role in polycystic ovary syndrome pathogenesis   137

Ayomide Michael Oshinjo, Olugbenga Samuel Michael, Lawrence Dayo Adedayo,
Charles Oluwaseun Adetunji, Bamidele Olubayode, Juliana Bunmi Adetunji,
Olaniyan Amos Morakinyo, Ebenezer Olusola Akinwale, Olulope Olufemi Ajayi,
Funmileyi Olubajo Awobajo, Ayodele Olufemi Soladoye and Oluwafemi Adebayo Oyewole

10. Introduction to epigenetic programming by gut microbiota   177

Olugbenga Samuel Michael, Oluwaseun Aremu Adeyanju, Charles Oluwaseun Adetunji,
Kehinde Samuel Olaniyi, Juliana Bunmi Adetunji and Ayodele Olufemi Soladoye

11. Gut microbiota, nutrition, and health: Fundamental and basic principle   **195**

Olugbenga Samuel Michael, Juliana Bunmi Adetunji, Oluwaseun Aremu Adeyanju,
Charles Oluwaseun Adetunji, Olufunto Olayinka Badmus, Kehinde Samuel Olaniyi and
Ayodele Olufemi Soladoye

Contributors

Lawrence Dayo Adedayo
Neurophysiology Research Unit, Department of Physiology, College of Health Sciences, Bowen University, Iwo, Osun State, Nigeria

Juliana Bunmi Adetunji
Nutritional and Toxicological Research Laboratory, Department of Biochemistry Sciences, Osun State University, Osogbo, Osun State, Nigeria; Department of Biochemistry, Edo State University, Uzairue, Auchi, Edo State, Nigeria

Charles Oluwaseun Adetunji
Applied Microbiology, Biotechnology and Nanotechnology Laboratory, Department of Microbiology, and Directorate of Research and Innovation, Edo State University Uzairue, Iyamho, Auchi, Edo State, Nigeria

Oluwaseun Aremu Adeyanju
Department of Physiology, College of Medicine and Health Sciences, Afe Babalola University, Ado-Ekiti, Ekiti, Nigeria; Department of Cellular and Molecular Biology, The University of Texas Health Science Center at Tyler, Tyler, TX, United States

Olulope Olufemi Ajayi
Department of Biochemistry, Edo State University Uzairue, Edo State, Nigeria; Department of Biochemistry, Federal University of Technology, Akure, Nigeria

Ebenezer Olusola Akinwale
Department of Physiology and Biomedical Science, Faculty of Science and Engineering, University of Wolverhampton, Wolverhampton, United Kingdom

Emmanuel Damilare Areola
Department of Physiology, College of Health Sciences, University of Ilorin, Ilorin, Kwara, Nigeria

Funmileyi Olubajo Awobajo
Department of Physiology, College of Medicine, University of Lagos, Idiaraba, Nigeria

Olufunto Olayinka Badmus
Department of Physiology and Biophysics, Cardiorenal, and Metabolic Diseases Research Center, University of Mississippi Medical Center, Jackson, MS, United States; Department of Public Health, Kwara State University, Malete, Kwara, Nigeria

Olugbenga Samuel Michael
Cardiometabolic, Microbiome and Applied Physiology Laboratory, Department of Physiology, College of Health Sciences, Bowen University, Iwo, Osun State, Nigeria; Department of Physiology, University of Tennessee Health Science Center, Memphis, TN, United States; Department of Medical Pharmacology and Physiology, University of Missouri, Columbia, MO, United States

Olaniyan Amos Morakinyo
Department of Anatomy, College of Health Sciences, Bowen University, Iwo, Nigeria

Frank Abimbola Ogundolie
Department of Biotechnology, Baze University, Abuja, Nigeria

Michael O. Okpara
Department of Science Laboratory Technology, Faculty of Science, Ekiti State University, Ado Ekiti, Nigeria

Kehinde Samuel Olaniyi
Department of Physiology, College of Medicine and Health Sciences, Afe Babalola University, Ado-Ekiti, Nigeria; Nutritional and Toxicological Research Laboratory, Department of Biochemistry Sciences, Osun State University, Osogbo, Osun State, Nigeria

Bamidele Olubayode
Cardiometabolic, Microbiome and Applied Physiology Laboratory, Department of Physiology, College of Health Sciences, Bowen University, Iwo, Osun State, Nigeria

Olufemi Idowu Oluranti
Applied and Environmental Research Unit, Department of Physiology, College of Health Sciences, Bowen University, Iwo, Osun State, Nigeria

Ayomide Michael Oshinjo
Cardiometabolic, Microbiome and Applied Physiology Laboratory, Department of Physiology, College of Health Sciences, Bowen University, Iwo, Osun State, Nigeria; Faculty of Biochemistry and Molecular Medicine, University of Oulu, Oulu, Finland

Oluwafemi Adebayo Oyewole
Department of Microbiology, Federal University of Technology, Minna, Nigeria

Ayodele Olufemi Soladoye
Cardiometabolic, Microbiome and Applied Physiology Laboratory, Department of Physiology, College of Health Sciences, Bowen University, Iwo, Osun State, Nigeria

Ohunayo Adeniyi Success
Nutritional and Toxicological Research Laboratory, Department of Biochemistry Sciences, Osun State University, Osogbo, Osun State, Nigeria

CHAPTER 1

Microbiome: Introduction and recent advances

Charles Oluwaseun Adetunji[1], Olugbenga Samuel Michael[3,4,7], Olulope Olufemi Ajayi[2], Frank Abimbola Ogundolie[5], Juliana Bunmi Adetunji[6] and Oluwafemi Adebayo Oyewole[8]

[1]Applied Microbiology, Biotechnology and Nanotechnology Laboratory, Department of Microbiology, and Directorate of Research and Innovation, Edo State University Uzairue, Iyamho, Auchi, Edo State, Nigeria; [2]Department of Biochemistry, Edo State University Uzairue, Edo State, Nigeria; [3]Cardiometabolic, Microbiome and Applied Physiology Laboratory, Department of Physiology, College of Health Sciences, Bowen University, Iwo, Osun State, Nigeria; [4]Department of Physiology, University of Tennessee Health Science Center, Memphis, TN, United States; [5]Department of Biotechnology, Baze University, Abuja, Nigeria; [6]Nutritional and Toxicological Research Laboratory, Department of Biochemistry Sciences, Osun State University, Osogbo, Osun State, Nigeria; [7]Department of Medical Pharmacology and Physiology, University of Missouri, Columbia, MO, United States; [8]Department of Microbiology, Federal University of Technology, Minna, Nigeria

Introduction

A variety of microorganisms are found in the gut relative to other parts of the body such as the skin (Liang et al., 2018). Bacteria constitute the prominent GIT microbial population in mammals. They include members of Proteobacteria, Firmicutes, Actinobacteria, and Bacteroidetes phyla (Das and Nair, 2019). Firmicutes and Bacteroidetes have been reported to account for a higher percentage (Liang et al., 2018). The presence of other bacterial phyla such as Tenericutes, Spirochetes, and Verrucomicrobia has also been reported (Caporaso et al., 2011).

Microbiomes are defined as an assemblage of genomes and gene products of microbiota in a host or environment (Breasalier and Chapkin, 2020). Human microbiome has more than 40 trillion microorganism per individual (Breasalier and Chapkin, 2020). Microbiomes are found in every part of the human body, including the plasma once seen as sterile environment (Liang et al., 2018). GIT microbiome composition and homeostasis are influenced by environmental factors and the host's genetic makeup (Das and Nair, 2019). A healthy microbiome is characterized by resilient ability to return to a state of equilibrium (Breasalier and Chapkin, 2020). The dynamism and variability of gut microbiota among individuals are dependent on gender, age, diet, lifestyle, antibiotics use, alcohol, and state of health (Das and Nair, 2019). Liang et al. (2018) reported that alcohol alters intestinal microbiota (Liang et al., 2018).

Gut microbiota enhances the digestibility of complex dietary polysaccharides; this results in the liberation of short-chain fatty acids (SCFAs) from indigestible dietary fibers, which provide energy for the intestinal mucosa and also play a significant role in immunity (Shreiner et al., 2015). Gut microbiota also boosts the synthesis of vitamins and essential

An Introduction to the Microbiome in Health and Diseases
ISBN 978-0-323-91190-0
https://doi.org/10.1016/B978-0-323-91190-0.00001-1

amino acids, alteration of xenobiotics' toxicities, and development of the immune system, thereby protecting against pathogenic agents among others (Das and Nair, 2019). Lipopolysaccharide (LPS), a product of microbial metabolism often produced by gram-negative bacteria, enhances innate immunity (Liang et al., 2018).

Diet, antibiotic use, and gender are important factors to consider while conducting microbiome study. The role of diet in microbiome research cannot be underestimated. Relationship between high-protein/animal fat diet and *Bacteroides* has been reported. Furthermore, the relationship between high-carbohydrate diet and *Prevotella* has also been observed (Kim et al., 2017).

The human microbiome constantly develops throughout life. At about age 3, the gut microbiome attains an anaerobic pattern, but gets altered in old age (Kim et al., 2017). It therefore becomes necessary to age-match control participants when conducting microbiome research.

Gender is another important factor to consider in microbiome research. It is assumed that the gut microbiome possesses endocrine abilities because of certain biomolecules it produces (Clarke et al., 2014). In a study, testosterone level was increased in male mice exposed early to microbes; this confers protection against type 1 diabetes mellitus (Markle et al., 2013). Similar protection against type 1 diabetes mellitus was observed in female mice to which the microbiota from the protected male mice was transplanted (Markle et al., 2013).

Microbiome and diseases

An imbalance in gut microbial ecology has been associated with pathologic conditions including cardiovascular diseases, colon cancer, inflammatory bowel disease, and neuro-degenerative diseases (Das and Nair, 2019). The association of microbiota with cardiovascular disease has been reported. Trimethylamine-*N*-oxide (TMAO), a product of dietary phosphatidylcholine metabolism, has been reported to be proatherosclerotic (Wang et al., 2011). There are also suspicions on the association of altered microbiota with irritable bowel syndrome. The pathogenesis of IBS is thought to involve microbiota—gut—brain axis (Shreiner et al., 2015). Altered gut microbiota has also been implicated in inflammatory bowel diseases, which is marked by inflammatory responses caused by both genetic and environmental factors in the gut (Shreiner et al., 2015).

Positive modulation of microbiota using prebiotics, probiotics, and synbiotics has proven to be effective in maintaining a healthy microbial ecology. *Lactobacillus* has been reported to prevent against antibiotic-resistant diarrhea in children (Goldenberg et al., 2015). Specifically, *Lactobacillus casei* prevented the growth of *Helicobacter pylori* (Sgouras et al., 2004). Certain strains of *Lactobacillus* have been used in treating diseases including type 2 diabetes mellitus, HIV infection, nonalcoholic fatty liver, among others (Liang et al., 2018).

The beneficial effects of Bifidobacteria as a probiotic have been reported. There are evidence that they offer great relief in respiratory diseases including asthma. The positive effect of Bifidobacteria on the cells of the intestines enhances the regulation of immunity as well as the expression of inflammatory genes (Liang et al., 2018).

The expression of TNF-α and IL-α was regulated by *Bifidobacterium longum* in individuals with ulcerative colitis (Furrie et al., 2005). In another study, tumor-specific immunity was enhanced upon the oral administration of *Bifidobacterium* (Sivan et al., 2015; Vetizou et al., 2015). A study showed the restoration of healthy gut microbiome via microbial transplantation as well as enhanced treatment of recurrent *Clostridium difficile* colitis (Bakken et al., 2011). Furthermore, the protective potential of probiotics (microbiome) against *Citrobacter rodentium* was reported in another study (Ivanov et al., 2009). Prebiotics, probiotics, synbiotics, or microbiome transplantation is promising in managing the effects of alcoholism (Liang et al., 2018)

The medical relevance of microbiome in the health sector

The microbiome, also known as the microbial population of a given habitat/biota, is very significant and plays a critical role in the health of humans, plants, and animals. An alteration in these microbiomes can result in several changes leading to several diseases; the microbiome composition of humans can be directly linked to their genetic makeup; today, advances in science have enabled scientists to identify several diseases that have been linked to the microbiome (Ursell et al., 2012; Gilbert et al., 2018; Michael et al., 2022a,b; Esiobu et al., 2022; Adetunji et al., 2022a,b,c,d,e,f,g,h,i; Olaniyan et al. 2022a,b; Oyedara et al., 2022). Naturally, microbiomes living on or in humans are not invasive but colonizers that are of great benefit to man and are essential for nutrition, immunity, and human development.

However, over time, the invasion of pathogenic microbes into the human system leads to the accumulation of toxic microbes, which eventually causes dysbacteriosis or imbalance in the microbiome. Intestinal microbiome imbalance, for instance, either informs of imbalance in the microbial population; metabolic or functional activities of the gut microbe have been associated with chronic inflammation of the gastrointestinal (GI) tract resulting in inflammatory bowel disease (IBD), a term for two disease conditions such as Crohn's disease and ulcerative colitis. Irritable bowel syndrome (IBS) is another common disease associated with an imbalance in the microbiome. In addition, it plays a significant role in the pathogenesis of other intestinal disorders, including metabolic syndrome, cardiovascular disease, asthma, allergies, obesity, and type 1 diabetes.

The emergence of metagenomics and gene profiling has given a better understanding of the influence of microbiota present in different organs of the body ranging from the urinary tract (Aragon et al., 2018), gut (Cénit, et al., 2014), oral (Wade, 2013), lungs, vaginal (Ma et al., 2012), brain—gut axis (Moloney et al., 2014), and intestine (Lynch and Pedersen, 2016) among others and the health state of humans. This has greatly led to a better understanding and diagnosis of different levels of diseases whether chronic or acute (Pflughoeft and Versalovic, 2012).

Metagenomics technology is important in understanding and gaining insight into the roles microbial composition plays in the health of both humans and animals. For instance, in dental health, perturbations leading to changes in the oral microbial composition has been associated with several dental disease conditions, such as tooth decay (Luo et al., 2016), gum disease, pyorrhea disease (Darveau et al., 1997; Burne et al., 2012; Luo et al., 2016), and inflammation of the gums (Huang et al. 2011, 2014; Luo et al., 2016).

References

Adetunji, C.O., Olaniyan, O.T., Adeyomoye, O., Dare, A., Adeniyi, M.J., Alex, E., Rebezov, M., Garipova, L., Ali Shariati, M., 2022a. eHealth, mHealth, and Telemedicine for COVID-19 Pandemic. In: Pani, S.K., Dash, S., dos Santos, W.P., Chan Bukhari, S.A., Flammini, F. (Eds.), Assessing COVID-19 and Other Pandemics and Epidemics Using Computational Modelling and Data Analysis. Springer, Cham. https://doi.org/10.1007/978-3-030-79753-9_10.

Adetunji, C.O., Olaniyan, O.T., Adeyomoye, O., Dare, A., Adeniyi, M.J., Enoch, A., Rebezov, M., Petukhova, E., Ali Shariati, M., 2022b. Machine learning approaches for COVID-19 pandemic. In: Pani, S.K., Dash, S., dos Santos, W.P., Chan Bukhari, S.A., Flammini, F. (Eds.), Assessing COVID-19 and Other Pandemics and Epidemics Using Computational Modelling and Data Analysis. Springer, Cham. https://doi.org/10.1007/978-3-030-79753-9_8.

Adetunji, C.O., Olaniyan, O.T., Adeyomoye, O., Dare, A., Adeniyi, M.J., Enoch, A., Rebezov, M., Isabekova, O., Ali Shariati, M., 2022c. Smart sensing for COVID-19 pandemic. In: Pani, S.K., Dash, S., dos Santos, W.P., Chan Bukhari, S.A., Flammini, F. (Eds.), Assessing COVID-19 and Other Pandemics and Epidemics Using Computational Modelling and Data Analysis. Springer, Cham. https://doi.org/10.1007/978-3-030-79753-9_9.

Adetunji, C.O., Olaniyan, O.T., Adeyomoye, O., Dare, A., Adeniyi, M.J., Enoch, A., Rebezov, M., Petukhova, E., Ali Shariati, M., 2022d. Internet of health things (IoHT) for COVID-19. In: Pani, S.K., Dash, S., dos Santos, W.P., Chan Bukhari, S.A., Flammini, F. (Eds.), Assessing COVID-19 and Other Pandemics and Epidemics Using Computational Modelling and Data Analysis. Springer, Cham. https://doi.org/10.1007/978-3-030-79753-9_5.

Adetunji, C.O., Olaniyan, O.T., Adeyomoye, O., Dare, A., Adeniyi, M.J., Enoch, A., Rebezov, M., Koriagina, N., Ali Shariati, M., 2022e. Diverse techniques applied for effective diagnosis of COVID-19. In: Pani, S.K., Dash, S., dos Santos, W.P., Chan Bukhari, S.A., Flammini, F. (Eds.), Assessing COVID-19 and Other Pandemics and Epidemics Using Computational Modelling and Data Analysis. Springer, Cham. https://doi.org/10.1007/978-3-030-79753-9_3.

Adetunji, C.O., Samuel, M.O., Adetunji, J.B., Oluranti, O.I., 2022f. Corn silk and health benefits. In: Medical Biotechnology, Biopharmaceutics, Forensic Science and Bioinformatics, first ed. Imprint CRC Press, ISBN 9781003178903, p. 12. https://doi.org/10.1201/9781003178903-11. First Published 2022.

Adetunji, C.O., Wilson, N., Olayinka, A.S., Olugbemi, O.T., Akram, M., Laila, U., Samuel, M.O., Oshinjo, A.M., Adetunji, J.B., Okotie, G.E., Esiobu, N.(D.), 2022g. Computational intelligence techniques for combating COVID-19. In: Medical Biotechnology, Biopharmaceutics, Forensic Science and Bioinformatics, first ed. Imprint CRC Press, ISBN 9781003178903, p. 12. https://doi.org/10.1201/9781003178903-16. First Published 2022.

Adetunji, C.O., Olugbemi, O.T., Akram, M., Laila, U., Samuel, M.O., Oshinjo, A.M., Adetunji, J.B., Okotie, G.E., Esiobu, N.(D.), Oyedara, O.O., Adeyemi, F.M., 2022h. Application of computational and bioinformatics techniques in drug repurposing for effective development of potential drug candidate for the management of COVID-19. In: Medical Biotechnology, Biopharmaceutics, Forensic Science and Bioinformatics, first ed. Imprint CRC Press, ISBN 9781003178903, p. 14. https://doi.org/10.1201/9781003178903-15. First Published 2022.

Adetunji, C.O., Wilson, N., Olayinka, A.S., Olugbemi, O.T., Akram, M., Laila, U., Olugbenga, M.S., Oshinjo, A.M., Adetunji, J.B., Okotie, G.E., Esiobu, N.(D.), 2022i. Machine learning and behaviour modification for COVID-19. In: Medical Biotechnology, Biopharmaceutics, Forensic Science and Bioinformatics, first ed. Imprint CRC Press, ISBN 9781003178903, p. 17. https://doi.org/10.1201/9781003178903-17. First Published 2022.

Aragon, I.M., Herrera-Imbroda, B., Queipo-Ortuño, M.I., Castillo, E., Del Moral, J.S.G., Gomez-Millan, J., Yucel, G., Lara, M.F., 2018. The urinary tract microbiome in health and disease. European urology focus 4 (1), 128—138.

Bakken, J.S., Borody, T., Brandt, L.J., Brill, J.V., Demarco, D.C., Franzos, M.A., Kelly, C., Khoruts, A., Louie, T., Martinelli, L.P., et al., 2011. Treating *Clostridium difficile* infection with fecal microbiota transplantation. Clin. Gastroenterol. Hepatol. 9 (12), 1044—1049.

Bresalier, R.S., Chapkin, R.S., 2020. Human microbiome in health and disease: the good, the bad, and the bugly. Dig. Dis. Sci. 65, 671—673. https://doi.org/10.1007/s10620-020-06059-y.

Burne, R.A., Zeng, L., Ahn, S.J., Palmer, S.R., Liu, Y., Lefebure, T., Stanhope, M.J., Nascimento, M.M., 2012. Progress dissecting the oral microbiome in caries and health. Adv. Dent. Res. 24, 77—80.

Caporaso, J.G., Lauber, C.L., Costello, E.K., Berg-Lyons, D., Gonzalez, A., Stombaugh, J., Knights, D., Gajer, P., et al., 2011. Moving pictures of the human microbiome. Genome Biol. 12, R50.

Cénit, M.C., Matzaraki, V., Tigchelaar, E.F., Zhernakova, A., 2014. Rapidly expanding knowledge on the role of the gut microbiome in health and disease. Biochim. Biophys. Acta, Mol. Basis Dis. 1842 (10), 1981—1992.

Clarke, G., Stilling, R.M., Kennedy, P.J., Stanton, C., Cryan, J.F., Dinan, T.G., 2014. Minireview: gut microbiota: the neglected endocrine organ. Mol. Endocrinol. 28 (8), 1221—1238. https://doi.org/10.1210/me.2014-1108.

Darveau, R.P., Tanner, A., Page, R.C., 1997. The microbial challenge in periodontitis. Periodontology 14, 12—32, 2000.

Das, B., Nair, G.B., 2019. Homeostasis and dysbiosis of the gut microbiome in health and disease. J. Biosci. 44, 117. https://doi.org/10.1007/s12038-019-9926-y.

Esiobu, N.(D.), Ogbonna, J.C., Adetunji, C.O., Obembe, O.O., Ezeonu, I.M., Ibrahim, A.B., Ubi, B.E., 2022. Microbiomes and Emerging Applications, first ed. Imprint CRC Press, Boca Raton, ISBN 9781003180241, p. 186. https://doi.org/10.1201/9781003180241. First Published 2022. eBook Published 11 May 2022. Subjects Bioscience, Engineering and Technology.

Furrie, E., Macfarlane, S., Kennedy, A., Cummings, J.H., Walsh, S.V., O'Neil, D.A., Macfarlane, G.T., 2005. Synbiotic therapy (*Bifidobacterium longum*/Synergy 1) initiates resolution of inflammation in patients with active ulcerative colitis: a randomised controlled pilot trial. Gut 54 (2), 242—249.

Gilbert, J.A., Blaser, M.J., Caporaso, J.G., Jansson, J.K., Lynch, S.V., Knight, R., 2018. Current understanding of the human microbiome. Nat. Med. 24 (4), 392—400. https://doi.org/10.1038/nm.4517.

Goldenberg, J.Z., Lytvyn, L., Steurich, J., Parkin, P., Mahant, S., Johnston, B.C., 2015. Probiotics for the prevention of pediatric antibiotic-associated diarrhea. Cochrane Database Syst. Rev. 12, CD004827.

Huang, S., Yang, F., Zeng, X., Chen, J., Li, R., Wen, T., Li, C., Wei, W., Liu, J., Chen, L., Davis, C., Xu, J., 2011. Preliminary characterization of the oral microbiota of Chinese adults with and without gingivitis. BMC Oral Health 11, 33.

Huang, S., Li, R., Zeng, X., He, T., Zhao, H., Chang, A., Bo, C., Chen, J., Yang, F., Knight, R., Liu, J., Davis, C., Xu, J., 2014. Predictive modeling of gingivitis severity and susceptibility via oral microbiota. ISME J. 8, 1768—1780.

Ivanov II, Atarashi, K., Manel, N., Brodie, E.L., Shima, T., Karaoz, U., Wei, D., Goldfarb, K.C., Santee, C.A., Lynch, S.V., et al., 2009. Induction of intestinal Th17 cells by segmented filamentous bacteria. Cell 139 (3), 485—498.

Kim, D., Hofstaedter, C.E., Zhao, C., Mattei, L., Tanes, C., Clarke, E., Lauder, A., Sherrill-Mix, S., Chehoud, C., Kelsen, J., Conrad, M., Collman, R.G., Baldassano, R., Bushman, F.D., Bittinger, K., 2017. Optimizing methods and dodging pitfalls in microbiome research. Microbiome 5, 52. https://doi.org/10.1186/s40168-017-0267-5.

Liang, D., Leung, R.K.-K., Guan, W., Au, W.W., 2018. Involvement of gut microbiome in human health and disease: brief overview, knowledge gaps and research opportunities. Gut Pathog. 10, 3. https://doi.org/10.1186/s13099-018-0230-4.

Luo, T., Srinivasan, U., Ramadugu, K., Shedden, K.A., Neiswanger, K., Trumble, E., Li, J.J., McNeil, D.W., Crout, R.J., Weyant, R.J., Marazita, M.L., 2016. Effects of specimen collection methodologies and storage conditions on the short-term stability of oral microbiome taxonomy. Appl. Environ. Microbiol. 82 (18), 5519–5529.

Lynch, S.V., Pedersen, O., 2016. The human intestinal microbiome in health and disease. N. Engl. J. Med. 375 (24), 2369–2379.

Ma, B., Forney, L.J., Ravel, J., 2012. Vaginal microbiome: rethinking health and disease. Annu. Rev. Microbiol. 66, 371–389.

Markle, J.G., Frank, D.N., Mortin-Toth, S., Robertson, C.E., Feazel, L.M., Rolle- Kampczyk, U., Danska, J.S., 2013. Sex differences in the gut microbiome drive hormone-dependent regulation of autoimmunity. Science 339 (6123), 1084–1088. https://doi.org/10.1126/science.1233521.

Michael, O.S., Oluranti, O.I., Oshinjo, A.M., Adetunji, C.O., Olaniyi, K.S., Adetunji, J.B., 2022a. Microbiome Reshaping and Epigenetic Regulation, first ed. Imprint CRC Press, ISBN 9781003180241, p. 22. https://doi.org/10.1201/9781003180241-6. First Published 2022.

Michael, O.S., Oluranti, O.I., Oshinjo, A.M., Adetunji, C.O., Adetunji, J.B., Esiobu, N.(D.), 2022b. Microbiota transplantation, health implications, and the way forward. In: Microbiomes and Emerging Applications, first ed. Imprint CRC Press, ISBN 9781003180241, p. 19. https://doi.org/10.1201/9781003180241-5. First Published 2022.

Moloney, R.D., Desbonnet, L., Clarke, G., Dinan, T.G., Cryan, J.F., 2014. The microbiome: stress, health and disease. Mamm. Genome 25 (1), 49–74.

Olaniyan, O.T., Adetunji, C.O., Adeniyi, M.J., Hefft, D.I., 2022a. Machine learning techniques for high-performance computing for IoT applications in healthcare. In: Deep Learning, Machine Learning and IoT in Biomedical and Health Informatics, first ed. Imprint CRC Press, ISBN 9780367548445, p. 13. https://doi.org/10.1201/9780367548445-20. First Published 2022.

Olaniyan, O.T., Adetunji, C.O., Adeniyi, M.J., Hefft, D.I., 2022b. Computational intelligence in IoT healthcare, 2022m. In: Deep Learning, Machine Learning and IoT in Biomedical and Health Informatics, first ed. Imprint CRC Press, ISBN 9780367548445, p. 13. https://doi.org/10.1201/9780367548445-19. First Published 2022.

Oyedara, O.O., Adeyemi, F.M., Adetunji, C.O., Elufisan, T.O., 2022. Repositioning antiviral drugs as a rapid and cost-effective approach to discover treatment against SARS-CoV-2 infection. In: Medical Biotechnology, Biopharmaceutics, Forensic Science and Bioinformatics, first ed. Imprint CRC Press, ISBN 9781003178903, p. 12. https://doi.org/10.1201/9781003178903-10. First Published 2022.

Pflughoeft, K.J., Versalovic, J., 2012. Human microbiome in health and disease. Annu. Rev. Pathol. 7, 99–122.

Sgouras, D., Maragkoudakis, P., Petraki, K., Martinez-Gonzalez, B., Eriotou, E., Michopoulos, S., Kalantzopoulos, G., Tsakalidou, E., Mentis, A., 2004. In vitro and in vivo inhibition of *Helicobacter pylori* by *Lactobacillus casei* strain Shirota. Appl. Environ. Microbiol. 70 (1), 518–526.

Shreiner, A.B., Kao, J.Y., Young, V.B., January 2015. The gut microbiome in health and in disease. Curr. Opin. Gastroenterol. 31 (1), 69–75. https://doi.org/10.1097/MOG.0000000000000139.

Sivan, A., Corrales, L., Hubert, N., Williams, J.B., Aquino-Michaels, K., Earley, Z.M., Benyamin, F.W., Lei, Y.M., Jabri, B., Alegre, M.L., et al., 2015. Commensal *Bifidobacterium* promotes antitumor immunity and facilitates anti-PD-L1 efficacy. Science 350 (6264), 1084–1089.

Ursell, L.K., Metcalf, J.L., Parfrey, L.W., Knight, R., 2012. Defining the human microbiome. Nutr. Rev. 70 (Suppl. 1), S38–S44. https://doi.org/10.1111/j.1753-4887.2012.00493.x. Suppl 1.

Vetizou, M., Pitt, J.M., Daillere, R., Lepage, P., Waldschmitt, N., Flament, C., Rusakiewicz, S., Routy, B., Roberti, M.P., Duong, C.P., et al., 2015. Anticancer immunotherapy by CTLA-4 blockade relies on the gut microbiota. Science 350 (6264), 1079–1084.

Wade, W.G., 2013. The oral microbiome in health and disease. Pharmacol. Res. 69 (1), 137–143.

Wang, Z., Klipfell, E., Bennett, B.J., et al., Apr 7; 2011. Gut flora metabolism of phosphatidylcholine promotes cardiovascular disease. Nature 472 (7341), 57–63. PubMed PMID: 21475195. Pubmed Central PMCID: 3086762.

CHAPTER 2

Why the need for microbiome? An updated perspective

Olugbenga Samuel Michael[1,5,6], Juliana Bunmi Adetunji[2], Ebenezer Olusola Akinwale[3], Charles Oluwaseun Adetunji[4] and Ayodele Olufemi Soladoye[1]

[1]Cardiometabolic, Microbiome and Applied Physiology Laboratory, Department of Physiology, College of Health Sciences, Bowen University, Iwo, Osun State, Nigeria; [2]Nutritional and Toxicological Research Laboratory, Department of Biochemistry Sciences, Osun State University, Osogbo, Osun State, Nigeria; [3]Department of Physiology and Biomedical Science, Faculty of Science and Engineering, University of Wolverhampton, Wolverhampton, United Kingdom; [4]Applied Microbiology, Biotechnology and Nanotechnology Laboratory, Department of Microbiology, and Directorate of Research and Innovation, Edo State University Uzairue, Iyamho, Auchi, Edo State, Nigeria; [5]Department of Physiology, University of Tennessee Health Science Center, Memphis, TN, United States; [6]Department of Medical Pharmacology and Physiology, University of Missouri, Columbia, MO, United States

Introduction

The need for adequate understanding and proper application of microbiome innovation has become very pertinent because of the sudden emergence of microbiome research as a field of global interest with enormous potential and untapped or unearthed bioresources. The relevance of the microbiome is not limited to humans' health and diseases because our world is a microbial world, and we are in constant interaction with microorganisms both the ones living outside and inside us. The microbiome has environmental influence, bioengineering, biotechnology, agricultural and food production, biomedical, nutritional, and industrial applications. Plants, animals, and environment also have their microbiomes. This makes understanding or deciphering the microbiome very complex and requires interdisciplinary approach. There has been numerous significant improvements in understanding the microbiome with the use of next-generation sequencing technology, omics-approach such as genomics, metagenomics, and metabolomics. These techniques have made it possible to quantify, estimate, and determine microbial diversity and functions (Zhang et al., 2019; Cullen et al., 2020; Ghebretatios et al., 2021).

Physiological and health relevance of the microbiome in the maintenance of human well-being is one of the major reasons why the microbiome has gained so much attention from the scientific community and the general public. The microbiome research has also been heavily funded in recent years because of the promise it holds in providing novel therapeutics through the use of microbes or their metabolites in the treatment of prevention of diseases (Hadrich, 2018; Berg et al., 2020). A healthy gut microbial composition has been shown to possess transformational and/or transcriptional potentials in reversing or managing disease conditions caused by distortion of the microbial homeostasis in the gut. This distortion in gut microbial homeostasis is termed dysbiosis. Microbiome

An Introduction to the Microbiome in Health and Diseases
ISBN 978-0-323-91190-0,
https://doi.org/10.1016/B978-0-323-91190-0.00002-3

dysbiosis has been connected to hypertension, metabolic syndrome, neurological disorders, polycystic ovary syndrome, infertility, ulcerative colitis, inflammatory bowel diseases, and clostridium difficile (Weiss and Hennet, 2017; Belizário et al., 2018). Therefore, application of microbiome as therapeutic intervention is gradually gaining momentum in the clinical practice showing the interesting and needed translation of findings from the bench to the bedside. An example of the popular application of microbiome is fecal microbiome transplantation for management of ulcerative colitis, and intractable clostridium difficile (Smits et al., 2013).

Although fecal microbiota transplantation (FMT) currently has investigational approval for therapy of many chronic gastrointestinal disorders, intensive research into fundamental genomics, ecological interactions, and culture of the members of human gut microbiome has become a global priority. Transplanting fecal microbial biomass can restore the dynamic community equilibrium, healthy microbiome composition, and diversity (Cho and Blaser, 2012; Wilson et al., 2019). Regardless of the successes achieved, the FMT application has not gained global acceptance due to various challenges such as complex regulatory and safety guidelines, inadequate donation facilities, donor recruitment difficulties, and adverse unintended versus host outcomes. Mitigation efforts include stool banking and stool encapsulation, which allows for consistent, suitable, and unbiased access to FMT for individuals with gastrointestinal disorders. The mechanisms of action of FMT are extremely complex and are reported to involve reduction of gut inflammation through modulatory activities on the inflammatory genes, mucins, and antimicrobial peptides. In addition, FMT helps to correct dysbiosis by replenishing the gut microbiota communities, which in turn mediates brain signaling, nutritional enhancement, and immune regulation to resolve colonic inflammatory disorders (Vendrik et al., 2020). Antimicrobial resistance is another prevalent condition facing the world, and microbiome research could be a solution through microbiome replacement therapy or enrichment by using probiotics and prebiotics as the case may be (Brinkac et al., 2017; Ghosh et al., 2019; Wieërs et al., 2020).

Plant microbiome possesses immense potentials that when fully harnessed will be beneficial for improving agricultural produce through enhancement of soil quality, plant growth, health, and production. Since the population of the world is constantly increasing, there is a serious challenge of hunger, famine, food insecurity, land degradation, and environmental pollution resulting in unfavorable climate change (Royal Agricultural Society of NSW, 2017; National Academies of Sciences and Medicine, 2018). This is due to the commercialization and industrialization of agriculture resulting in intensive farming and crop production activities through large quantities of fertilizers and pesticides with detrimental consequences on the plant and soil microbes, natural resources, human health, and the environment. Hence, there is an urgent need to stop the dependence on the use of chemicals and fertilizers for increased crop productivity because this method is no longer sustainable for humans, plants, soil, environment, and the climate. Taking advantage of the plant and soil microbiome can bring numerous benefits to agricultural practices through the promotion of plant growth, efficient nutrient usage, and pest control (Ray et al., 2020). The soil and plant microbiome are vital for ensuring global food security through the improvement of agricultural productivity using a purely organic and sustainable approach without any damaging consequences on the environment and soil (Qiu et al., 2019). Essentially, microorganisms are important for driving the enhancement of soil quality and function and improved agricultural production (Nazaries et al., 2013; Singh and Trivedi, 2017).

Microbiomes can be innovatively manipulated or engineered using biotechnological approaches. Microbiome engineering holds huge promise in enhancing the revolution that the world needs in the areas of agriculture, food production, environmental rejuvenation, climate change, human, and animal health management. Chemical fertilizers and pesticides that have been reported to cause negative effect on human well-being, soil microbiome, environmental pollution, and land degradation can be eradicated with the use of engineered microbiomes that have the potential to impact the soil microbiome, crop resilience, environmental microbiome, and climate change in a sustainable way. Engineered microbiomes could also offer interesting health-enhancing benefits in the area of precision medicine by personalizing the human microbiomes. The marine environment possesses a great deposit of microbiome that can be engineered for drug discovery, bioremediation, and waste management. Hence, applications of biotechnology to manipulate or engineered microbiomes have immense benefits that could bring global transformation (Lee et al., 2020).

Microbiome and Metabolism

Metabolism is critical to human survival; it determines the amount of energy, water, and carbon dioxide that could be generated from macromolecules. Carbohydrates, proteins, and lipids are key nutritional constituents such as simple sugar, fatty acids, and amino acids for humans, which are metabolized by human microbiomes such as gut microbiota. Meanwhile, the enzymes present in humans only have the capability of hydrolyzing the disaccharides and starches to form simple sugar while there is a minimal capability to hydrolyze other polysaccharides. As a result, the bulk of the undigested carbohydrates such as cellulose, pectin, and xylan and some unmetabolized starch get to the distal gut where the microbial communities then act on them to be digested (Devaraj et al., 2013). In the gut, there are communities of microbiota that are active metabolically and could degrade complex carbohydrates; as a result, mammals avoid synthesizing complex enzymes that could degrade several polysaccharides present in the diet (Devaraj et al., 2013). Cantarel et al. (2012) documented that microbe has numerous genes that encode a large number of bioactive enzymes in the human microbiome that could act on carbohydrates.

Moreover, the microbial carbohydrate-active enzymes (MCAzymes) such as carbohydrate esterases, glycosyl transferases, polysaccharide lyases, and glycoside hydrolases can be found in mammalian host (Cantarel et al., 2012), make use of carbohydrate fermentation to generate substrates for sustaining viable, functional microbial communities, and also produce bioactive signals, which can alter metabolic process in mammals. It was documented that intestinal bacterial taxonomy varies concerning their abilities to use the carbohydrate derived from nutrients or the host, e.g., mucus components (Sonnenburg et al., 2005, 2010).

Also, *Bacteriodetes* was reported to demonstrate simpler dietary carbohydrate assimilation because it is a phylum that possesses numerous routes for utilization of carbohydrates. Though in the absence of carbohydrates, the intestinal microbes stimulate degradation of mucin, which is a part of intestinal epithelial lining as a substitute to generate carbohydrate resulting in loss of gut mucosal barrier integrity. Garrido and coworkers established the capability of gut microbiota to metabolize several plants and host-derived glycoconjugates (glycans) as well as glycosaminoglycans such as hyaluronic acid, cellulose, chondroitin sulfate, mucins, and heparin. Moreover, endoglycosidase enzymes act on dietary substrates to produce from dairy sources and human milk complex *N*-glycans (Garrido et al., 2012).

Lipid metabolism

Fatty acids with promising potential are known to be produced from numerous intestinal bacteria, mostly probiotics such as intestinal bifidobacteria responsible for synthesizing conjugated linoleic acid that could modulate adipose tissue and liver fatty acid composition in murine animal models (OShea et al., 2012). Intestinal bacteria generate short-chain fatty acids (SCFAs) such as acetate, butyric, and propionic acids through the fermentation of fiber generated from undigested carbohydrate in humans. Martin et al. (2008) reported that germ-free mice lack SCFAs which explains the relevance of gut microbes in intestinal SCFA production.

Moreover, acetic acids serve as the most dominant of the SCFA present in mammals, and it appears to be involved in the modulation of the activity of 5αAMP-activated protein kinase as well as adipose tissue macrophage infiltration (Carvaho et al., 2012). However, propionic acid is known to be involved in lipid and glucose de novo synthesis and serves as a host energy source. SCFAs act as microbe-derived signals, which initiate carbohydrate degradation, and gut physiology and stimulate mammalian peptide release, which serves as gut epithelial cells' energy sources. In Ducastel et al. (2020) documented that gut microbes are partly involved in energy homeostasis control via dietary fiber fermentation, thereby releasing SCFAs that help to promote incretin secretion via FFAR2 and FFAR3 receptor binding on enteroendocrine L cells of the SCFA. Consequently, diets high in fat replaced with butyrate have improved insulin sensitivity and enhanced glucose regulation in obese mice. Also, reduction in some carbohydrates results in low concentration of fecal butyric acid and butyric acid—producing bacteria (Gao et al., 2009). Interestingly, Cho and Blaser (2012) demonstrated that alteration in gut microbiome population structure as well as its metabolic capabilities results from subtherapeutic administration of antibiotics. The antibiotics caused increased adiposity when administered to young mice and increased incretin GIP-1 level.

Protein metabolism

The microbiota such as bifidobacteria and lactobacilli present in the gut have the potential to produce biologically active compounds from amino acids such as numerous biogenic amines. Subsequently, the derived proteins and peptides have been hydrolyzed to produce amino acids from the action of proteinases and peptidases. These amino acids are substrates for luminal bioconversion via the action of gut microbiome (Devaraj et al., 2013). The commonly found enzymes in the gut microbiota are amino acid decarboxylases, which link amino acids with dietary compounds to signaling and microbial metabolism in the gut mucosa. Moreover, gut microbiota—derived bioactive molecules via liquid chromatography hydrophilic interaction HPLC (LCHI-HPLC) were isolated and screened as antiinflammatory compounds, and fractions of LCHI-HPLC were evaluated through mass spectrophotometry as well as nuclear magnetic resonance (Devaraj et al., 2013). Gut microbiome—derived histamine possesses antiinflammatory activity through suppression of tumor necrosis factor via its modulatory action on protein kinase A (Thomas et al., 2012).

Microbiome and the environment

In the environment, there are numerous microbes that get attached to the gastrointestinal tract, the skin, and the mucosa, which outweigh those in the genes and cells (Chatterjee et al., 2017). Authors also reported that the variety of microbes in the gut rely on the competition existing between microorganism in the environment, their sieving effects, and elimination. The gut of insects has numerous but unexplored microbes that are critical in the uptake of diverse chemicals useful in nitrogen cycle, pathogen prevention, production of pheromone, and breakdown of pesticides (Fig. 2.1) (Reeson et al., 2003; Mrazek et al., 2008). It was documented that microbes colonize the gastrointestinal tract of infants right from birth; this was revealed within 20 minutes postbirth for vaginally delivered infants that carry an identical microbiome present in the birth canal. However, a study found in humans that infant skin has a different microbial composition after delivery by caesarean section compared with vagina delivery (Dominguez-Bello et al., 2010). Moreover, these microbes keep on changing during development with an increase in infant gut microbes while major variation occurs at the onset of breastfeeding, cereal introduction, immune system development, and at the point of feed formulation (Ursell et al., 2012; Turnbaugh et al., 2009a,b). Chatterjee et al. (2017) documented that

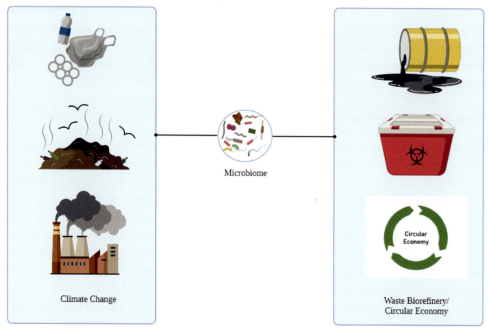

Figure 2.1 Involvement of microbiome in climate change and circular economy. Microbiomes are important regulators of climate change through their ability to consume and produce the greenhouse gases. Global warming is having devastating effects on microbial diversity, but environmental microbiome engineering can alleviate the deterioration caused by climate change. The use of microbiome in waste and pollution control or recycling for production of valuable materials and products for the society, e.g., biofuel, biogas, bioplastic, etc. *(Created with Biorender.com.)*

environmental as well as human microbial interaction is dynamic, leading to constant transfer of the communities of microbes within individuals and surfaces.

Gut microbiome and its metabolites

In mammals, the intestinal microbes secrete bioactive compounds such as ammonia, bile acids, phenols, endotoxins, SCFAs, and so on, stimulated by dietary macronutrients. Schroeder and Backhed in 2016 documented that microbes-derived metabolites could serve as agents for microorganism—host communication that are critical in maintaining host physiology.

Moreover, a robust technology was developed to identify and quantify the metabolite of interest through the characterization of the microbes and hence outline the metabolite biochemical routes (Vernocchi et al., 2016). One such metabolite is the primary bile acid with liver cholesterol as the precursor; also glycine and taurine conjugates are secreted into the guts where they can be transmuted via gut microbiome-derived bile salt hydrolase into secondary bile acids (Matsubara et al., 2013). Bile acids function via stimulating the pregnane X receptor, farnesoid X receptor (FXR), and G protein—coupled (GPC) receptors, resulting in alteration of metabolic processes. However, lithocholic acid and deoxycholic acid serve as the most available secondary bile acids present in the gut microbiota and then modulate host energy homeostasis and metabolism (Duboc et al., 2014).

Moreover, microbes in the gut also modulate the bile acid metabolic processes. The report of Swann et al. (2011) revealed that there is an increase in the available bile acids conjugated with taurine that is present in antibiotic and germ-free treated rats. Also, Gu et al. (2017) established that the antidiabetic medication acarbose causes alterations in intestinal microbes responsible for bile acid metabolism in diabetic patients and hence alters the bile acid composition ((Islam et al., 2011) by triggering result of type 2 diabetes patients. In another report, taurodeoxycholic acid derivative treatment improved the sensitivity of muscle and hepatic insulin (Kars et al., 2010). Furthermore, CYP7B1 was upregulated through antibiotic treatment and also caused an increase in hamster of tauro-β-muricholate (TβMC) with a decline in intestinal FXR signaling, hence decreasing glucose tolerance and liver damage (Sun et al., 2019). Consequently, targeting FXR and TGR5 bile acid receptors through alteration of intestinal bile acids composition via gut microbiome could be beneficial therapeutic interventions to ameliorate metabolic diseases.

Canfora et al. (2019) documented that colonic microbiota has the potential to add amino acids and proteins to their biomass via enzymatic mechanisms on fermentable and poorly absorbed carbohydrates, probiotics, and resistant starch for the generation of metabolites such as succinic acid and SCFAs. Microbiome-derived SCFAs are made through fermentation of resistant dietary fiber and starch by gut microbes (Kau et al., 2011).

Also, fermentable carbohydrate is not available for the distal colon microbiome, and the microbes turn to ferment the protein and thereby produce some detrimental metabolites, which are phenols, ammonia, and branched-chain fatty acids (BCFAs) (Canfora et al., 2019). Meanwhile, Cummings et al. (1987) established that the key microbial fermentation products of SCFAs in the guts are acetate, propionic acid, and butyrate. Moreover, the action of some gut bacteria such as *Blautia hydrogenotrophica* on pyruvate through acetyl-CoA or the Wood-Ljungdahl route can produce acetate (Ragsdale and Pierce, 2008). However, *Dialister* spp., *Phascolarctobacterium succinatutens*, and *Veillonella* spp.) of the *Bacteroidetes* and *Firmicutes* can produce propionic acid through the succinic acid route (Louis et al., 2014).

Larraufie et al. (2018) and Psichas et al. (2015) documented that SCFAs alter energy consumption as well as secretion of insulin through the production of satiety hormone via stimulation that releases peptide YY (PYY) and glucagon-like peptide (GLP)-1 in enteroendocrine cells via GPR41 (FFAR3) and GPR43 (FFAR2) (Li et al., 2018). SCFAs were documented to be responsible for inflammation conditions and immune responses.

However, Brown et al. (2011) reported that clusters of lactate- and butyric acid–producing gut microbes can impose a substantial quantity of mucin metabolism for the gut homeostasis to be maintained and also suppress autoimmune disease development such as type 1 diabetes.

The polyphenols derived from the diet have been reported to be xenobiotics used by humans, but their polymerization and structural complexity sometimes affect their absorption in the small intestine (Appeldoorn et al., 2009). However, their formulation into therapeutic agents is due to poor bioavailability, and this serves as a major concern. Consequently, unabsorbed polyphenols gathered in large intestines alongside bile conjugate, which thereafter undergoes enzymatic reactions by the gut microbes (Cardona et al., 2013). Interestingly, polyphenol metabolism in humans deserves a critical investigation because the gut microbes contribute to the synthesis of potentially active compounds through de novo. One of such is the synthesis of a low-molecular-weight compound, e.g., phenolic acid by the action of the gut microbiome on glycosylated polyphenols (Bode et al., 2013). Thus, the protective potential of the polyphenols also relies on microbiome-mediated metabolic processes.

Report also revealed that synthesis of piceid from resveratrol is due to the action of gut microbiota (Cichewicz and Kouzi, 1998), an active substance that serves as a more potent bioavailable antioxidant than resveratrol (Fabris et al., 2008). Oxidative stress and inflammation are induced by oxidized-LDL through NF-κB route can be attenuated by the derivatives from proanthocyanidins and 3-hydroxyphenylpyruvic acid (3-HPPA) through the inhibition of macrophage conversion into foam cells and then regulate in vitro cellular lipid metabolism (Zhang et al., 2018).

Microbiome and health

Microbiome constitutes the microbes that are both beneficial and potentially harmful. The microbes have symbiotic (health-promoting) and pathogenic (disease-promoting) relationship with the host. The pathogenic and symbiotic microbes live together without issues in a healthy body but a distortion of the balance or harmony between these microbes caused by diets, antibiotics, infections, or other xenobiotics resulting in dysbiosis involved in microbiome-related disorders such as gastrointestinal disorders, diabetes, hypertension, obesity, etc. Microbiome of a healthy individual is highly diverse and possesses beneficial microbial agents with capacity to offer protection against stress and disease-causing organisms through the stimulation of the immunological system of the body. However, microbiome diversity is low with reduction in the amount of beneficial microorganism in nonhealthy individuals resulting in diseases in the presence of stress or inflammation. Hence, adequate elucidation of the differences in microbiome diversity and properties between healthy and nonhealthy individuals may increase the possibility of detecting and identifying microbiome-associated with disease conditions (Sekirov et al., 2010; Kelly et al., 2016).

Therefore, intestinal microbiome has modulatory roles in the treatment, management, prevention, pathobiology, and progression of neurodegenerative diseases, metabolic dysregulations, cardiovascular diseases, and gastrointestinal disorders (Ley et al., 2006; Cho and Blaser, 2012; Le Chatelier et al., 2013). Hence, microbiome can be reshaped to achieve a desired therapeutic outcome or an unintended pathological outcome. For instance, microbiome can be manipulated to treat a disease condition as found in the case of using FMT for treatment of ulcerative colitis or *Clostridium difficile* or prevent diseases through the fortification and enrichment of the microbiome diversity by taking diets rich in probiotics, prebiotics, and postbiotics (Zeevi, et al., 2015; Zhao et al., 2018; Zmora et al., 2018). Likewise, microbiome can be altered to cause disease as seen in the case of diabetes, hypertension, and obesity due to long-term consumption of Western diets, and sedentary lifestyle (Turnbaugh et al., 2006; Turnbaugh et al., 2009a,b; Bouter et al., 2017; Kootte et al., 2017). Microbiome function and composition changes from early neonatal years by maternal and environmental microbiomes colonization, which has long lasting consequences in influencing microbiome later in life (Koenig et al., 2011; Gomez de Aguero et al., 2016).

Maternal microbiome and fetal health

The first few years of life of an infant are a very critical window for microbiome colonization. This is very essential as it has long-term health consequences in adulthood. Microbiome is transmitted to the fetus in utero at the time of delivery when the fetus passes through the birth canal where the fetus is bathe in the vaginal microbiome. Studies have reported this has a role in enhancing the health of the infant. Babies delivered by caesarean (C-) section lack the vaginal microbiome baptism that happens in birth canal during labor and delivery. Neonates also acquire microbiome from the environment of delivery, and this is the case with babies delivered through C-section. The benefits offered by the vaginal microbiome have been reported to be compromised in the babies delivered by C-section, and this has been associated with disease pathogenesis later in adulthood. Hence, the mode of delivery, surrounding environment, and diet are important factors influencing neonatal microbiome colonization or acquisition (Yatsunenko et al., 2012; Arrieta et al., 2014; Laursen et al., 2017).

Microbiome from mothers is involved in the preservation of maternal health, and fetal development during pregnancy cannot be overemphasized. Microbiome dysbiosis during pregnancy has been associated with stimulation of inflammatory signals (Afkham et al., 2019) causing abnormal placentation resulting in poor pregnancy outcome such as preterm delivery, preeclampsia, periodontal diseases, gestational weight gain, and cardiometabolic dysregulations. Interestingly, impaired vaginal microbiota homeostasis affects fertility (Xu et al., 2020) before conception or decreased success rate of in vitro fertilization (Kong et al., 2020; Schoenmakers et al., 2020), suggesting microbiome significance in pregnancy. A dysbiotic vaginal microbiome-associated with adverse pregnancy outcome has been shown to have a very low abundance of *Lactobacillus* (*Firmicutes*) prevent urinary tract infections as well as sexually transmitted infections (Klebanoff et al., 1991; Donders et al., 2000; Wiesenfeld et al., 2003). The following microbes are associated with healthy pregnancies Bacteroides, *Lactobacillus*, Clostridiales, and Actinobacteria (Prince et al., 2015). The vaginal microbiome has *Lactobacillus* as the dominant microbe with approximately 70% abundance (Huttenhower et al., 2012). Maternal microbiome has been demonstrated to influence fetal neural development (Vuong et al., 2020). Placental microbiome also contributes to neural development during pregnancy through tryptophan metabolism (Goeden et al., 2013). Interestingly, oral microbiome community has been shown to have close similarities with the placental microbiome. Hence, oral microbiome dysbiosis as a result of periodontitis, dental caries, and gingivitis has been linked to preterm delivery and preeclampsia (Bearfield et al., 2002; Boggess et al., 2003; Cobb et al., 2017).

Microbiome and metabolic health

Microbiome has been reported to have regulatory influences on the processes of metabolism in the gastrointestinal system. Metabolic disorders are linked to microbiome dysbiosis (Bouter et al., 2017). This is obvious from the observed microbiome properties, functional and diversity alteration reported in prediabetes and overt type 2 diabetes patients (Qin et al., 2012; Allin et al., 2018). However, mechanistic processes involved have not been fully elucidated, impaired bile acid metabolism, compromised generation of SCFAs and BCAAs, impaired permeability of the gut, and endotoxemia (Utzschneider et al., 2016). Glucometabolic regulation and insulin sensitivity have been demonstrated to be enhanced by transplanting microbiome from hale and hearty donors to individuals with gastrointestinal dysregulations using FMT (Kootte et al., 2017). Likewise, acetate, butyrate, and propionate have been reported in many animal models to improve glucometabolic regulation and bioenergetics (Lin et al., 2012; De Vadder et al., 2014; Michael et al., 2020; Olaniyi et al., 2021). Psichas and colleagues have also reported that SCFAs stimulate production of glucagon-like peptide-1 and peptide YY hormones, which are crucial regulators of glucose metabolism and cellular energy utilization (Psichas et al., 2015). In addition, consumption of diets rich in fibers offers favorable metabolic action in insulin-resistant and diabetic subjects through increased gut microbiome-mediated generation of SCFAs and enhanced glucagon-like peptide-1 secretion (Upadhyaya et al., 2016; Zhao et al., 2018). Wu and coworkers revealed that antidiabetic effect of metformin in regulating glucose homeostasis involves the enhancement of microbiome-mediated SCFAs generation (Wu et al., 2017).

Microbiome-derived metabolites have been shown to have epigenetic modification properties. Several studies have associated acetate, butyrate, and propionate with histone modification, DNA methylation, and noncoding RNAs regulation. SCFAs-mediated inhibition of histone deacetylase improves metabolic regulation resulting in enhanced glucose regulation and improved energy utilization (Sharma and Tripathi, 2019; Olaniyi et al., 2021). Epigenetic alterations induced by gut microbes influence neurological disorders. Intestinal microbiome interconnects with the brain and influences neurobehavioral processes through the mechanistic pathway of gut—brain axis, endocrine regulation, stress hormones, and neuroimmunological pathways (Vuong et al., 2017; Lynch and Hsiao, 2019; Kundu et al., 2021).

Microbiome and cardiovascular health

Cardiovascular disease (CVD) is the foremost cause of death worldwide (Hulten et al., 2017). Hypertension, a major CVD and intestinal microbiome, is implicated in its pathogenesis. Dietary plus lifestyle factors have capability to reshape and modify the microbiome, which further increases the risk of developing hypertension. Essentially, diverse studies using rodent models of gut dysbiosis reported to escalate progression of CVD (Lau et al., 2017; Avery et al., 2021). In addition, the levels of microbiome-derived bile acids, SCFA, and trimethylamine-N-oxide have been correlated with CVD development (Tang et al., 2017; Kitai and Tang, 2018).

Trimethylamine-N-oxide is a proatherogenic microbiome-derived metabolite that has been linked to CVD development through vascular inflammation, endothelial dysfunction, dyslipidemia, platelet aggregation, and leukocyte adhesion (Seldin et al., 2016; Ma et al., 2017; Li et al., 2017a,b). Regardless of the deleterious effect of trimethylamine-N-oxide on the cardiovascular health, SCFA is another microbiome-derived metabolite consisting largely of acetate, butyrate, and propionate display positive cardiovascular and cardiometabolic effects such as antiinflammatory, glucoregulatory, hypolipidemic, cardiorenal protective, antidiabetic, and hypotensive effects (Andrade-Oliveira et al., 2015; Natarajan et al., 2016; Bartolomaeus et al., 2020; Michael et al., 2020; Olaniyi et al., 2021).

Fecal microbiome transplantation to a germ-free mouse from human donor with hypertension has been documented to result in increased blood pressure after 8 weeks while those that received FMT from normotensive human donors are normotensive (Li et al., 2017a,b). Similarly, normotensive Wistar rats become hypertensive when infused with microbiome from stroke-prone spontaneously hypertensive rats (Adnan et al., 2017). Beneficial effect of microbiome to the cardiovascular health was also demonstrated by Kaye and colleagues who revealed that prebiotic fiber cardioprotective effects when

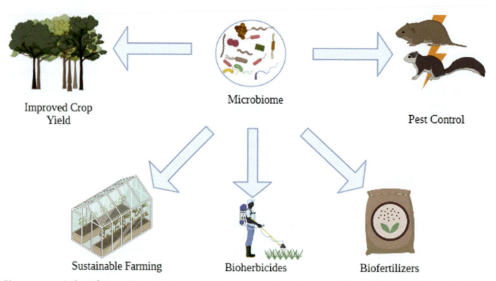

Figure 2.2 Role of microbiome in agriculture. Agricultural application of microbiome includes increased crop production for high yield, pest control, biofertilizers, bioherbicides, and sustainable farming. Nitrogen-producing bacteria can decrease or eliminate the use of chemical fertilizers in the crop cultivation. Chemical nitrogen fertilizers are not cost-effective for farmers; hence, using naturally occurring beneficial plant-associated microbiome for sustainable farming is very promising. *(Created with Biorender.com.)*

present in diets stimulate hypertension and cardiovascular disease development when deficient in diets (Kaye et al., 2020). Interestingly, microbiome can be a therapeutic target or pathogenic influence in CVD formation.

Microbiome in agriculture and food industry

Worldwide, the demand for food and dietary fiber is constantly on the rise, with a forecast of 70% increase by 2050. This pressure has been addressed with the use of arable land and chemicals (fertilizers and pesticides), which is attended with negative impacts on the environment. The result of this is depleted soil and water quality and climate change (Fig. 2.2). Apart from environmental impact, indulgent use of these chemicals has resulted in food contamination and even pesticide resistance, which has led to pest crises and lost crop yield; the annual loss of crops to pests in the United States now stands at 18%–26%, valued at over $470million (Oliveira et al., 2014; Sharma et al., 2017). The proposed solution to these issues is the use of associated microbiome as a natural, albeit effective, sustainable, and safer way of increasing agricultural output and productivity (Fig. 2.2) (Oliveira et al., 2014; Sharma et al., 2017; Singh and Trivedi, 2017).

Relationship between plants and microbiome

All through evolution, plants and microbiome have shared a symbiotic relationship (Singh and Trivedi, 2017). Microbiome in the rhizosphere functions similarly to human gut microbiota, both being inhabited by various bacteria, viruses, archaea, and eukaryotes. These functions include nutrient uptake, defense against diseases, and aiding growth and general health. And, just as in humans, pathogenesis and diseased states in plants correlate with microbiome population and diversity (Martin-Gallausiaux et al., 2020; Ramírez-puebla et al., 2013; Rowland et al., 2018; Sanna et al., 2019; Xue et al., 2019; Zhang et al., 2016). Roots have folds, root hair, and microvilli, which contribute to a large surface area for effective absorption of nutrients and water. Microbiome can lower pH in the rhizosphere because of the various acids that they produce and hence further provide a wide range of metabolic benefits (Singh and Trivedi, 2017).

Furthermore, by producing enzymes such as 1-aminocyclopropane-1-carboxy-late (ACC) deaminase, microbiome can ameliorate the impact of abiotic stress and increase resilience to biotic stress. Microbiome also competes with pathogens and hence fights off infections in plants (Ramírez-Pérez et al., 2017; Ramírez-puebla et al., 2013; Rodriguez et al., 2004; Wu et al., 2020; Zhang et al., 2016). In addition to aiding uptake of nutrients from the soil, some plant-associated microbiota have been shown to impact hormone-level modulation through the synthesis of polyamines and phytohormones, such as abscisic acid, auxins, cytokinins, ethylene, and gibberellins among others (Ding et al., 2013; Podlešáková et al., 2019; Ramírez-Pérez et al., 2017). This hormone signaling is effective regulator of plant growth and crop yield (Ding et al., 2013; Santner et al., 2009).

Nonetheless, the effective use of microbiota in inoculating plants against pathogens still eludes the agricultural and food production sector due to existing knowledge gaps. Current technique involves either modification of practices such as tillage and agrochemical use or addition of microbial inoculates. The addition of microbial inoculates to the soil have had little success mainly due to competition from other undesired native plants (Singh and Trivedi, 2017).

Efficient in situ application of microbiome is the recommended ideal means for tackling the problems of crop loss, low yield, pest resistance, food contamination, and environmental hazards that attend the current widespread use of chemical fertilizers and pesticides. And, to better harness the potential of microbiome for agriculture and food industries, microbiomes could undergo specie- and function-specific engineering for even higher yields. More research needs to focus on this area so as to provide better knowledge on its application in the food industry. If plant microbiome is properly harnessed, it can be a tool to match agricultural productivity with increasing food demand.

Microbiome as a source of novel therapeutics

The understanding of connection of microbiome to disease pathogenesis through dysbiosis has led to the concept of using fecal microbiome transplantation, microbiome metabolites, probiotics, prebiotics, and postbiotics for the purpose of replenishing or rejuvenating lost microbiome involved in restoration of health and microbiome diversity. Postbiotics, meaning microbial-derived metabolites supplements, have also been used for restoration of lost microbiome or inhibition of some detrimental metabolites. This proof of concept has been used in the management of gut disorders. Retractable Clostridium *difficile* where administrations of SCFAs exogenously have produced beneficial outcomes (Maslowski et al., 2009; Lee et al., 2015; Song et al., 2018).

Microbiome has connection to diseases such as diabetes, hypertension, neurological disorder, and psychological diseases. Interestingly, microbiome has also been shown to hold huge promise in the treatment of these diseases. For instance, cardiometabolic disorders/diabetes have been demonstrated to be ameliorated by microbiome-derived metabolites (acetate and butyrate) (Olaniyi et al., 2021; Michael et al., 2020), neurological disorders (Grochowska et al., 2019), hypertension (Bartolomaeus et al., 2019; Vallianou et al., 2020; Avery et al., 2021), metabolic syndrome (Maifeld et al., 2021), kidney diseases (Ramezani and Raj, 2014; Yang et al., 2018), liver diseases (Son et al., 2010), etc. Reduction of obesity-associated weight gain has been achieved by microbial-derived flavonoid supplements (Thaiss et al., 2016). Furthermore, postbiotic-mediated inhibition of microbial-derived trimethylamine N-oxide from L-carnitine is associated with amelioration of stroke, atherosclerosis, and myocardial infarction (Koeth et al., 2013).

Probiotics represent another microbiome therapy that is currently adopted in treatment of disease condition because of their ability to restore depleted microbiome population. Probiotics is the administration of live microorganisms with the capacity to offer health benefits when taken in appropriate quantities. Probiotics are called live biopharmaceutical products (LBPs) by the FDA. No LBPs have been approved by FDA for clinical application due to the fact that studies are ongoing to ascertain their interactions with resident host microbiome, safety profile, and mucosal colonization. Probiotic therapeutics is not a new practice, it dates back to centuries, it is still under review because microbiome diversity varies across individuals and communities (Kristensen et al., 2016; Walter et al., 2018). Reports exist that probiotics are efficacious against some disease conditions (Thompson et al., 2017; Veronese et al., 2018). Furthermore, probiotics result in the restoration of microbiome diversity following dysbiosis induced by antibiotics even though it takes time for the microbiome reconstruction to occur (Ekmekciu et al., 2017; Suez et al., 2018; Taur et al., 2018).

Numerous human research studies using probiotic therapies have not produced consistent results because of interindividual microbiome variability; hence, there is need for further studies on the compatibility or suitability of probiotics in diverse populations. The understanding of interindividual differences in microbiome is still evolving; however, studies have implicated environmental factors as critical influencer of microbiome diversity or population suggesting that genetics does not play any significant role in this variability (Rothschild et al., 2018). The knowledge of the microbiome differences in individuals can be used for patient's stratification before any administration of microbiome-based therapeutics to enhance the likelihood of positive response to the treatment. In cancer chemotherapy; microbiome is involved in determination of patient response to treatment or adverse reaction (Ma et al., 2019).

The world is the middle of antimicrobial resistance where known antibiotics are not effective against infections that they used to treat. This has led to the search of different remedy through natural means to combat the microbial infections. Interestingly, the microbiome contains a huge reserve and unexploited microbial community with immense ability to generate peptides and metabolites. The production of these metabolites and peptides by the microbiome represents the crucial step in finding novel and effective agents for combating antimicrobial resistance (Relman et al., 2018; Garcia-Gutierrez et al., 2019; Pilmis et al., 2020). Microorganisms from marine as well as soils have been demonstrated to possess metabolites with varied therapeutic applications (Molinski et al., 2009; Ling et al., 2015; Grzelak et al., 2019). Technological advancement using genomic sequencing and metabolomics approaches has allowed production of novel natural therapeutics from the gut microbiome mining (Donia et al., 2014; Donia and Fischbach 2015).

Microbiome-derived metabolites are small molecule end products of fermentation of indigestible dietary fibers. These small molecules have signaling capabilities responsible for the modulation of physiological, cellular, and organ-wide activities of the host (Blacher et al., 2017; Levy et al., 2017; Franzosa et al., 2019; Skelly et al., 2019). Microbiome—host cross-talk takes place effectively through the utilization of these small molecules metabolites. These microbiome-derived metabolites or postbiotics largely act on microbiome-associated downstream signaling mechanisms and alleviate the detrimental consequences of over-, lack of, or impaired production of metabolites involved in the microbiome signaling processes. Hence, metabolites blockade or administration exogenously has the ability to mitigate the dysbiosis-induced negative actions. SCFAs (acetate, butyrate, propionate) are examples of the microbiome-derived metabolites that have been shown to possess huge capabilities such as antiinflammatory effect, glucose regulatory effects, energy production, gut integrity, and immune regulation (Morrison and Preston, 2016). SCFAs quantities have been shown to be decreased in inflammatory bowel disease and ulcerative colitis (Butzner et al., 1996; Scheppach et al., 1996; Macia et al., 2015). Taurine and flavonoids are also microbiome-derived metabolites that have shown promise against intestinal inflammation and metabolic dysregulation, respectively (Levy et al., 2015; Thaiss et al., 2016).

Microbiome-derived metabolites as therapeutic intervention are promising because these metabolites are abundant in large quantities in the human body under physiological conditions making them less toxic compared with application of foreign entity or live microbes. Metabolites can be administered through diverse routes because they are found in most parts of the body. Also, microbiome-derived metabolites are quite stable in the circulatory system of humans as such their concentration can be manipulated to achieve desired physiological outcome. The main concern is that microbiome-derived metabolites have short half-life when compared with live microbes administration; thus, treatment of dysbiosis-induced pathology may require repeated administration of microbiome-derived metabolites to achieve desired results. Furthermore, some microbiome-derived metabolites have several effects and have cell type specificity. Very few microbiome-derived metabolites have characterized, cellular receptors, tissue of interest, mechanistic pathways, and physiological result of utility (Zheng et al., 2011; Levy et al., 2015; Wong and Levy, 2019). Next-generation sequencing technology, metabolomics, metagenomic, and proteomics characterization approach will be needed to be applied systematically to be able to provide a comprehensive blueprint of microbiome—metabolites—host interactions. Discovery of microbiome metabolite sensing technique will enhance direct or targeted modulation signaling mechanistic machinery involved in the tackling the detrimental actions of dysbiosis on the host (Wong and Levy, 2019).

Interestingly, much microbiome-associated therapeutics (Fig. 2.3) have been highly successful in animal models, but translational application from bench to bedside is still in the ongoing phase, and many microbiome-based therapeutics are in phase 3 of clinical trials to determine their efficacy and safety profiles in humans.

Conclusion

Microbiome relevance across various fields and spheres of life is increasing or unfolding at a very fast rate. The application of microbiome is changing various industries positively such as agriculture and food, biomedical and healthcare, engineering, cosmetics and beauty, environment, etc. Hence, adequate knowledge or research to uncover the full potentials of microbiome will have significant societal and economic impacts. The microbiome forms an integral part of human beings, which has physiological and pathological significance. Investigation of individual differences in microbiome composition will further advance the application of microbiome in disease management. The crucial roles played by microbiome span the various parts or system of the human body such as reproductive, cardiovascular, gastrointestinal, endocrine, and nervous systems.

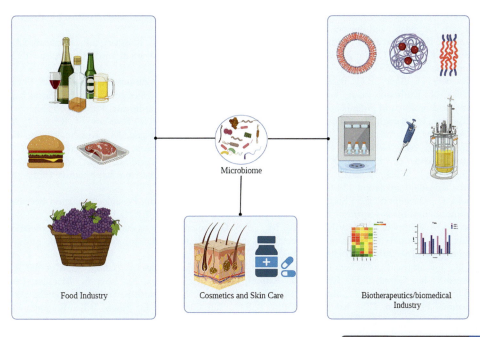

Created in **BioRender.com** bio

Figure 2.3 Biotechnological uses of microbiome. Microbiome use spans across various industries such as food and beverage industry, cosmetic and skin care industry, and biotherapeutic and biomedical industry. Diet has a major role in modulating the microbiome, food has diverse microbiome communities, and food waste is a major problem around the world due largely to microbial spoilage. Microbial fermentation is the process used in the food and beverage industry for production of bread, alcoholic drinks, yogurt, etc. Skincare and cosmetic industry are now developing products using the knowledge of skin microbiome. Microbiome-based biotherapeutics uses microbiome-derived metabolites and live microbes to modulate the microbiome for the treatment of diseases. *(Created with Biorender.com.)*

References

Adnan, S., Nelson, J.W., Ajami, N.J., Venna, V.R., Petrosino, J.F., Bryan Jr., R.M., Durgan, D.J., 2017. Alterations in the gut microbiota can elicit hypertension in rats. Physiol. Genom. 49, 96—104.

Afkham, A., Eghbal-Fard, S., Heydarlou, H., Azizi, R., Aghebati-Maleki, L., Yousefi, M., 2019. Toll-like receptors signaling network in pre-eclampsia: an updated review. J. Cell. Physiol. 234, 2229—2240.

Allin, K.H., Tremaroli, V., Caesar, R., Jensen, B.A.H., Damgaard, M.T.F., Bahl, M.I., Licht, T.R., Hansen, T.H., Nielsen, T., Dantoft, T.M., et al., 2018. Aberrant intestinal microbiota in individuals with prediabetes. Diabetologia 61, 810—820.

Andrade-Oliveira, V., Amano, M.T., Correa-Costa, M., Castoldi, A., Felizardo, R.J., de Almeida, D.C., Bassi, E.J., Moraes-Vieira, P.M., Hiyane, M.I., Rodas, A.C., et al., 2015. Gut bacteria products prevent AKI induced by ischemia-reperfusion. J. Am. Soc. Nephrol. 26, 1877—1888.

Appeldoorn, M.M., Vincken, J.P., Gruppen, H., Hollman, P.C., 2009. Procyanidin dimers A1, A2, and B2 are absorbed without conjugation or methylation from the small intestine of rats. J. Nutr. 139, 1469—1473.

Arrieta, M.C., Stiemsma, L.T., Amenyogbe, N., Brown, E.M., Finlay, B., 2014. The intestinal microbiome in early life: health and disease. Front. Immunol. 5, 427.

Avery, E.G., Bartolomaeus, H., Maifeld, A., Marko, L., Wiig, H., Wilck, N., Rosshart, S.P., Forslund, S.K., Müller, D.N., 2021. The gut microbiome in hypertension: recent advances and future perspectives. Circ. Res. 128 (7), 934—950.

Bartolomaeus, H., Avery, E.G., Bartolomaeus, T.U.P., Kozhakhmetov, S., Zhumadilov, Z., Müller, D.N., Wilck, N., Kushugulova, A., Forslund, S.K., 2020. Blood pressure changes correlate with short-chain fatty acid production potential shifts under a synbiotic intervention. Cardiovasc. Res. 116, 1252.

Bartolomaeus, H., Balogh, A., Yakoub, M., Homann, S., Markó, L., Höges, S., Tsvetkov, D., Krannich, A., Wundersitz, S., Avery, E.G., Haase, N., Kräker, K., Hering, L., Maase, M., Kusche-Vihrog, K., Grandoch, M., Fielitz, J., Kempa, S., Gollasch, M., Zhumadilov, Z., Kozhakhmetov, S., Kushugulova, A., Eckardt, K.U., Dechend, R., Rump, L.C., Forslund, S.K., Müller, D.N., Stegbauer, J., Wilck, N., 2019. Short-chain fatty acid propionate protects from hypertensive cardiovascular damage. Circulation 139 (11), 1407—1421.

Bearfield, C., Davenport, E.S., Sivapathasundaram, V., Allaker, R.P., 2002. Possible association between amniotic fluid micro-organism infection and microflora in the mouth. BJOG 109, 527—533.

Belizário, J.E., Faintuch, J., Garay-Malpartida, M., 2018. Gut microbiome dysbiosis and immunometabolism: new frontiers for treatment of metabolic diseases. Mediat. Inflamm. 2018, 2037838. https://doi.org/10.1155/2018/2037838.

Berg, G., Rybakova, D., Fischer, D., Cernava, T., Vergès, M.C., Charles, T., Chen, X., Cocolin, L., Eversole, K., Corral, G.H., Kazou, M., Kinkel, L., Lange, L., Lima, N., Loy, A., Macklin, J.A., Maguin, E., Mauchline, T., McClure, R., Mitter, B., Ryan, M., Sarand, I., Smidt, H., Schelkle, B., Roume, H., Kiran, G.S., Selvin, J., Souza, R.S.C., van Overbeek, L., Singh, B.K., Wagner, M., Walsh, A., Sessitsch, A., Schloter, M., 2020. Microbiome definition re-visited: old concepts and new challenges. Microbiome 8 (1), 103. https://doi.org/10.1186/s40168-020-00875-0.

Blacher, E., Levy, M., Tatirovsky, E., Elinav, E., 2017. Microbiome-modulated metabolites at the interface of host immunity. J. Immunol. 198, 572—580.

Bode, L.M., Bunzel, D., Huch, M., Cho, G.S., Ruhland, D., Bunzel, M., Bub, A., Franz, C.M., Kulling, S.E., 2013. In vivo and in vitro metabolism of trans-resveratrol by human gut microbiota. Am. J. Clin. Nutr. 97, 295—309.

Boggess, K.A., Lieff, S., Murtha, A.P., Moss, K., Beck, J., Offenbacher, S., 2003. Maternal periodontal disease is associated with an increased risk for preeclampsia. Obstet. Gynecol. 101, 227—231.

Bouter, K.E., van Raalte, D.H., Groen, A.K., Nieuwdorp, M., 2017. Role of the gut microbiome in the pathogenesis of obesity and obesity-related metabolic dysfunction. Gastroenterology 152, 1671—1678.

Brinkac, L., Voorhies, A., Gomez, A., Nelson, K.E., November 2017. The threat of antimicrobial resistance on the human microbiome. Microb. Ecol. 74 (4), 1001—1008. https://doi.org/10.1007/s00248-017-0985-z.

Brown, C.T., Davis-Richardson, A.G., Giongo, A., Gano, K.A., Crabb, D.B., Mukherjee, N., Casella, G., Drew, J.C., Ilonen, J., Knip, M., et al., 2011. Gut microbiome metagenomics analysis suggests a functional model for the development of autoimmunity for type 1 diabetes. PLoS One 6, e25792.

Butzner, J.D., Parmar, R., Bell, C.J., Dalal, V., 1996. Butyrate enema therapy stimulates mucosal repair in experimental colitis in the rat. Gut 38, 568—573.

Canfora, E.E., Meex, R.C.R., Venema, K., Blaak, E.E., 2019. Gut microbial metabolites in obesity, NAFLD and T2DM. Nat. Rev. Endocrinol. 15, 261—273.

Cantarel, B.L., Lombard, V., Henrissat, B., 2012. Complex carbohydrate utilization by the healthy human microbiome. PLoS One 7, e28742.

Cardona, F., Andres-Lacueva, C., Tulipani, S., Tinahones, F.J., Queipo-Ortuno, M.I., 2013. Benefits of polyphenols on gut microbiota and implications in human health. J. Nutr. Biochem. 24, 1415—1422.

Carvalho, B.M., Guadagnini, D., Tsukumo, D.M., Schenka, A.A., Latuf-Filho, P., Vassallo, J., et al., 2012. Modulation of gut microbiota by antibiotics improves insulin signalling in high-fat fed mice. Diabetologia 55, 2823—2834.

Chatterjee, S., Datta, S., Sharma, S., Tiwari, S., Gupta, D.K., 2017. Health and environmental applications of gut microbiome: a review. Ecol. Chem. Eng. S 24 (3), 467—482.

Cho, I., Blaser, M.J., 2012. The human microbiome: at the interface of health and disease. Nat. Rev. Genet. 13, 260–270.

Cichewicz, R.H., Kouzi, S.A., 1998. Biotransformation of resveratrol to piceid by Bacillus cereus. J. Nat. Prod. 61, 1313–1314.

Cobb, C.M., Kelly, P.J., Williams, K.B., Babbar, S., Angolkar, M., Derman, R.J., 2017. The oral microbiome and adverse pregnancy outcomes. Int J Womens Health 9, 551–559.

Cullen, C.M., Aneja, K.K., Beyhan, S., Cho, C.E., Woloszynek, S., Convertino, M., McCoy, S.J., Zhang, Y., Anderson, M.Z., Alvarez-Ponce, D., Smirnova, E., Karstens, L., Dorrestein, P.C., Li, H., Sen Gupta, A., Cheung, K., Powers, J.G., Zhao, Z., Rosen, G.L., 2020. Emerging priorities for microbiome research. Front. Microbiol. 11, 136. https://doi.org/10.3389/fmicb.2020.00136 eCollection 2020.

Cummings, J.H., Pomare, E.W., Branch, W.J., Naylor, C.P., Macfarlane, G.T., 1987. Short chain fatty acids in human large intestine, portal, hepatic and venous blood. Gut 28, 1221–1227.

De Vadder, F., Kovatcheva-Datchary, P., Goncalves, D., Vinera, J., Zitoun, C., Duchampt, A., B€ackhed, F., Mithieux, G., 2014. Microbiota-generated metabolites promote metabolic benefits via gut-brain neural circuits. Cell 156, 84–96.

Devaraj, S., Hemarajata, P., Versalovic, J., 2013. The human gut microbiome and body metabolism: implications for obesity and diabetes. Clin. Chem. 59 (4), 617–628. https://doi.org/10.1373/clinchem.2012.187617.

Ding, J., Chen, B., Xia, X., Mao, W., Shi, K., Zhou, Y., Yu, J., 2013. Cytokinin-induced parthenocarpic fruit development in tomato is partly dependent on enhanced gibberellin and auxin biosynthesis. PLoS One 8, 70080. https://doi.org/10.1371/journal.pone.0070080.

Dominguez-Bello, M.G., Costello, E.K., Contreras, M., Magris, M., Hidalgo, G., Fierer, N., et al., 2010. Delivery mode shapes the acquisition and structure of the initial microbiota across multiple body habitats in newborns. Proc. Natl. Acad. Sci. U.S.A. 107, 11971–11975. https://doi.org/10.1073/pnas.1002601107.

Donders, G.G., Bosmans, E., Dekeersmaecker, A., Vereecken, A., Van Bulck, B., Spitz, B., 2000. Pathogenesis of abnormal vaginal bacterial flora. Am. J. Obstet. Gynecol. 182, 872–878.

Donia, M.S., Cimermancic, P., Schulze, C.J., Wieland Brown, L.C., Martin, J., Mitreva, M., Clardy, J., Linington, R.G., Fischbach, M.A., 2014. A systematic analysis of biosynthetic gene clusters in the human microbiome reveals a common family of antibiotics. Cell 158 (6), 1402–1414.

Donia, M.S., Fischbach, M.A., 2015. Human microbiota. Small molecules from the human microbiota. Science 349 (6246), 1254766.

Duboc, H., Taché, Y., Hofmann, A.F., 2014. The bile acid TGR5 membrane receptor: from basic research to clinical application. Dig. Liver Dis. 46, 302–312.

Ducastel, S., Touche, V., Trabelsi, M.S., et al., 2020. The nuclear receptor FXR inhibits Glucagon-Like Peptide-1 secretion in response to microbiota-derived Short-Chain Fatty Acids. Sci. Rep. 10, 174. https://doi.org/10.1038/s41598-019-56743-x.

Ekmekciu, I., von Klitzing, E., Fiebiger, U., Neumann, C., Bacher, P., Scheffold, A., Bereswill, S., Heimesaat, M.M., 2017. The probiotic compound VSL#3 modulates mucosal, peripheral, and systemic immunity following murine broad-spectrum antibiotic treatment. Front. Cell. Infect. Microbiol. 7, 167. https://doi.org/10.3389/fcimb.2017.00167 eCollection 2017.

Fabris, S., Momo, F., Ravagnan, G., Stevanato, R., 2008. Antioxidant properties of resveratrol and piceid on lipid peroxidation in micelles and monolamellar liposomes. Biophys. Chem. 135 (1–3), 76–83. https://doi.org/10.1016/j.bpc.2008.03.005.

Franzosa, E.A., Sirota-Madi, A., Avila-Pacheco, J., Fornelos, N., Haiser, H.J., Reinker, S., Vatanen, T., Hall, A.B., Mallick, H., McIver, L.J., Sauk, J.S., Wilson, R.G., Stevens, B.W., Scott, J.M., Pierce, K., Deik, A.A., Bullock, K., Imhann, F., Porter, J.A., Zhernakova, A., Fu, J., Weersma, R.K., Wijmenga, C., Clish, C.B., Vlamakis, H., Huttenhower, C., Xavier, R.J., 2019. Gut microbiome structure and metabolic activity in inflammatory bowel disease. Nat. Microbiol. 4, 293–305.

Gao, Z., Yin, J., Zhang, J., Ward, R.E., Martin, R.J., Lefevre, M., et al., 2009. Butyrate improves insulin sensitivity and increases energy expenditure in mice. Diabetes 58, 1509–1517.

Garcia-Gutierrez, E., Mayer, M.J., Cotter, P.D., Narbad, A., 2019. Gut microbiota as a source of novel antimicrobials. Gut Microb. 10 (1), 1–21.

Garrido, D., Nwosu, C., Ruiz-Moyano, S., Aldredge, D., German, J.B., Lebrilla, C.B., Mills, D.A., 2012. Endo-beta-Nacetylglucosaminidases from infant-gut associated bifidobacteria release complex N-glycans from human milk glycoproteins. Mol. Cell. Proteomics 11, 775–785, 2012.

Ghebretatios, M., Schaly, S., Prakash, S., 2021. Nanoparticles in the food industry and their impact on human gut microbiome and diseases. Int. J. Mol. Sci. 22 (4), 1942. https://doi.org/10.3390/ijms22041942.

Ghosh, C., Sarkar, P., Issa, R., Haldar, J., 2019. Alternatives to conventional antibiotics in the era of antimicrobial resistance. Trends Microbiol. 27 (4), 323–338.

Goeden, N., Velasquez, J.C., Bonnin, A., 2013. Placental tryptophan metabolism as a potential novel pathway for the developmental origins of mental diseases. Transl. Dev. Psychiatry 1, 20593.

Gomez de Agüero, M., Ganal-Vonarburg, S.C., Fuhrer, T., Rupp, S., Uchimura, Y., Li, H., Steinert, A., Heikenwalder, M., Hapfelmeier, S., Sauer, U., McCoy, K.D., Macpherson, A.J., 2016. The maternal microbiota drives early postnatal innate immune development. Science 351 (6279), 1296–1302. https://doi.org/10.1126/science.aad2571.

Grochowska, M., Laskus, T., Radkowski, M., 2019. Gut microbiota in neurological disorders. Arch. Immunol. Ther. Exp. 67, 375–383.

Grzelak, E.M., Choules, M.P., Gao, W., Cai, G., Wan, B., Wang, Y., McAlpine, J.B., Cheng, J., Jin, Y., Lee, H., Suh, J.W., Pauli, G.F., Franzblau, S.G., Jaki, B.U., Cho, S., 2019. Strategies in anti-Mycobacterium tuberculosis drug discovery based on phenotypic screening. J. Antibiot. (Tokyo). https://doi.org/10.1038/s41429-019-0205-9.

Gu, Y., Wang, X., Li, J., Zhang, Y., Zhong, H., Liu, R., Zhang, D., Feng, Q., Xie, X., Hong, J., et al., 2017. Analyses of gut microbiota and plasma bile acids enable stratification of patients for antidiabetic treatment. Nat. Commun. 8, 1785.

Hadrich, D., June 13, 2018. Microbiome research is becoming the key to better understanding health and nutrition. Front. Genet. 9, 212. https://doi.org/10.3389/fgene.2018.00212 eCollection 2018.

Hulten, E.A., Bittencourt, M.S., Preston, R., Singh, A., Romagnolli, C., Ghoshhajra, B., Shah, R., Abbasi, S., Abbara, S., Nasir, K., Blaha, M., Hoffmann, U., Di Carli, M.F., Blankstein, R., 2017. Obesity, metabolic syndrome and cardiovascular prognosis: from the Partners coronary computed tomography angiography registry. Cardiovasc. Diabetol. 16, 14.

Huttenhower, C., Gevers, D., Knight, R., Abubucker, S., Badger, J.H., Chinwalla, A.T., Creasy, H.H., Earl, A.M., FitzGerald, M.G., Fulton, R.S., 2012. Structure, function and diversity of the healthy human microbiome. Nature 486, 207.

Islam, K.B., Fukiya, S., Hagio, M., Fujii, N., Ishizuka, S., Ooka, T., Ogura, Y., Hayashi, T., Yokota, A., 2011. Bile acid is a host factor that regulates the composition of the cecal microbiota in rats. Gastroenterology 141, 1773–1781.

Kars, M., Yang, L., Gregor, M.F., Mohammed, B.S., Pietka, T.A., Finck, B.N., Patterson, B.W., Horton, J.D., Mittendorfer, B., Hotamisligil, G.S., et al., 2010. Tauroursodeoxycholic acid may improve liver and muscle but not adipose tissue insulin sensitivity in obese men and women. Diabetes 59, 1899–1905.

Kau, A.L., Ahern, P.P., Gri_n, N.W., Goodman, A.L., Gordon, J.I., 2011. Human nutrition, the gut microbiome and the immune system. Nature 474, 327–336.

Kaye, D.M., Shihata, W.A., Jama, H.A., Tsyganov, K., Ziemann, M., Kiriazis, H., Horlock, D., Vijay, A., Giam, B., Vinh, A., Johnson, C., Fiedler, A., Donner, D., Snelson, M., Coughlan, M.T., Phillips, S., Du, X.J., El-Osta, A., Drummond, G., Lambert, G.W., Spector, T.D., Valdes, A.M., Mackay, C.R., Marques, F.Z., 2020. Deficiency of prebiotic fiber and insufficient signaling through gut metabolite-sensing receptors leads to cardiovascular disease. Circulation 141 (17), 1393–1403. https://doi.org/10.1161/CIRCULATIONAHA.119.043081.

Kelly, B.J., Imai, I., Bittinger, K., Laughlin, A., Fuchs, B.D., Bushman, F.D., Collman, R.G., 2016. Composition and dynamics of the respiratory tract microbiome in intubated patients. Microbiome 4, 7.

Kitai, T., Tang, W.H.W., 2018. Gut microbiota in cardiovascular disease and heart failure. Clin. Sci. (Lond.) 132 (85–91).

Klebanoff, S.J., Hillier, S.L., Eschenbach, D.A., Waltersdorph, A.M., 1991. Control of the microbial flora of the vagina by H2O2-generating lactobacilli. J. Infect. Dis. 164, 94–100.

Koenig, J.E., et al., 2011. Succession of microbial consortia in the developing infant gut microbiome. Proc. Natl. Acad. Sci. U. S. A. 108, 4578–4585.

Koeth, R.A., et al., 2013. Intestinal microbiota metabolism of L-carnitine, a nutrient in red meat, promotes atherosclerosis. Nat. Med. 19, 576—585.

Kong, Y., Liu, Z., Shang, Q., Gao, Y., Li, X., Zheng, C., Deng, X., Chen, T., 2020. The disordered vaginal microbiota is a potential indicator for a higher failure of in vitro fertilization. Front. Med. 7, 217.

Kootte, R.S., Levin, E., Saloj€arvi, J., Smits, L.P., Hartstra, A.V., Udayappan, S.D., Hermes, G., Bouter, K.E., Koopen, A.M., Holst, J.J., et al., 2017. Improvement of insulin sensitivity after lean donor feces in metabolic syndrome is driven by baseline intestinal microbiota composition. Cell Metabol. 26, 611—619.e6.

Kristensen, N.B., et al., 2016. Alterations in fecal microbiota composition by probiotic supplementation in healthy adults: a systematic review of randomized controlled trials. Genome Med. 8, 52.

Kundu, P., Torres, E.R.S., Stagaman, K., et al., 2021. Integrated analysis of behavioral, epigenetic, and gut microbiome analyses in AppNL-G-F, AppNL-F, and wild type mice. Sci. Rep. 11, 4678. https://doi.org/10.1038/s41598-021-83851-4.

Larraufie, P., Martin-Gallausiaux, C., Lapaque, N., Dore, J., Gribble, F.M., Reimann, F., Blottiere, H.M., 2018. SCFAs strongly stimulate PYY production in human enteroendocrine cells. Sci. Rep. 8, 74.

Lau, K., Srivatsav, V., Rizwan, A., Nashed, A., Liu, R., Shen, R., Akhtar, M., 2017. Bridging the gap between gut microbial dysbiosis and cardiovascular diseases. Nutrients 9, 859—874.

Laursen, M.F., Bahl, M.I., Michaelsen, K.F., Licht, T.R., 2017. First foods and gut microbes. Front. Microbiol. 8, 356.

Le Chatelier, E., et al., 2013. Richness of human gut microbiome correlates with metabolic markers. Nature 500, 541—546.

Lee, E.D., Aurand, E.R., Friedman, D.C., December 18, 2020. Engineering biology research consortium microbiomes roadmapping working group. Engineering microbiomes-looking ahead. ACS Synth. Biol. 9 (12), 3181—3183.

Lee, W.J., Lattimer, L.D.N., Stephen, S., et al., 2015. Fecal microbiota transplantation: a review of emerging indications beyond relapsing *Clostridium difficile* toxin colitis. Gastroenterol. Hepatol. 11 (1), 24—32.

Levy, M., Blacher, E., Elinav, E., 2017. Microbiome, metabolites and host immunity. Curr. Opin. Microbiol. 35, 8—15.

Levy, M., Thaiss, C.A., Zeevi, D., Dohnalová, L., Zilberman-Schapira, G., Mahdi, J.A., David, E., Savidor, A., Korem, T., Herzig, Y., Pevsner-Fischer, M., Shapiro, H., Christ, A., Harmelin, A., Halpern, Z., Latz, E., Flavell, R.A., Amit, I., Segal, E., Elinav, E., 2015. Microbiota-modulated metabolites shape the intestinal microenvironment by regulating NLRP6 inflammasome signaling. Cell 163, 1428—1443.

Ley, R.E., Turnbaugh, P.J., Klein, S., Gordon, J.I., 2006. Microbial ecology: human gut microbes associated with obesity. Nature 444, 1022—1023.

Li, J., Zhao, F., Wang, Y., Chen, J., Tao, J., Tian, G., Wu, S., Liu, W., Cui, Q., Geng, B., et al., 2017b. Gut microbiota dysbiosis contributes to the development of hypertension. Microbiome 5, 14. https://doi.org/10.1186/s40168-016-0222-x64.

Li, X.S., Obeid, S., Klingenberg, R., Gencer, B., Mach, F., Räber, L., Windecker, S., Rodondi, N., Nanchen, D., Muller, O., Miranda, M.X., Matter, C.M., Wu, Y., Li, L., Wang, Z., Alamri, H.S., Gogonea, V., Chung, Y.M., Tang, W.H., Hazen, S.L., Lüscher, T.F., 2017a. Gut microbiota-dependent trimethylamine N-oxide in acute coronary syndromes: a prognostic marker for incident cardiovascular events beyond traditional risk factors. Eur. Heart J. 38, 814—824.

Li, M., van Esch, B.C., Henricks, P.A., Folkerts, G., Garssen, J., 2018. The anti-inflammatory effects of short chain fatty acids on lipopolysaccharide-or tumor necrosis factor-stimulated endothelial cells via activation of GPR41/43 and inhibition of HDACs. Front. Pharmacol. 9, 533.

Lin, H.V., Frassetto, A., Kowalik Jr., E.J., Nawrocki, A.R., Lu, M.M., Kosinski, J.R., Hubert, J.A., Szeto, D., Yao, X., Forrest, G., Marsh, D.J., 2012. Butyrate and propionate protect against diet-induced obesity and regulate gut hormones via free fatty acid receptor 3-independent mechanisms. PLoS One 7, e35240.

Ling, L.L., Schneider, T., Peoples, A.J., Spoering, A.L., Engels, I., Conlon, B.P., Mueller, A., Schaberle, T.F., Hughes, D.E., Epstein, S., Jones, M., Lazarides, L., Steadman, V.A., Cohen, D.R., Felix, C.R., Fetterman, K.A., Millett, W.P., Nitti, A.G., Zullo, A.M., Chen, C., Lewis, K., 2015. A new antibiotic kills pathogens without detectable resistance. Nature 517 (7535), 455—459.

Louis, P., Hold, G.L., Flint, H.J., 2014. The gut microbiota, bacterial metabolites and colorectal cancer. Nat. Rev. Microbiol. 12, 661—672.

Lynch, J., Hsiao, E., 2019. Microbiomes as sources of emergent host phenotypes. Science 365, 1405—1409.

Ma, G., Pan, B., Chen, Y., Guo, C., Zhao, M., Zheng, L., Chen, B., 2017. Trimethylamine N-oxide in atherogenesis: impairing endothelial self-repair capacity and enhancing monocyte adhesion. Biosci. Rep. 37, BSR20160244.

Ma, W., Mao, Q., Xia, W., Dong, G., Yu, C., Jiang, F., 2019. Gut microbiota shapes the efficiency of cancer therapy. Front. Microbiol. 10, 1050.

Macia, L., Tan, J., Vieira, A.T., Leach, K., Stanley, D., Luong, S., Maruya, M., McKenzie, C.I., Hijikata, A., Wong, C., Binge, L., Thorburn, A.N., Chevalier, N., Ang, C., Marino, E., Robert, R., Offermanns, S., Teixeira, M.M., Moore, R.J., Flavell, R.A., Fagarasan, S., Mackay, C.R., 2015. Metabolite-sensing receptors GPR43 and GPR109A facilitate dietary fibre-induced gut homeostasis through regulation of the inflammasome. Nat. Commun. 6, 6734.

Maifeld, A., Bartolomaeus, H., Löber, U., Avery, E.G., Steckhan, N., Markó, L., Wilck, N., Hamad, I., Šušnjar, U., Mähler, A., Hohmann, C., Chen, C.Y., Cramer, H., Dobos, G., Lesker, T.R., Strowig, T., Dechend, R., Bzdok, D., Kleinewietfeld, M., Michalsen, A., Müller, D.N., Forslund, S.K., 2021. Fasting alters the gut microbiome reducing blood pressure and body weight in metabolic syndrome patients. Nat. Commun. 12 (1), 1970.

Martin, F.P., Wang, Y., Sprenger, N., Yap, I.K., Lundstedt, T., Lek, P., et al., 2008. Probiotic modulation of symbiotic gut microbial-host metabolic interactions in a humanized microbiome mouse model. Mol. Syst. Biol. 4, 157.

Martin-Gallausiaux, C., Marinelli, L., Blottière, H.M., Larraufie, P., Lapaque, N., 2020. SCFA: mechanisms and functional importance in the gut. Proc. Nutr. Soc. https://doi.org/10.1017/S0029665120006916.

Maslowski, K.M., et al., 2009. Regulation of inflammatory responses by gut microbiota and chemoattractant receptor GPR43. Nature 461, 1282—1286.

Matsubara, T., Li, F., Gonzalez, F.J., 2013. FXR signaling in the enterohepatic system. Mol. Cell. Endocrinol. 368, 17—29.

Michael, O.S., Dibia, C.L., Soetan, O.A., Adeyanju, O.A., Oyewole, A.L., Badmus, O.O., Adetunji, C.O., Soladoye, A.O., 2020. Sodium acetate prevents nicotine-induced cardiorenal dysmetabolism through uric acid/creatine kinase-dependent pathway. Life Sci. 257, 118127. https://doi.org/10.1016/j.lfs.2020.118127.

Molinski, T.F., Dalisay, D.S., Lievens, S.L., Saludes, J.P., 2009. Drug development from marine natural products. Nat. Rev. Drug Discov. 8 (1), 69—85.

Morrison, D.J., Preston, T., 2016. Formation of short chain fatty acids by the gut microbiota and their impact on human metabolism. Gut Microb. 7, 189—200.

Mrázek, J., Strosová, L., Fliegerová, K., Kott, T., Kopecný, J., 2008. Diversity of insect intestinal microflora. Folia Microbiol. Praha 53, 229—233. https://doi.org/10.1007/s12223-008-0032-z.

Natarajan, N., Hori, D., Flavahan, S., Steppan, J., Flavahan, N.A., Berkowitz, D.E., Pluznick, J.L., 2016. Microbial short chain fatty acid metabolites lower blood pressure via endothelial G protein-coupled receptor 41. Physiol. Genom. 48, 826—834.

National Academies of Sciences, Engineers and Medicine, 2018. Science Breakthroughs to Advance Food and Agricultural Research by 2030. National Academies Press. (Accessed 1 February 2019).

Nazaries, L., Pan, Y., Bodrossy, L., Baggs, E.M., Millard, P., et al., 2013. Microbial regulation of biogeochemcial cycles: evidence from a stidy on methane flux and land-use change. Appl. Environ. Microbiol. 79, 4031—4040.

O'Shea, E.F., Cotter, P.D., Stanton, C., Ross, R.P., Hill, C., 2012. Production of bioactive substances by intestinal bacteria as a basis for explaining probiotic mechanisms: bacteriocins and conjugated linoleic acid. Int. J. Food Microbiol. 152, 189—205, 2012.

Olaniyi, K.S., Amusa, O.A., Akinnagbe, N.T., Ajadi, I.O., Ajadi, M.B., Agunbiade, T.B., Michael, O.S., 2021. Acetate ameliorates nephrotoxicity in streptozotocin-nicotinamide-induced diabetic rats: involvement of xanthine oxidase activity. Cytokine 142, 155501. https://doi.org/10.1016/j.cyto.2021.155501.

Oliveira, C.M., Auad, A.M., Mendes, S.M., Frizzas, M.R., 2014. Crop losses and the economic impact of insect pests on Brazilian agriculture. Crop Protect. 56, 50—54. https://doi.org/10.1016/j.cropro.2013.10.022.

Pilmis, B., Le Monnier, A., Zahar, J.R., 2020. Gut microbiota, antibiotic therapy and antimicrobial resistance: a narrative review. Microorganisms 8 (2), 269.

Podlešáková, K., Ugena, L., Spíchal, L., Doležal, K., de Diego, N., 2019. Phytohormones and polyamines regulate plant stress responses by altering GABA pathway. N. Biotech. https://doi.org/10.1016/j.nbt.2018.07.003.

Prince, A.L., Chu, D.M., Seferovic, M.D., Antony, K.M., Ma, J., Aagaard, K.M., 2015. The perinatal microbiome and pregnancy: moving beyond the vaginal microbiome. Cold Spring Harb. Perspect Med. 5, a023051.

Psichas, A., Sleeth, M.L., Murphy, K.G., Brooks, L., Bewick, G.A., Hanyaloglu, A.C., Ghatei, M.A., Bloom, S.R., Frost, G., 2015. The short chain fatty acid propionate stimulates GLP-1 and PYY secretion via free fatty acid receptor 2 in rodents. Int. J. Obes. 39, 424–429.

Qin, J., Li, Y., Cai, Z., Li, S., Zhu, J., Zhang, F., Liang, S., Zhang, W., Guan, Y., Shen, D., et al., 2012. A metagenome-wide association study of gut microbiota in type 2 diabetes. Nature 490, 55–60.

Qiua, Z., Egidia, E., Liua, H., Kaura, S., Singh, B.K., 2019. New frontiers in agriculture productivity: optimised microbial inoculants and in situ microbiome engineering. Biotechnol. Adv. 37, 107371.

Ragsdale, S.W., Pierce, E., 2008. Acetogenesis and the wood-ljungdahl pathway of CO(2) fixation. Biochem. Biophys. Acta 1784, 1873–1898.

Ramezani, A., Raj, D.S., April 2014. The gut microbiome, kidney disease, and targeted interventions. J. Am. Soc. Nephrol. 25 (4), 657–670.

Ramírez-Pérez, O., Cruz-Ramón, V., Chinchilla-López, P., Méndez-Sánchez, N., 2017. The role of the gut microbiota in bile acid metabolism. Ann. Hepatol. 16, S21–S26. https://doi.org/10.5604/01.3001.0010.5672.

Ramírez-puebla, S.T., Servín-Garcidueñas, L.E., Jiménez-marín, B., Bolaños, L.M., Rosenblueth, M., Martínez, J., Rogel, M.A., Ormeño-orrillo, E., Martínez-romero, E., 2013. Gut and root microbiota commonalities. Appl. Environ. Microbiol. https://doi.org/10.1128/AEM.02553-12.

Ray, P., Lakshmanan, V., Labbé, J.L., Craven, K.D., 2020. Microbe to microbiome: a paradigm shift in the application of microorganisms for sustainable agriculture. Front. Microbiol. 11, 622926. https://doi.org/10.3389/fmicb.2020.622926, 2020 Dec 21.

Reeson, A.F., Jankovic, T., Kasper, M.L., Rogers, S., Austin, A.D., 2003. Application of 16S rDNA-DGGE to examine the microbial ecology associated with a social wasp Vespulagermanica. Insect Mol. Biol. 12, 85–91. https://link.springer.com/article/10.1007/s00284-008-9243-4.

Relman, D.A., Lipsitch, M., 2018. Microbiome as a tool and a target in the effort to address antimicrobial resistance. Proc. Natl. Acad. Sci. U. S. A. 115 (51), 12902–12910.

Rodriguez, H., Gonzalez, T., Goire, I., Bashan, Y., 2004. Gluconic acid production and phosphate solubilization by the plant growth-promoting bacterium Azospirillum spp. Naturwissenschaften 91, 552–555. https://doi.org/10.1007/s00114-004-0566-0.

Rothschild, D., Weissbrod, O., Barkan, E., Kurilshikov, A., Korem, T., Zeevi, D., Costea, P.I., Godneva, A., Kalka, I.N., Bar, N., 2018. Environment dominates over host genetics in shaping human gut microbiota. Nature 555, 210.

Rowland, I., Gibson, G., Heinken, A., Scott, K., Swann, J., Thiele, I., Tuohy, K., 2018. Gut microbiota functions: metabolism of nutrients and other food components. Eur. J. Nutr. https://doi.org/10.1007/s00394-017-1445-8.

Royal Agricultural Society of NSW, 2017. Annual Report 2016/2017. https://www.rasnsw.com.au/globalassets/document-library/annual-report/2016-2017_Annual_Report. (Accessed 18 February 2019).

Sanna, S., van Zuydam, N.R., Mahajan, A., Kurilshikov, A., Vich Vila, A., Võsa, U., Mujagic, Z., Masclee, A.A.M., Jonkers, D.M.A.E., Oosting, M., Joosten, L.A.B., Netea, M.G., Franke, L., Zhernakova, A., Fu, J., Wijmenga, C., McCarthy, M.I., 2019. Causal relationships among the gut microbiome, short-chain fatty acids and metabolic diseases. Nat. Genet. https://doi.org/10.1038/s41588-019-0350-x.

Santner, A., Calderon-Villalobos, L.I.A., Estelle, M., 2009. Plant hormones are versatile chemical regulators of plant growth. Nat. Chem. Biol. https://doi.org/10.1038/nchembio.165.

Scheppach, W., Bartram, H., Richter, F., Muller, J., 1996. Treatment of distal ulcerative colitis with short-chain fatty acid enemas—a placebo controlled trial. Dig. Dis. Sci. 41, 2254—2259.

Schoenmakers, S., Laven, J., 2020. The vaginal microbiome as a tool to predict IVF success. Curr. Opin. Obstet. Gynecol. 32, 169—178.

Schroeder, B.O., Backhed, F., 2016. Signals from the gut microbiota to distant organs in physiology and disease. Nat. Med. 22, 1079—1089.

Sekirov, I., Russell, S.L., Antunes, L.C., Finlay, B.B., 2010. Gut microbiota in health and disease. Physiol. Rev. 90, 859—904.

Seldin, M.M., Meng, Y., Qi, H., Zhu, W., Wang, Z., Hazen, S.L., Lusis, A.J., Shih, D.M., 2016. Trimethyl-amine-N-oxide promotes vascular inflammation through signaling of mitogen-activated protein kinase and nuclear factor-B. J. Am. Heart Assoc. 5, e002767.

Sharma, S., Kooner, R., Arora, R., 2017. Insect pests and crop losses. In: Breeding Insect Resistant Crops for Sustainable Agriculture. Springer Singapore, pp. 45—66. https://doi.org/10.1007/978-981-10-6056-4_2.

Sharma, S., Tripathi, P., 2019. Gut microbiome and type 2 diabetes: where we are and where to go? J. Nutr. Biochem. 63, 101—108. https://doi.org/10.1016/j.jnutbio.2018.10.003.

Singh, B.K., Trivedi, P., 2017. Microbiome and the future for food and nutrient security. Microb. Bio-technol. 10, 50—53. https://doi.org/10.1111/1751-7915.12592.

Skelly, A.N., Sato, Y., Kearney, S., Honda, K., 2019. Mining the microbiota for microbial and metabolite-based immunotherapies. Nat. Rev. Immunol.

Smits, L.P., Bouter, K.E., de Vos, W.M., Borody, T.J., Nieuwdorp, M., November 2013. Therapeutic potential of fecal microbiota transplantation. Gastroenterology 145 (5), 946—953. https://doi.org/10.1053/j.gastro.2013.08.058.

Son, G., Kremer, M., Hines, I.N., 2010. Contribution of gut bacteria to liver pathobiology. Gastroenterol. Res. Pract. 2010, 1—13.

Song, Y., Wang, P., Parian, A., et al., 2018. Where will fecal microbiota transplantation fit in the treatment algorithms for Crohn's disease and ulcerative colitis: a synthesis of completed, ongoing and future trials. Gastroenterology 154 (1, Suppl. p), S90.

Sonnenburg, J.L., Xu, J., Leip, D.D., Chen, C.H., Westover, B.P., et al., 2005. Glycan foraging in vivo by an intestine-adapted bacterial symbiont. Science 307, 1955—1959.

Sonnenburg, E.D., Zheng, H., Joglekar, P., Higginbottom, S.K., Firbank, S.J., Bolam, D.N., Sonnenburg, J.L., 2010. Specificity of polysaccharide use in intestinal bacteroides species determines diet-induced microbiota alterations. Cell 141, 1241—1252.

Suez, J., Zmora, N., Zilberman-Schapira, G., Mor, U., Dori-Bachash, M., Bashiardes, S., Zur, M., Regev-Lehavi, D., Ben-Zeev Brik, R., Federici, S., Horn, M., Cohen, Y., Moor, A.E., Zeevi, D., Korem, T., Kotler, E., Harmelin, A., Itzkovitz, S., Maharshak, N., Shibolet, O., Pevsner-Fischer, M., Shapiro, H., Sharon, I., Halpern, Z., Segal, E., Elinav, E., September 6, 2018. Post-antibiotic gut mucosal micro-biome reconstitution is impaired by probiotics and improved by autologous FMT. Cell 174 (6), 1406—1423.e16. https://doi.org/10.1016/j.cell.2018.08.047.

Sun, L., Pang, Y., Wang, X., Wu, Q., Liu, H., Liu, B., Liu, G., Ye, M., Kong, W., Jiang, C., 2019. Ablation of gut microbiota alleviates obesity induced hepatic steatosis and glucose intolerance by modulating bile acid metabolism in hamsters. Acta Pharm. Sin. B 9, 702—710.

Swann, J.R., Want, E.J., Geier, F.M., Spagou, K., Wilson, I.D., Sidaway, J.E., Nicholson, J.K., Holmes, E., 2011. Systemic gut microbial modulation of bile acid metabolism in host tissue compartments. Proc. Natl. Acad. Sci. U.S.A. 108 (Suppl. 1), 4523—4530.

Tang, W.H., Kitai, T., Hazen, S.L., 2017. Gut microbiota in cardiovascular health and disease. Circ. Res. 120, 1183—1196.

Taur, Y., Coyte, K., Schluter, J., Robilotti, E., Figueroa, C., Gjonbalaj, M., Littmann, E.R., Ling, L., Miller, L., Gyaltshen, Y., Fontana, E., Morjaria, S., Gyurkocza, B., Perales, M.A., Castro-Malaspina, H., Tamari, R., Ponce, D., Koehne, G., Barker, J., Jakubowski, A., Papadopoulos, E., Dahi, P., Sauter, C., Shaffer, B., Young, J.W., Peled, J., Meagher, R.C., Jenq, R.R., van den Brink, M.R.M., Giralt, S.A., Pamer, E.G., Xavier, J.B., September 26, 2018. Reconstitution of the gut microbiota of antibiotic-treated patients by autologous fecal microbiota transplant. Sci. Transl. Med. 10 (460), eaap9489. https://doi.org/10.1126/scitranslmed.aap9489.

Thaiss, C.A., Itav, S., Rothschild, D., Meijer, M.T., Levy, M., Moresi, C., Dohnalová, L., Braverman, S., Rozin, S., Malitsky, S., Dori-Bachash, M., Kuperman, Y., Biton, I., Gertler, A., Harmelin, A., Shapiro, H., Halpern, Z., Aharoni, A., Segal, E., Elinav, E., 2016. Persistent microbiome alterations modulate the rate of postdieting weight regain. Nature 540, 544.

Thomas, C.M., Hong, T., van Pijkeren, J.P., Hemarajata, P., Trinh, D.V., Hu, W., et al., 2012. Histamine derived from probiotic lactobacillus reuteri suppresses TNF via modulation of PKA and ERK signaling. PLoS One 7, e31951.

Thompson, S.V., Hannon, B.A., An, R., Holscher, H.D., 2017. Effects of isolated soluble fiber supplementation on body weight, glycemia, and insulinemia in adults with overweight and obesity: a systematic review and meta-analysis of randomized controlled trials. Am. J. Clin. Nutr. 106, 1514–1528.

Turnbaugh, P.J., Hamady, M., Yatsunenko, T., Cantarel, B.L., Duncan, A., Ley, R.E., 2009a. A core gut microbiome in obese and lean twins. Nature 457, 480–484. https://doi.org/10.1038/nature07540.

Turnbaugh, P.J., Ley, R.E., Mahowald, M.A., Magrini, V., Mardis, E.R., Gordon, J.I., 2006. An obesity-associated gut microbiome with increased capacity for energy harvest. Nature 444, 1027–1031.

Turnbaugh, P.J., Ridaura, V.K., Faith, J.J., Rey, F.E., Knight, R., Gordon, J.I., 2009b. The effect of diet on the human gut microbiome: a metagenomic analysis in humanized gnotobiotic mice. Sci. Transl. Med. 1, 6ra14.

Upadhyaya, B., McCormack, L., Fardin-Kia, A.R., Juenemann, R., Nichenametla, S., Clapper, J., Specker, B., Dey, M., 2016. Impact of dietary resistant starch type 4 on human gut microbiota and immunometabolic functions. Sci. Rep. 6, 28797.

Ursell, L.K., Metcalf, J.L., Parfrey, L.W., Knight, R., 2012. Defining the human microbiome. Nutr. Rev. 70, S38–S44. https://doi.org/10.1111/j.1753-4887.2012.00493.x.

Utzschneider, K.M., Kratz, M., Damman, C.J., Hullar, M., 2016. Mechanisms linking the gut microbiome and glucose metabolism. J. Clin. Endocrinol. Metab. 101, 1445–1454.

Vallianou, N.G., Geladari, E., Kounatidis, D., 2020. J Microbiome and hypertension: where are we now? Cardiovasc. Med. 21 (2), 83–88.

Vendrik, K.E.W., Ooijevaar, R.E., de Jong, P.R.C., Laman, J.D., van Oosten, B.W., van Hilten, J.J., Ducarmon, Q.R., Keller, J.J., Kuijper, E.J., Contarino, M.F., March 24, 2020. Fecal microbiota transplantation in neurological disorders. Front. Cell. Infect. Microbiol. 10, 98. https://doi.org/10.3389/fcimb.2020.00098.

Vernocchi, P., Del Chierico, F., Putignani, L., 2016. Gut microbiotaprofiling: metabolomics based approach to unravel compounds affecting human health. Front. Microbiol. 7, 1144.

Veronese, N., et al., 2018. Dietary fiber and health outcomes: an umbrella review of systematic reviews and meta-analyses. Am. J. Clin. Nutr. 107, 436–444.

Vuong, H.E., Pronovost, G.N., Williams, D.W., Coley, E.J.L., Siegler, E.L., Qiu, A., Kazantsev, M., Wilson, C.J., Rendon, T., Hsiao, E.Y., 2020. The maternal microbiome modulates fetal neurodevelopment in mice. Nature 586, 281–286.

Vuong, H., Yano, J., Fung, T., Hsiao, E., 2017. The microbiome and host behavior. Annu. Rev. Neurosci. 40, 21–49.

Walter, J., Maldonado-Gómez, M.X., Martínez, I., 2018. To engraft or not to engraft: an ecological framework for gut microbiome modulation with live microbes. Curr. Opin. Biotechnol. 49, 129–139.

Weiss, G.A., Hennet, T., 2017. Mechanisms and consequences of intestinal dysbiosis. Cell. Mol. Life Sci. 74 (16), 2959–2977.

Wieërs, G., Belkhir, L., Enaud, R., Leclercq, S., Philippart de Foy, J.M., Dequenne, I., de Timary, P., Cani, P.D., January 15, 2020. How probiotics affect the microbiota. Front. Cell. Infect. Microbiol. 9, 454. https://doi.org/10.3389/fcimb.2019.00454.

Wiesenfeld, H.C., Hillier, S.L., Krohn, M.A., Landers, D.V., Sweet, R.L., 2003. Bacterial vaginosis is a strong predictor of Neisseria gonorrhoeae and Chlamydia trachomatis infection. Clin. Infect. Dis. 36, 663–668.

Wilson, B.C., Vatanen, T., Cutfield, W.S., O'Sullivan, J.M., January 21, 2019. The super-donor phenomenon in fecal microbiota transplantation. Front. Cell. Infect. Microbiol. 9, 2. https://doi.org/10.3389/fcimb.2019.00002.

Wong, A.C., Levy, M., 2019. New approaches to microbiome-based therapies. mSystems 4, 001222-19.

Wu, H., Esteve, E., Tremaroli, V., Khan, M.T., Caesar, R., Mannera° s-Holm, L., Sta° hlman, M., Olsson, L.M., Serino, M., Planas-Fe` lix, M., et al., 2017. Metformin alters the gut microbiome of individuals with treatment-naive type 2 diabetes, contributing to the therapeutic effects of the drug. Nat. Med. 23, 850–858.

Wu, J., Wang, K., Wang, X., Pang, Y., Jiang, C., 2020. The role of the gut microbiome and its metabolites in metabolic diseases. Protein and Cell. https://doi.org/10.1007/s13238-020-00814-7.

Xu, J., Bian, G., Zheng, M., Lu, G., Chan, W.Y., Li, W., Yang, K., Chen, Z.J., Du, Y., 2020. Fertility factors affect the vaginal microbiome in women of reproductive age. Am. J. Reprod. Immunol. 83, e13220.

Xue, J., Li, X., Liu, P., Li, K., Sha, L., Yang, X., Zhu, L., Wang, Z., Dong, Y., Zhang, L., Lei, H., Zhang, X., Dong, X., Wang, H., 2019. Inulin and metformin ameliorate polycystic ovary syndrome via anti-inflammation and modulating gut microbiota in mice. Endocr. J. 66, 859–870. https://doi.org/10.1507/endocrj.EJ18-0567.

Yang, T., Richards, E.M., Pepine, C.J., Raizada, M.K., 2018. The gut microbiota and the brain-gut-kidney axis in hypertension and chronic kidney disease. Nat. Rev. Nephrol. 14 (7), 442–456.

Yatsunenko, T., et al., 2012. Human gut microbiome viewed across age and geography. Nature 486, 222–227.

Zeevi, D., Korem, T., Zmora, N., Israeli, D., Rothschild, D., Weinberger, A., Ben-Yacov, O., Lador, D., Avnit-Sagi, T., Lotan-Pompan, M., Suez, J., Mahdi, J.A., Matot, E., Malka, G., Kosower, N., Rein, M., Zilberman-Schapira, G., Dohnalová, L., Pevsner-Fischer, M., Bikovsky, R., Halpern, Z., Elinav, E., Segal, E., November 19, 2015. Personalized nutrition by prediction of glycemic responses. Cell 163 (5), 1079–1094. https://doi.org/10.1016/j.cell.2015.11.001.

Zhang, X., Li, L., Butcher, J., Stintzi, A., Figeys, D., December 6, 2019. Advancing functional and translational microbiome research using meta-omics approaches. Microbiome 7 (1), 154. https://doi.org/10.1186/s40168-019-0767-6.

Zhang, Y.Y., Li, X.L., Li, T.Y., Li, M.Y., Huang, R.M., Li, W., Yang, R.L., 2018. 3-(4-Hydroxyphenyl) propionic acid, a major microbial metabolite of procyanidin A2, shows similar suppression of macrophage foam cell formation as its parent molecule. RSC Adv. 8, 6242–6250. https://doi.org/10.1039/C7RA13729J.

Zhang, L., Xie, C., Nichols, R.G., Chan, S.H.J., Jiang, C., Hao, R., Smith, P.B., Cai, J., Simons, M.N., Hatzakis, E., Maranas, C.D., Gonzalez, F.J., Patterson, A.D., 2016. Farnesoid X receptor signaling shapes the gut microbiota and controls hepatic lipid metabolism. mSystems 1. https://doi.org/10.1128/msystems.00070-16.

Zhao, L., Zhang, F., Ding, X., Wu, G., Lam, Y.Y., Wang, X., Fu, H., Xue, X., Lu, C., Ma, J., Yu, L., Xu, C., Ren, Z., Xu, Y., Xu, S., Shen, H., Zhu, X., Shi, Y., Shen, Q., Dong, W., Liu, R., Ling, Y., Zeng, Y., Wang, X., Zhang, Q., Wang, J., Wang, L., Wu, Y., Zeng, B., Wei, H., Zhang, M., Peng, Y., Zhang, C., March 9, 2018. Gut bacteria selectively promoted by dietary fibers alleviate type 2 diabetes. Science 359 (6380), 1151–1156. https://doi.org/10.1126/science.aao5774.

Zheng, X., Xie, G., Zhao, A., Zhao, L., Yao, C., Chiu, N.H.L., Zhou, Z., Bao, Y., Jia, W., Nicholson, J.K., Jia, W., 2011. The footprints of gut microbial–mammalian co-metabolism. J. Proteome Res. 10, 5512–5522.

Zmora, N., Zilberman-Schapira, G., Suez, J., Mor, U., Dori-Bachash, M., Bashiardes, S., Kotler, E., Zur, M., Regev-Lehavi, D., Brik, R.B., Federici, S., Cohen, Y., Linevsky, R., Rothschild, D., Moor, A.E., Ben-Moshe, S., Harmelin, A., Itzkovitz, S., Maharshak, N., Shibolet, O., Shapiro, H., Pevsner-Fischer, M., Sharon, I., Halpern, Z., Segal, E., Elinav, E., September 6, 2018. Personalized gut mucosal colonization resistance to empiric probiotics is associated with unique host and microbiome features. Cell 174 (6), 1388–1405.e21. https://doi.org/10.1016/j.cell.2018.08.041.

CHAPTER 3

Procedures for sampling of small and larger samples of microbiome

Juliana Bunmi Adetunji[1], Olugbenga Samuel Michael[2,3,7], Charles Oluwaseun Adetunji[4], Olulope Olufemi Ajayi[5] and Frank Abimbola Ogundolie[6]

[1]Nutritional and Toxicological Research Laboratory, Department of Biochemistry Sciences, Osun State University, Osogbo, Osun State, Nigeria; [2]Cardiometabolic, Microbiome and Applied Physiology Laboratory, Department of Physiology, College of Health Sciences, Bowen University, Iwo, Osun State, Nigeria; [3]Department of Physiology, University of Tennessee Health Science Center, Memphis, TN, United States; [4]Applied Microbiology, Biotechnology and Nanotechnology Laboratory, Department of Microbiology, and Directorate of Research and Innovation, Edo State University Uzairue, Iyamho, Auchi, Edo State, Nigeria; [5]Department of Biochemistry, Edo State University Uzairue, Edo State, Nigeria; [6]Department of Biotechnology, Baze University, Abuja, Nigeria; [7]Department of Medical Pharmacology and Physiology, University of Missouri, Columbia, MO, United States

Introduction

There is increasing evidence of the contribution of the microbiome in regulation of a wide array of physiological activities resulting in the maintenance of health (Foster, 2013; Shreiner et al., 2015). The relevance of microbiome to human existence as an integral part of human system is gaining center stage with many innovative approaches and technological advancement driving the microbiome research discoveries and translational applications from the bench to the bed side (Michael et al. 2022a,b; Esiobu et al., 2022, Cullen et al., 2020). Also the alteration in the balance of the gut microbiome homeostasis has been reported to be connected with pathogenesis of diverse disease conditions such as diabetes, hypertension, obesity, gastrointestinal disorders, and autoimmune condition (Morgan et al., 2012; Lynch and Pedersen, 2016; Sharma and Tripathi, 2019; O'Donnell et al., 2023). Regardless of technological innovations, increasing scientific findings, and understanding of some microbiome functionalities, the causes of individual microbiome variability and its evolution over the course of life are yet to be fully clarified or elucidated (Berg et al., 2020). However, studies have implicated genetics and epigenetical modifications to be possibly responsible for a large part of the microbiome variation indicating the significant role of the environment as a main component influencing the microbiome community and function (Rothschild et al., 2018).

Effective microbiome sampling techniques are significant for identification of microbiome in various parts of the body (Bokulich et al., 2020; Bharti and Grimm, 2021). Microbiomes are located in different areas of the body, and microbiome compositions are different across various sites (Kennedy and Chang, 2020). The skin, mouth, vagina, intestine, etc. have their varying microbiome composition (Kennedy and Chang, 2020). Furthermore, since no two individuals have similar microbiome, and the

An Introduction to the Microbiome in Health and Diseases
ISBN 978-0-323-91190-0,
https://doi.org/10.1016/B978-0-323-91190-0.00003-5

differences among individuals are large compared with the typical biochemical differences within a person over time (Integrative HMP (iHMP) Research Network Consortium, 2014; Lax et al., 2014). Identical twins are barely more similar to one another in microbial composition and structure than are nonidentical twins (Goodrich et al., 2014). Effective microbiome sampling will enhance detection and determination of the microbiome compositions and functions using next-generation sequencing techniques such as metagenomic and metabolomics sequencing (Costea et al., 2017). Metagenomic sequencing has been a very useful technique for studying microbiome communities associated with a host, whereas metabolomics analyses provide downstream information on the metabolic activities of microbiome generating data for valid clinical interpretation and translational purposes (Mallick et al., 2019; Yin et al., 2020; Liu et al., 2021).

Fecal microbiome sampling provided the foundational information on the structural and functional properties of microbiome in human (Jones et al., 2021). However, this method provides access to minor portion of gut microbes for culturing or sequencing (Lau et al., 2016). Hence, to gain access to further samples, direct microbiome sample collection methods from the gut using colonoscopy and gastroscopy have been developed (Kim et al., 2021). Interestingly, direct sample collection also has its challenges, which include inability to reach entire length of the gut due to its extensive length and diameter variations, and pain and discomfort to patients due to the invasive nature of this method resulting in low compliance (Moglia et al., 2009; Valdastri et al., 2012).

The microbiome composition or diversity is very multifaceted and dynamic with variations in different healthy persons (Human Microbiome Project Consortium, 2012). Interestingly, microbiome composition can change in the same individual when measured over a long period of time. These changes in microbiome composition may be due largely to disease conditions or environmental influence (David et al., 2014; Flores et al., 2014; Fukuyama et al., 2017). Young and older adults have been reported to have differences in microbiome composition due to its continuing alterations with time (O'Toole and Jeffery, 2015). Genetic and dietary manipulation have been reported to cause alteration in microbiome composition (Maier et al., 2017) as such microbiome composition can serve as a predictor or biomarkers of disorders or impending disease conditions. Korem et al. demonstrated that microbiome composition can serve as a predictor of biomarkers of systemic glucoregulation (Korem et al., 2017). In addition, microbiome composition across various parts of the gut such as small intestine and colon experience some physiological alterations, which may be due to immunological activity of the host, nutritional, and biochemical gradients (Donaldson et al., 2016).

The relevance of the microbiome in the regulation of physiological activities in the human body is tremendous. Therefore, analysis and explanation of the connection of the gut microbiome and its changes to physiological dysregulation resulting in diseases such as metabolic, neurological, gastrointestinal, and cardiovascular disorders is warranted.

General overview

A variety of procedures are currently in use for sampling microbiome. This includes fecal sampling, samples from endoscopy, samples from aspirated intestinal fluid, surgical samples, and use of ingestible sampling technique. Fecal sampling is usually used to obtain information about gut microbiota. Its noninvasive nature allows it to be used frequently (Tang et al., 2020).

There are emerging evidence that variation exists between microbial composition in intestinal mucosa and feces. Fecal samples do not represent both the composition and metagenomic utility of mucosa-related microbiota dispersed along the intestinal wall (Zmora et al., 2018). Additionally, fecal microbiota are not evenly distributed in feces (Swidsinski et al., 2008). For small-scale sampling, the gold standard for profiling gut microbiota is fecal material frozen at $-80°C$. This is because microbial integrity is maintained at this temperature in the absence of preservatives (Fouhy et al., 2015).

Certain methods have been developed for large-scale sampling. This involves appropriate sample collection, storage, and transportation. Sample storage and transportation at $4°C$ has been reported to significantly reduce alteration in microbial composition in the absence of ultralow temperature storage facility (Choo et al., 2015). There are other effective methods that do not require the use of preservatives. These include OMNIgene gut kit, fecal occult blood test cards, and FTA cards, which have been reported to maintain microbial integrity for some days at room temperature (Tang et al., 2020).

Gut microbiome sampling can also be done using endoscopy. The use of endoscopic biopsy allows assessment of mucosal microbiota at diverse anatomical locations of the gut (Tang et al., 2020). Sampling using this technique has a number of demerits including its invasive nature, which is unpleasant to patients; the gut microflora may be affected by bowel preparation, which could lead to contamination and the limitation of the technique in reaching certain areas of the small intestine, particularly the distal small intestine (Tang et al., 2020).

The use of protected specimen brush (PSB) is another sampling technique for gut microbiome. It is a germ-free disposable sheath brush. There are indications that this technique offers a better representative sample of the mucosal surface than mucosa biopsy, which is characterized with bleeding and risk of infection. Another advantage it has is its ability to provide large ratio of bacterial to host DNA (Huse et al., 2014). Just like biopsies, it is invasive and prone to contamination.

The use of laser capture microdissection offers better outcomes in comparison with other tissue microdissection methods. It offers a better and more efficient way of obtaining mucosal bacteria in other to assess their interactions with the host (Tang et al., 2020). This technique is, however, tedious; this is a major constraint in using it for large-scale investigations.

Surgery is another gut microbiome technique. This includes biopsy of mucosal samples and needle aspiration. It allows the sampling of the distal ileum. Samples taken by surgery are not prone to contamination and may represent the gut microbiota. It must also be known that presurgical preparations including abstinence from food, use of antibiotics, and bowel cleansing could alter gut microbiota (Tang et al., 2020). Obtaining samples from healthy patients using this technique appears impossible; it is also not suitable to determine the association between bacterial flora and diseases in diverse population (Tang et al., 2020).

The development of ingestible sampling devices is currently being explored. Some of the ingestible systems have not proven to be effective because of their susceptibility of samples to contamination with intestinal fluid from areas where sampling is not required.

Sampling of microbiomes

The microbiome of either plants or animals in any particular environment has a lot of importance; it is closely related to the well-being of the main organism. The biota can either aid nutrient breakdown and easy adsorption or can result in various disease conditions. During diagnosis also, the microbiome of a particular tissue can give a better understanding of possible causes of disease and management strategies. It has been observed that clinical diagnosis of several diseases using the microbiota and subsequent treatment is quite difficult to use as the sole guide to the management or treatment of such disease because the microbial population analysis relies on not only advances in high-throughput DNA sequencing methods as reported by Bassiouni et al. (2015) but also greatly the sampling methods, which influence the prevalence of microbes to present per sample obtained (Tang et al., 2020). Effective sampling methods for microbiome research are expected to be based on scientific questions to be addressed, and the pain threshold when collecting such sample.

Over the years, sample collection methods frequently used for microbiome analysis can either be invasive or noninvasive. Either method has a significant role to play in the microbiota after sequencing. The common methods of swabs, fluids, and biopsy. Storage and means of transportation have been observed to play a significant role in the characteristics of the microbiota present at a given time with several reports on the use of different storage conditions and transportation conditions being reported to observe their impacts on the microbial population (Swidsinski et al., 2008; Choo et al., 2015; Fouhy et al., 2015; Tedjo et al., 2015; Song et al., 2016; Vogtmann et al., 2017; Wang et al., 2018; Tang et al., 2020).

In a 2017 study by Santigli et al., they reported that the microbiome of the oral cavity of children changes as they grow, and this microbial population can be attributed to several medical conditions across the age bracket of children (Santigli et al., 2017). The use of oral cavity biofilms for sampling for microbiome was said to be of great importance because they are made up of microbes that are both commensals and bene-ficial (Santigli et al., 2017). They further stated that the sampling methods can either be parallel where samples are collected from parallel cavities separately, which means samples are obtained using a single cavity at a time as the source of starting material, and finally, the sampling can be done in a combined mixture.

Johnson et al., (2019) in their study argued that despite the importance of the micro-biome to health, the microbial population at a given time does not depend on conven-tional nutrients present, but on the choice of food taken at a time. According to David et al. (2014), the composition of microorganisms present in the human gastrointestinal tract depends largely on the kind of diet taken as they observed that the composition has large variations when switching from foods that are plant-based and animal-based. However, Flores et al. (2014) and Fukuyama et al., (2017) suggested sampling based on environmental change and conditions. Johnson et al., (2019) further stated that the best way of sampling is daily sample collections, which are usual factors in the environ-ment and dietary change on daily bases.

In microbiome sampling, the sample collection method techniques are an important factor to consider because it sometimes determines the quality of microbial biomass to be obtained (Kong et al., 2017). An important factor that also should be considered is that any method used should have a low probability of cross-contamination (Tang et al., 2020).

Although sample collection methods, genomic DNA isolation methods, purification, preparation of the libraries, sequencing, and genomic data analysis contribute to the composition of the microbiota, variations observed in microbial communities/composi-tion observed during metagenomics can also depend on other factors. This showed that standardizing sample collections can be challenging, as storage conditions, temperature, durations, and collection may be also influenced by environmental factors (Flores et al., 2015; Vogtmann et al., 2017; Poulsen et al., 2021).

Poulsen et al. (2021) observed that although sample storage in preservatives is some-what usual practice, it might actually interfere with the result of other analyses such as metabolomics, transcriptomics, and proteomics. However, deep storage in -20°C or -80°C for biospecimen is encouraged.

Ingala et al. (2018) compared fecal sampling with intestinal sampling in bats and observed sample variation between the sampling methods. They also inferred that sam-ples from the sampling methods are not interchangeable and that each gives distinct in-formation regarding the host.

Zmora et al. (2018) showed that fecal sample does not represent the composition and metagenomic utility of membrane-associated microbiota in various sites of the intestine. An association between stool consistency and microbiota richness has been reported. This association also relates to community composition and abundance of specific enterotypes (Vandeputte et al., 2016).

In a multiple sampling carried out by Wu et al. (2010), about 35% of low-abundance microbial taxa that made up of between 0.2% and 0.4% of total microbiome in a replicate were absent in second fecal sample. The detection of microbial taxa by qPCR was uneven in a study involving subsampling of fecal sample (Gorzelak et al., 2015).

A study that focused on the upper respiratory tract microbiome was carried out (Kumpitsch et al., 2019). The authors recommended recent technologies including Oxford Nanopore, Pacific Bioscience technology to analyze sinonasal microbiome (Earl et al., 2018). They further suggested the use of shotgun metagenomics, which gives details of mycobiome, virobiome, and archaeome in the upper respiratory tract.

The metatranscriptome of oral microbiome was assessed. This is an opportunity for the recognition of the modulation of oral microbiota by certain environmental signals. In dental carries, sugar metabolism distinguishes between health and disease, and more recently, sorbitol supplies carbon by *Streptococcus mutans* in dental carries (Solbiati et al., 2018).

A study aimed at determining the impact of sampling device and anatomical sampling site on outcomes regarding vaginal microbiota was conducted using 16S rRNA gene analysis. It was observed that DNA yield depends strongly on the sampling device relative to the anatomical site. Varying vaginal microbiota composition was also observed (Virtanen et al., 2017). In another study, it was observed that alterations to cystic fibrosis lung microbiome occur outside of acute pulmonary episodes and are specific to patients (Whelan et al., 2017).

In a study, the microbiota of lower urinary tract of apparently healthy premenopausal women was determined. This was via midstream-voided urine samples with the intention to establish the relationship between microbial dynamics and personal factors. Lower urinary tract microbiota correlated with symptoms of lower urinary tract disorders (Price et al., 2020).

Emerging evidence suggest a link between human intestinal microbiota alteration and systemic cancer treatment. The different kinds of cancer treatment, hormonal therapy, chemotherapy, and immunotherapy are reported to alter gut microbiota (Aarnoutse et al., 2019).

Fecal microbiome sample collection

Oftentimes, the fecal samples are used as substitutions for gut microbiomes; the fecal samples are noninvasive, naturally collected, and can be sampled repetitively and thus serve as a source of specimens employed in the assessment of most microbiomes in the intestine (Tang et al., 2020). Carroll et al. (2010) reported that there seems to be a significant distinction in the microbial composition of the feces and the mucosa. Also, it was demonstrated that there is a distinct niche between fecal and mucosal-associated microbiomes (Tap et al., 2017). However, feces were considered a substitute of the gastrointestinal lumen contents, but its constituents revealed it is directly linked with the mucosa. Zmora et al. (2018) established that fecal specimens are not indicators of the composition and metagenomic function existing between the association of microbiota and the mucosa lining of numerous sites of the intestine.

Moreover, the involvement of the next-generation sequencing (NGS) technology helps to broaden the understating of the gut microbiome. However, the sequencing technologies employed are based on the specimen of feces collected, intestinal fluid, etc. The current method of sampling by proxy suggests that fecal samples could be used for intestinal microbiota, but for patients, the biopsies are invasive methods and are not suitable. Moreover, Tang et al. (2020) reported that there are numerous modes of obtaining gut microbiome samples through endoscopy; this includes biopsy (it gives insight to composition of mucosal microbiota in several anatomical site of the gastrointestinal tract, which can be done via Brisbane Aseptic Biopsy Device), luminal brushing (this involves the use of protected sterile brush, which is plugged to the distal top of a sheath and sealed when retracted or inserted via the colonoscopic channel), and laser capture microdissection (this method selectively attached the materials of interest over the tissue section via infrared laser pulse to a tiny transparent film) (Emmert-Buck et al., 1996). Reports have also shown that fecal samples can be gotten from aspirated intestinal fluid through an invented Shiner stainless steel that is fitted using a cap at distal end with hollow connection toward the center (Shiner, 1963). Fecal samples are also collected from surgery via mucosal biopsy of sample or through direct needle aspiration (Lavelle et al., 2015).

Fecal microbiome sampling can be done using stool collection, rectal swab, and glove tip method. Stool collection is the gold standard for fecal sample collection. When stool sample is collected, they have to be analyzed and used immediately or stored frozen quickly such that microbiome composition in the sample is not altered. Short and co-workers reported that microbiome samples collected using the glove tip method are not different from those collected with other techniques, basically stool collection and rectal swab microbiome sample collection. The glove tip method of fecal microbiome sampling is very easy to use especially within office setting. Therefore, this sampling technique interestingly will positively impact microbiome research and clinical practice due to the ease of usage and practicability of implementing the sample collection process (Short et al., 2021).

Fecal microbiome sampling using stool collection technique in microbiome-based research may sometimes be unattractive yet unavoidable. Hence, provision of hygienic and easy sampling method with assurance of improvement in the integrity of sample would enhance compliance by participants. Microbiome composition alteration has been reported when using some commercially available tubes for stool collection, even though they provided simple collection and little period of storage in atmospheric temperature for approximately 14 days (Penington et al., 2018). It is pertinent to note the variability in microbiome across different individuals when fecal microbiome sample is collected from day to day or over 1 week period suggesting that the time and periods of sample collection may affect the accuracy or alteration in microbiome composition (Caporaso et al., 2011; Flores et al., 2014).

Jones et al. recommended homogenization of sample before analysis or subsample collection due to variability in microbiome composition and microbiome-derived metabolites they observed from spot stool sample collection. When stool samples are collected twice within 25 hours, alterations in microbiome composition were observed. Therefore, to ensure reliability, consistency, and reproducibility of microbiome-related studies especially for translational purposes, sample collection and storage must be done in such a way as to avoid microbiome compositional alterations (Jones et al., 2021). Therefore, fecal sample collection should be done following the first movement of a full bowel in a day as well as immediate freezing of sample at −80°C after collection is advised. This approach will be very useful for microbiome-related studies with multiple analyses (Jones et al., 2021).

Endoscopy

Microbiome sample collection can be done using endoscopic procedures. This method is invasive and causes discomfort to individuals undergoing the procedure, but it gives more widespread microbiome composition information. Endoscopy is associated with many challenges such as contamination of samples, discomfort, and inability to sample across the whole areas of the gut. Endoscopy makes use of the following tools for sample collection; luminal brushes, and mucosal biopsy. Mucosal biopsy is very efficient for sampling of microbiome in distinct areas of the gut such as the colon, cecum, and small intestine. The microbiota located in the mucosal are unique and have very crucial roles in the regulation of the host immunological processes because they are in constant contact with the lymphoid tissues of the intestine (Heinsen et al., 2015).

Endoscopy procedure for sampling microbiome is associated with contamination, and this is a major concern because, before sampling, endoscopy tube may be contaminated especially during its passage either through the oral cavity or anus due to attraction of microbes from other areas of the gut to the site of sample collection. Mucosal biopsy sample collection method is limited because it cannot cover all sections of the gut. Large sample collection consisting of proteins, ribonucleic acid, and deoxyribonucleic acid is needed for the next-generation sequencing approach for characterization of microbiome, and mucosal biopsy cannot provide the needed quantity of microbiome samples for the metagenomic and metabolomics sequencing analyses due to its limitation or restriction to a certain site of sample collection within the gut (Huse et al., 2014). Therefore, Watt and coworkers reported another sampling technique called colonic lavage that is capable of collection of large microbiome samples from the gut with enough genetic materials for the metagenomic analysis (Watt et al., 2016).

Colonic lavage fluid

Colonic microbiota was surveyed by a group of researchers to examine its link with the host and the effect of diseases such as obesity, inflammatory bowel disease, pouchitis, and antibiotic-associated colitis (Manichanh et al., 2006; Ley et al., 2005; Komanduri et al., 2007; Young and Schmidt, 2004; Mai et al., 2006). Almost all studies of the mucosal-associated gut microbiota have sampled the colonic microbiota of cleansed/purged bowel after laxation, this was used for colonoscopic examination of the colon. Several studies conducted have considered the confounding potential of bowel cleansing gut microbiota preparation that could result in misleading or artifactual information of colonic microbiota at the natural state.

Mai et al. (2006) address the potentials of bowel preparation and colonoscopy on the intestinal microbiota. The authors investigated changes in microbiota fecal in about five persons under colonoscopy screening. However, about three subjects revealed that the stool samples collected after microbial profiles via denaturation of the gel electrophoresis gradient have different colonoscopy when compared with that collected before. This signifies that bowel preparations cause a critical consequence on the fecal gut microbiota.

Gut microbiota has been considered the special communities with several microbe populations that work together bringing about key physiological functions (Harrell et al., 2012). Eckburg et al. (2005) reported that gut microbiota is, however, considered for the study of microbial species available in the lumen and colonic mucosa supporter.

Zoetendal et al. (1998) documented that alteration in diet and luminal content contributes to luminal microbiota fluctuation, while mucosa-associated microbiota maintain stability in individuals for a lifetime. Moreover, attachment of the microbes to the mucosa influences the stability of microbiota and as such established a niche via biofilms formation and creation in the microbial communities through expansion inhibition. Mucosa microbes associate itself to the host to form a stable and intimate association, which influences the physiology of the host and the disease development critically (Backhed et al., 2005).

Harrell et al. (2012) sampled microbial communities for mucosal—host association in their natural state because it is paramount to understanding the health and disease relationship of host microbial and enteric microbiota in the human colon. It was observed that the lavage solution seems to affect overlying microbes mucosa-associated thereby affecting the colonic epithelium mucus integrity and abundance. Report also reveals that mucus changes automatically affect the structure and diversity of ecosystems associated a microbe that relies on it for nutrient source, attachment, and stability (Harrel et al., 2012). Authors also documented that communities of mucosa association are not discrete in the distal colon of human when compared with colon of mouse. It was concluded in the authors' report that colonic lavage may impact significantly the distal human colon microbiota of the associated mucosa.

Swallowable sampling devices

Many ingestible devices have been developed by researchers for the purpose of microbiome sample collection and site-specific delivery of drugs. Ingestible devices are noninvasive, and they are becoming popular collection of microbiome sample from the gut.
Cui et al. developed a swallowable capsule using microelectromechanical systems for
collection of sample from the intestine and drug delivery (Cui et al., 2008). The ingestible
device with its encapsulated feature uses wireless communication system for enhancement of large microbiome sample collection and passage through the gut making seam-
less contact with the intestine resulting in collection of fluids from the intestine, but the
downside to this is that the device becomes contaminated by the liquid below the area of
sample collection.

In addition, some smart swallowable capsule with inbuilt temperature sensor, pH,
batteries, motors, communication units, and microcomputers. This device is able to
collect sample from the small intestine. The capsule possesses the capacity for microbiome
preservation in a qualitative and quantitative manner especially through its gut pH measurement activity (Koziolek et al., 2015). The mechanism used by the swallowable
capsule at desired section of the gut for the collection of intestinal fluids is aspiration.
The fluid collected from the intestine through aspiration by the capsule is expelled
from the body along with the swallowable device.

Conclusion

Reliable, adequate, accurate, and efficient microbiome sampling methods are needed for
detailed sample collection in microbiome-related studies to give exact picture of the
contribution of the various microbial communities in disease pathogenesis and regulation
of host physiological processes. Sampling technique that does not inflict pain or cause
discomfort and psychological distress will encourage compliance. Development of effective sampling method is crucial to providing information about the microbiome—host
interactions, which will advance microbiome research tremendously.

References

Aarnoutse, R., Ziemons, J., Penders, J., Rensen, S.S., de Vos-Geelen, J., Smidt, M.L., 2019. The clinical link
between human intestinal microbiota and systemic cancer therapy. Int. J. Mol. Sci. 20, 4145. https://
doi.org/10.3390/ijms20174145.
Backhed, F., Ley, R.E., Sonnenburg, J.L., Peterson, D.A., Gordon, J.I., 2005. Host bacterial mutualism in
the human intestine. Science 307, 1915—1920.
Bassiouni, A., Cleland, E.J., Psaltis, A.J., Vreugde, S., Wormald, P.J., 2015. Sinonasal microbiome sampling:
a comparison of techniques. PLoS One 10 (4), e0123216.
Berg, G., Rybakova, D., Fischer, D., Cernava, T., Vergès, M.C., Charles, T., Chen, X., Cocolin, L.,
Eversole, K., Corral, G.H., Kazou, M., Kinkel, L., Lange, L., Lima, N., Loy, A., Macklin, J.A.,
Maguin, E., Mauchline, T., McClure, R., Mitter, B., Ryan, M., Sarand, I., Smidt, H., Schelkle, B.,

Roume, H., Kiran, G.S., Selvin, J., Souza, R.S.C., van Overbeek, L., Singh, B.K., Wagner, M., Walsh, A., Sessitsch, A., Schloter, M., June 30, 2020. Microbiome definition re-visited: old concepts and new challenges. Microbiome 8 (1), 103. https://doi.org/10.1186/s40168-020-00875-0.

Bharti, R., Grimm, D.G., January 18, 2021. Current challenges and best-practice protocols for microbiome analysis. Briefings Bioinf. 22 (1), 178–193. https://doi.org/10.1093/bib/bbz155.

Bokulich, N.A., Ziemski, M., Robeson 2nd, M.S., Kaehler, B.D., December 3, 2020. Measuring the microbiome: best practices for developing and benchmarking microbiomics methods. Comput. Struct. Biotechnol. J. 18, 4048–4062. https://doi.org/10.1016/j.csbj.2020.11.049.

Caporaso, J.G., Lauber, C.L., Costello, E.K., Berg-Lyons, D., Gonzalez, A., Stombaugh, J., Knights, D., Gajer, P., Ravel, J., Fierer, N., Gordon, J.I., Knight, R., 2011. Moving pictures of the human microbiome. Genome Biol. 12 (5), R50. https://doi.org/10.1186/gb-2011-12-5-r50.

Carroll, I.M., Chang, Y.-H., Park, J., Sartor, R.B., Ringel, Y., 2010. Luminal and mucosal-associated intestinal microbiota in patients with diarrhea-predominant irritable bowel syndrome. Gut Pathog. 2, 19. https://doi.org/10.1186/1757-4749-2-19.

Choo, J.M., Leong, L.E.X., Rogers, G.B., 2015. Sample storage conditions significantly influence faecal microbiome profiles. Sci. Rep. 5, 1–10. https://doi.org/10.1038/srep16350.

Costea, P.I., Zeller, G., Sunagawa, S., Pelletier, E., Alberti, A., Levenez, F., Tramontano, M., Driessen, M., Hercog, R., Jung, F.E., Kultima, J.R., Hayward, M.R., Coelho, L.P., Allen-Vercoe, E., Bertrand, L., Blaut, M., Brown, J.R.M., Carton, T., Cools-Portier, S., Daigneault, M., Derrien, M., Druesne, A., de Vos, W.M., Finlay, B.B., Flint, H.J., Guarner, F., Hattori, M., Heilig, H., Luna, R.A., van Hylckama Vlieg, J., Junick, J., Klymiuk, I., Langella, P., Le Chatelier, E., Mai, V., Manichanh, C., Martin, J.C., Mery, C., Morita, H., O'Toole, P.W., Orvain, C., Patil, K.R., Penders, J., Persson, S., Pons, N., Popova, M., Salonen, A., Saulnier, D., Scott, K.P., Singh, B., Slezak, K., Veiga, P., Versalovic, J., Zhao, L., Zoetendal, E.G., Ehrlich, S.D., Dore, J., Bork, P., November 2017. Towards standards for human fecal sample processing in metagenomic studies. Nat. Biotechnol. 35 (11), 1069–1076. https://doi.org/10.1038/nbt.3960.

Cui, J., Zheng, X., Hou, W., Zhuang, Y., Pi, X., Yang, J., 2008. The study of a remote-controlled gastrointestinal drug delivery and sampling system. Telemed. J. E. Health. 14, 715–719. https://doi.org/10.1089/tmj.2007.0118.

Cullen, C.M., Aneja, K.K., Beyhan, S., Cho, C.E., Woloszynek, S., Convertino, M., McCoy, S.J., Zhang, Y., Anderson, M.Z., Alvarez-Ponce, D., Smirnova, E., Karstens, L., Dorrestein, P.C., Li, H., Sen Gupta, A., Cheung, K., Powers, J.G., Zhao, Z., Rosen, G.L., February 19, 2020. Emerging priorities for microbiome research. Front. Microbiol. 11, 136. https://doi.org/10.3389/fmicb.2020.00136 eCollection 2020.

David, L.A., Maurice, C.F., Carmody, R.N., Gootenberg, D.B., Button, J.E., Wolfe, B.E., Ling, A.V., Devlin, A.S., Varma, Y., Fischbach, M.A., Biddinger, S.B., January 2014. Diet rapidly and reproducibly alters the human gut microbiome. Nature 505 (7484), 559–563.

Donaldson, G.P., Lee, S.M., Mazmanian, S.K., 2016. Gut biogeography of the bacterial microbiota. Nat. Rev. Microbiol. 14, 20–32.

Earl, J., Adappa, N., Krol, J., Bhat, A., Balashov, S., Ehrlich, R., et al., 2018. Species-level bacterial community profiling of the healthy sinonasal microbiome using Pacific Biosciences sequencing of full-length 16S rRNA genes. Microbiome 6, 190.

Eckburg, P.B., Bik, E.M., Bernstein, C.N., Purdom, E., Dethlefsen, L., et al., 2005. Diversity of the human intestinal microbial flora. Science 308, 1635–1638.

Emmert-Buck, M.R., Bonner, R.F., Smith, P.D., Chuaqui, R.F., Zhuang, Z., Goldstein, S.R., et al., 1996. Laser capture microdissection. Science 274, 998–1001. https://doi.org/10.1126/science.274.5289.998.

Esiobu, N.D., Ogbonna, J.C., Adetunji, C.O., Obembe, O.O., Ezeonu, I.M., Ibrahim, A.B., Ubi, B.E., 2022. Microbiomes and Emerging Applications, first ed. Imprint CRC Press, Boca Raton, ISBN 9781003180241, p. 186. https://doi.org/10.1201/9781003180241. First Published 2022. eBook Published 11 May 2022. eBook ISBN 9781003180241. Subjects Bioscience, Engineering and Technology.

Flores, G.E., Caporaso, J.G., Henley, J.B., Rideout, J.R., Domogala, D., Chase, J., Leff, J.W., Vázquez-Baeza, Y., Gonzalez, A., Knight, R., Dunn, R.R., December 2014. Temporal variability is a personalized feature of the human microbiome. Genome Biol. 15 (12), 1–3.

Flores, R., Shi, J., Yu, G., Ma, B., Ravel, J., Goedert, J.J., Sinha, R., 2015. Collection media and delayed freezing effects on microbial composition of human stool. Microbiome 3, 33.

Foster, J.A., McVey Neufeld, K.-A., 2013. Gut-brain axis: how the microbiome influences anxiety and depression. Trends Neurosci. 36, 305–312 (CrossRef) (PubMed).

Fouhy, F., Deane, J., Rea, M.C., O'Sullivan, Ó., Ross, R.P., O'Callaghan, G., et al., 2015. The effects of freezing on faecal microbiota as determined using MiSeq sequencing and culture-based investigations. PLoS One 10, e0119355. https://doi.org/10.1371/journal.pone.0119355.

Fukuyama, J., Rumker, L., Sankaran, K., Jeganathan, P., Dethlefsen, L., Relman, D.A., Holmes, S.P., 2017. Multidomain analyses of a longitudinal human microbiome intestinal cleanout perturbation experiment. PLoS Comput. Biol. 13 (8), e1005706.

Goodrich, J.K., Waters, J.L., Poole, A.C., Sutter, J.L., Koren, O., Blekhman, R., Beaumont, M., Van Treuren, W., Knight, R., Bell, J.T., Spector, T.D., Clark, A.G., Ley, R.E., November 6, 2014. Human genetics shape the gut microbiome. Cell 159 (4), 789–799. https://doi.org/10.1016/j.cell.2014.09.053.

Gorzelak, M.A., Gill, S.K., Tasnim, N., Ahmadi-Vand, Z., Jay, M., Gibson, D.L., 2015. Methods for improving human gut microbiome data by reducing variability through sample processing and storage of stool. PLoS One 10, e0134802. https://doi.org/10.1371/journal.pone.0134802.

Harrell, L., Wang, Y., Antonopoulos, D., Young, V., Lichtenstein, L., et al., 2012. Standard colonic lavage alters the natural state of mucosal-associated microbiota in the human colon. PLoS One 7 (2), e32545. https://doi.org/10.1371/journal.pone.0032545.

Heinsen, F.-A., Knecht, H., Neulinger, S.C., Schmitz, R.A., Knecht, C., Kühbacher, T., et al., 2015. Dynamic changes of the luminal and mucosa-associated gut microbiota during and after antibiotic therapy with paromomycin. Gut Microb. 6, 243–254. https://doi.org/10.1080/19490976.2015.1062959.

Human Microbiome Project Consortium, 2012. Structure, function and diversity of the healthy human microbiome. Nature 486, 207–214.

Huse, S.M., Young, V.B., Morrison, H.G., Antonopoulos, D.A., Kwon, J., Dalal, S., et al., 2014. Comparison of brush and biopsy sampling methods of the ileal pouch for assessment of mucosa-associated microbiota of human subjects. Microbiome 2, 5. https://doi.org/10.1186/2049-2618-2-5.

Ingala, M.R., Simmons, N.B., Wultsch, C., Krampis, K., Speer, K.A., Perkins, S.L., 2018. Comparing microbiome sampling methods in a wild mammal: fecal and intestinal samples record different signals of host ecology, evolution. Front. Microbiol. 9, 803. https://doi.org/10.3389/fmicb.2018.00803.

Integrative HMP (iHMP) Research Network Consortium, September 10, 2014. The Integrative Human Microbiome Project: dynamic analysis of microbiome-host omics profiles during periods of human health and disease. Cell Host Microbe 16 (3), 276–289. https://doi.org/10.1016/j.chom.2014.08.014.

Johnson, A.J., Vangay, P., Al-Ghalith, G.A., Hillmann, B.M., Ward, T.L., Shields-Cutler, R.R., Kim, A.D., Shmagel, A.K., Syed, A.N., Students, P.M., Walter, J., 2019. Daily sampling reveals personalized diet-microbiome associations in humans. Cell Host Microbe 25 (6), 789–802, 2019 Jun 12.

Jones, J., Reinke, S.N., Ali, A., Palmer, D.J., Christophersen, C.T., July 7, 2021. Fecal sample collection methods and time of day impact microbiome composition and short chain fatty acid concentrations. Sci. Rep. 11 (1), 13964. https://doi.org/10.1038/s41598-021-93031-z.

Kennedy, M.S., Chang, E.B., 2020. The microbiome: composition and locations. Prog. Mol. Biol. Transl. Sci. 176, 1–42. https://doi.org/10.1016/bs.pmbts.2020.08.013.

Kim, D., Jung, J.Y., Oh, H.S., Jee, S.R., Park, S.J., Lee, S.H., Yoon, J.S., Yu, S.J., Yoon, I.C., Lee, H.S., October 22, 2021. Comparison of sampling methods in assessing the microbiome from patients with ulcerative colitis. BMC Gastroenterol. 21 (1), 396. https://doi.org/10.1186/s12876-021-01975-3.

Komanduri, S., Gillevet, P.M., Sikaroodi, M., Mutlu, E., Keshavarzian, A., 2007. Dysbiosis in pouchitis: evidence of unique microfloral patterns in pouch inflammation. Clin. Gastroenterol. Hepatol. 5, 352–360.

Kong, H.H., Andersson, B., Clavel, T., Common, J.E., Jackson, S.A., Olson, N.D., Segre, J.A., Traidl-Hoffmann, C., March 1, 2017. Performing skin microbiome research: a method to the madness. J. Invest. Dermatol. 137 (3), 561–568.

Korem, T., Zeevi, D., Zmora, N., Weissbrod, O., Bar, N., Lotan-Pompan, M., Avnit-Sagi, T., Kosower, N., Malka, G., Rein, M., et al., 2017. Bread affects clinical parameters and induces gut microbiome-associated personal glycemic responses. Cell Metabol. 25, 1243–1253.

Koziolek, M., Grimm, M., Becker, D., Iordanov, V., Zou, H., Shimizu, J., et al., 2015. Investigation of pH and temperature profiles in the GI tract of fasted human subjects using the intellicap® system. J. Pharmaceut. Sci. 104, 2855–2863. https://doi.org/10.1002/jps.24274.

Kumpitsch, C., Koskinen, K., Schöp, V., Moissl-Eichinger, C., 2019. The microbiome of the upper respiratory tract in health and disease. BMC Biol. 17, 87. https://doi.org/10.1186/s12915-019-0703-z.

Lau, J.T., Whelan, F.J., Herath, I., Lee, C.H., Collins, S.M., Bercik, P., Surette, M.G., July 1, 2016. Capturing the diversity of the human gut microbiota through culture-enriched molecular profiling. Genome Med. 8 (1), 72. https://doi.org/10.1186/s13073-016-0327-7.

Lavelle, A., Lennon, G., O'Sullivan, O., Docherty, N., Balfe, A., Maguire, A., Mulcahy, H.E., Doherty, G., O'Donoghue, D., Hyland, J., Ross, R.P., Coffey, J.C., Sheahan, K., Cotter, P.D., Shanahan, F., Winter, D.C., O'Connell, P.R., October 2015. Spatial variation of the colonic microbiota in patients with ulcerative colitis and control volunteers. Gut 64 (10), 1553–1561. https://doi.org/10.1136/gutjnl-2014-307873.

Lax, S., Smith, D.P., Hampton-Marcell, J., Owens, S.M., Handley, K.M., Scott, N.M., Gibbons, S.M., Larsen, P., Shogan, B.D., Weiss, S., Metcalf, J.L., Ursell, L.K., Vázquez-Baeza, Y., Van Treuren, W., Hasan, N.A., Gibson, M.K., Colwell, R., Dantas, G., Knight, R., Gilbert, J.A., August 29, 2014. Longitudinal analysis of microbial interaction between humans and the indoor environment. Science 345 (6200), 1048–1052. https://doi.org/10.1126/science.1254529.

Ley, R.E., Backhed, F., Turnbaugh, P., Lozupone, C.A., Knight, R.D., et al., 2005. Obesity alters gut microbial ecology. Proc. Natl. Acad. Sci. U. S. A. 102, 11070–11075.

Liu, Z., Ma, A., Mathé, E., Merling, M., Ma, Q., Liu, B., March 22, 2021. Network analyses in microbiome based on high-throughput multi-omics data. Briefings Bioinf. 22 (2), 1639–1655. https://doi.org/10.1093/bib/bbaa005.

Lynch, S.V., Pedersen, O.N., December 15, 2016. The human intestinal microbiome in health and disease. Engl. J. Med. 375 (24), 2369–2379. https://doi.org/10.1056/NEJMra1600266.

Mai, V., Greenwald, B., Morris Jr., J.G., Raufman, J.P., Stine, O.C., 2006. Effect of bowel preparation and colonoscopy on post-procedure intestinal microbiota composition. Gut 55, 1822–1823.

Maier, T.V., Lucio, M., Lee, L.H., VerBerkmoes, N.C., Brislawn, C.J., Bernhardt, J., Lamendella, R., McDermott, J.E., Bergeron, N., Heinzmann, S.S., Morton, J.T., González, A., Ackermann, G., Knight, R., Riedel, K., Krauss, R.M., Schmitt-Kopplin, P., Jansson, J.K., October 17, 2017. Impact of dietary resistant starch on the human gut microbiome, metaproteome, and metabolome. mBio 8 (5), e01343-17. https://doi.org/10.1128/mBio.01343-17.

Mallick, H., Franzosa, E.A., McIver, L.J., Banerjee, S., Sirota-Madi, A., Kostic, A.D., Clish, C.B., Vlamakis, H., Xavier, R.J., Huttenhower, C., July 17, 2019. Predictive metabolomics profiling of microbial communities using amplicon or metagenomic sequences. Nat. Commun. 10 (1), 3136. https://doi.org/10.1038/s41467-019-10927-1.

Manichanh, C., Rigottier-Gois, L., Bonnaud, E., Gloux, K., Pelletier, E., et al., 2006. Reduced diversity of faecal microbiota in Crohn's disease revealed by a metagenomic approach. Gut 55, 205–211.

Michael, O.S., Oluranti, O.I., Oshinjo, A.M., Adetunji, C.O., Olaniyi, K.S., Adetunji, J.B., 2022a. Microbiome Reshaping and Epigenetic Regulation, first ed.st Edition. Imprint CRC Press, ISBN 9781003180241, p. 22. https://doi.org/10.1201/9781003180241-6. First Published 2022.

Michael, O.S., Oluranti, O.I., Oshinjo, A.M., Adetunji, C.O., Adetunji, J.B., Esiobu, N.D., 2022b. Microbiota transplantation, health implications, and the way forward. In: Microbiomes and Emerging Applications, first ed.st Edition. Imprint CRC Press, ISBN 9781003180241, p. 19. https://doi.org/10.1201/9781003180241-5. First Published 2022.

Moglia, A., Menciassi, A., Dario, P., Cuschieri, A., 2009. Capsule endoscopy: progress update and challenges ahead. Nat. Rev. Gastroenterol. Hepatol. 6, 353–361. https://doi.org/10.1038/nrgastro.2009.69.

Morgan, X.C., Kabakchiev, B., Waldron, L., Tyler, A.D., Tickle, T.L., Milgrom, R., et al., 2012. Dysfunction of the intestinal microbiome in inflammatory bowel disease and treatment. Genome Biol. 13, R79. https://doi.org/10.1186/gb-2012-13-9-r79.

O'Donnell, J.A., Zheng, T., Meric, G., Marques, F.Z., March 2023. The gut microbiome and hypertension. Nat. Rev. Nephrol. 19 (3), 153–167. https://doi.org/10.1038/s41581-022-00654-0.

O'Toole, P.W., Jeffery, I.B., 2015. Gut microbiota and aging. Science 350, 1214–1215.

Penington, J.S., Penno, M.A.S., Ngui, K.M., Ajami, N.J., Roth-Schulze, A.J., Wilcox, S.A., Bandala-Sanchez, E., Wentworth, J.M., Barry, S.C., Brown, C.Y., Couper, J.J., Petrosino, J.F., Papenfuss, A.T., Harrison, L.C., ENDIA Study Group, March 12, 2018. Influence of fecal collection conditions and 16S rRNA gene sequencing at two centers on human gut microbiota analysis. Sci. Rep. 8 (1), 4386. https://doi.org/10.1038/s41598-018-22491-7.

Poulsen, C.S., Kaas, R.S., Aarestrup, F.M., Pamp, S.J., 2021. Standard sample storage conditions have an impact on inferred microbiome composition and antimicrobial resistance patterns. Microbiol. Spectr. 9 (2), e01387-21.

Price, T.K., Wolff, B., Halverson, T., Limeira, R., Brubaker, L., Dong, Q., Mueller, E.R., Wolfe, A.J., 2020. Temporal dynamics of the adult female lower urinary tract microbiota. mBio 11, e00475-20. https://doi.org/10.1128/mBio.00475-20.

Rothschild, D., Weissbrod, O., Barkan, E., Kurilshikov, A., Korem, T., Zeevi, D., Costea, P.I., Godneva, A., Kalka, I.N., Bar, N., Shilo, S., Lador, D., Vila, A.V., Zmora, N., Pevsner-Fischer, M., Israeli, D., Kosower, N., Malka, G., Wolf, B.C., Avnit-Sagi, T., Lotan-Pompan, M., Weinberger, A., Halpern, Z., Carmi, S., Fu, J., Wijmenga, C., Zhernakova, A., Elinav, E., Segal, E., March 8, 2018. Environment dominates over host genetics in shaping human gut microbiota. Nature 555 (7695), 210–215. https://doi.org/10.1038/nature25973.

Santigli, E., Koller, M., Klug, B., 2017. Oral biofilm sampling for microbiome analysis in healthy children. JoVE 130, e56320.

Sharma, S., Tripathi, P., January 2019. Gut microbiome and type 2 diabetes: where we are and where to go? J. Nutr. Biochem. 63, 101–108. https://doi.org/10.1016/j.jnutbio.2018.10.003.

Shiner, M., 1963. A capsule for obtaining sterile samples of gastrointestinal fluids. Lancet 1, 532–533. https://doi.org/10.1016/s0140-6736(63)91328-x.

Short, M.I., Hudson, R., Besasie, B.D., Reveles, K.R., Shah, D.P., Nicholson, S., Johnson-Pais, T.L., Weldon, K., Lai, Z., Leach, R.J., Fongang, B., Liss, M.A., 2021. Comparison of rectal swab, glove tip, and participant-collected stool techniques for gut microbiome sampling. BMC Microbiol. 21 (1), 26. https://doi.org/10.1186/s12866-020-02080-3.

Shreiner, A.B., Kao, J.Y., Young, V.B., 2015. The gut microbiome in health and in disease. Curr. Opin. Gastroenterol. 31, 69 (CrossRef) (PubMed).

Solbiati, J., Frias-Lopez, J., 2018. Metatranscriptome of the oral microbiome in health and disease. J. Dent. Res. 97 (5), 492–500.

Song, S.J., Amir, A., Metcalf, J.L., Amato, K.R., Xu, Z.Z., Humphrey, G., Knight, R., 2016. Preservation methods differ in fecal microbiome stability, affecting suitability for field studies. mSystems 1 (3), 1–12. https://doi.org/10.1128/mSystems.00021-16.

Swidsinski, A., Loening-Baucke, V., Verstraelen, H., Osowska, S., Doerffel, Y., 2008. Biostructure of fecal microbiota in healthy subjects and patients with chronic idiopathic diarrhea. Gastroenterology 135, 568–579. https://doi.org/10.1053/j.gastro.2008.04.017.

Tang, Q., Jin, G., Wang, G., Liu, T., Liu, X., Wang, B., Cao, H., 2020. Current sampling methods for gut microbiota: a call for more precise devices. Front. Cell. Infect. Microbiol. 10, 151. https://doi.org/10.3389/fcimb.2020.00151.

Tap, J., Derrien, M., Törnblom, H., Brazeilles, R., Cools-Portier, S., Dor,é, J., et al., 2017. Identification of an intestinal microbiota signature associated with severity of irritable bowel syndrome. Gastroenterology 152, 111–123.e8. https://doi.org/10.1053/j.gastro.2016.09.049.

Tedjo, D.I., Jonkers, D.M.A.E., Savelkoul, P.H., Masclee, A.A., van Best, N., Pierik, M.J., et al., 2015. The effect of sampling and storage on the fecal microbiota composition in healthy and diseased subjects. PLoS One 10, e0126685. https://doi.org/10.1371/journal.pone.0126685.

Valdastri, P., Simi, M., Webster, R.J., 2012. Advanced technologies for gastrointestinal endoscopy. Annu. Rev. Biomed. Eng. 14, 397–429.

Vandeputte, D., Falony, G., Vieira-Silva, S., Tito, R.Y., Joossens, M., Raes, J., 2016. Stool consistency is strongly associated with gut microbiota richness and composition, enterotypes and bacterial growth rates. Gut 65, 57–62.

Virtanen, S., Kalliala, I., Nieminen, P., Salonen, A., 2017. Comparative analysis of vaginal microbiota sampling using 16S rRNA gene analysis. PLoS One 12 (7), e0181477. https://doi.org/10.1371/journal.pone.0181477.

Vogtmann, E., Chen, J., Kibriya, M.G., Chen, Y., Islam, T., Eunes, M., et al., 2017. Comparison of fecal collection methods for microbiota studies in Bangladesh. Appl. Environ. Microbiol. 83, 17. https://doi.org/10.1128/AEM.00361-17.

Wang, Z., Zolnik, C.P., Qiu, Y., Usyk, M., Wang, T., Strickler, H.D., et al., 2018. Comparison of fecal collection methods for microbiome and metabolomics studies. Front. Cell. Infect. Microbiol. 8, 301. https://doi.org/10.3389/fcimb.2018.00301.

Watt, E., Gemmell, M.R., Berry, S., Glaire, M., Farquharson, F., Louis, P., et al., 2016. Extending colonic mucosal microbiome analysis-assessment of colonic lavage as a proxy for endoscopic colonic biopsies. Microbiome 4, 61. https://doi.org/10.1186/s40168-016-0207-9.

Whelan, F.J., Heirali, A.A., Rossi, L., Rabin, H.R., Parkins, M.D., Surette, M.G., 2017. Longitudinal sampling of the lung microbiota in individuals with cystic fibrosis. PLoS One 12 (3), e0172811. https://doi.org/10.1371/journal.pone.0172811.

Wu, G.D., Lewis, J.D., Hoffmann, C., Chen, Y.-Y., Knight, R., Bittinger, K., et al., 2010. Sampling and pyrosequencing methods for characterizing bacterial communities in the human gut using 16S sequence tags. BMC Microbiol. 10, 206. https://doi.org/10.1186/1471-2180-10-206.

Yin, X., Altman, T., Rutherford, E., West, K.A., Wu, Y., Choi, J., Beck, P.L., Kaplan, G.G., Dabbagh, K., DeSantis, T.Z., Iwai, S., December 4, 2020. A comparative evaluation of tools to predict metabolite profiles from microbiome sequencing data. Front. Microbiol. 11, 595910. https://doi.org/10.3389/fmicb.2020.595910.

Young, V.B., Schmidt, T.M., 2004. Antibiotic-associated diarrhea accompanied by large-scale alterations in the composition of the fecal microbiota. J. Clin. Microbiol. 42, 1203–1206.

Zmora, N., Zilberman-Schapira, G., Suez, J., Mor, U., Dori-Bachash, M., Bashiardes, S., et al., 2018. Personalized gut mucosal colonization resistance to empiric probiotics is associated with unique host and microbiome features. Cell 174, 1388–1405.e21. https://doi.org/10.1016/j.cell.2018.08.041.

Zoetendal, E.G., Akkermans, A.D., De Vos, W.M., 1998. Temperature gradient gel electrophoresis analysis of 16S rRNA from human fecal samples reveals stable and host-specific communities of active bacteria. Appl. Environ. Microbiol. 64, 3854–3859.

CHAPTER 4

Microbiome characterization and identification: key emphasis on molecular approaches

Frank Abimbola Ogundolie[1], Charles Oluwaseun Adetunji[2], Olulope Olufemi Ajayi[3], Michael O. Okpara[4], Olugbenga Samuel Michael[5,6,10], Juliana Bunmi Adetunji[7], Ohunayo Adeniyi Success[8] and Oluwafemi Adebayo Oyewole[9]

[1]Department of Biotechnology, Baze University, Abuja, Nigeria; [2]Applied Microbiology, Biotechnology and Nanotechnology Laboratory, Department of Microbiology, and Directorate of Research and Innovation, Edo State University Uzairue, Iyamho, Auchi, Edo State, Nigeria; [3]Department of Biochemistry, Federal University of Technology, Akure, Nigeria; [4]Department of Science Laboratory Technology, Faculty of Science, Ekiti State University, Ado Ekiti, Nigeria; [5]Cardiometabolic, Microbiome and Applied Physiology Laboratory, Department of Physiology, College of Health Sciences, Bowen University, Iwo, Osun State, Nigeria; [6]Department of Physiology, University of Tennessee Health Science Center, Memphis, TN, United States; [7]Department of Biochemistry, Edo State University, Uzairue, Auchi, Edo State, Nigeria; [8]Nutritional and Toxicological Research Laboratory, Department of Biochemistry Sciences, Osun State University, Osogbo, Osun State, Nigeria; [9]Department of Microbiology, Federal University of Technology, Minna, Nigeria; [10]Department of Medical Pharmacology and Physiology, University of Missouri, Columbia, MO, United States

Introduction

The complexity of microbiota in human gut coupled with the challenges of growing a high quantity of different gut microbes requires effective methods of identification and characterization. To this end, meta-omics techniques are used to determine the phylogenetic indicators, proteins, genes, and associated markers of a samples (Marchesi and Ravel, 2015). These techniques include metaproteomics, metatranscriptomics, metagenomics, metabolomics, and 16S rRNA sequencing (Zhang et al., 2019). These advanced techniques have been used for the relationship between human gut microbiome and chronic diseases including cancers, diabetes mellitus, and cardiovascular diseases (Cho and Blaser, 2012, Clemente et al., 2012; Michael et al., 2022a,b; Esiobu et al., 2022; Adetunji et al., 2022a,b,c,d,e,f,g,h,i; Olaniyan et al., 2022a,b; Oyedara et al., 2022).

Advanced sequencing techniques including shotgun metagenomics and amplicon sequencing have been used in characterizing the functional composition of human microbiomes. Shotgun metagenomics gives details about the microbial strain; metagenomic sequencing has also been used in determining the role of human microbiome in the early onset of type 1 diabetes mellitus and prediabetic conditions (Stewart et al., 2018; Vatanen et al., 2018; Zhou et al., 2019).

The analytical process of metatranscriptomics and metagenomics is similar. This is of significance in the ease of adaptability during data processing (Abu-Ali et al., 2018). The

An Introduction to the Microbiome in Health and Diseases
ISBN 978-0-323-91190-0,
https://doi.org/10.1016/B978-0-323-91190-0.00004-7

synergy of these techniques enhances microbial genomic stability and gene prediction as well as detection of induced and repressed genes (Zhang et al., 2019).

The use development of metaproteomic data processing devices appears as a step forward for metagenomics and metatranscriptomics. This offers detailed identification and characterization of microbiome proteins composition, i.e., proteins from all organisms (including the host) present in the microbiome irrespective of their origin (Zhang et al., 2019). Galaxy-P, MetaLab, and MetaProteome Analyzer are good examples of these metaproteomic data processors.

Intestinal metabolites are assessed using metabolomics. This involves techniques such as mass spectrometry and nuclear magnetic resonance (NMR). It is of note that the sensitivity of MS-based metabolomics is higher than that of NMR (Emwas, 2015). Fecal metabolome is regarded as the final readout of biological events that emanated from gut microbiome (Zierer et al., 2018). Products of fecal metabolomics could also comprise metabolites from gut microbiota or from the host; butyrate and microbial peptides, respectively. There are evidence that these metabolites are involved in signal transduction between the microbiome and the host (Zhang et al., 2019).

Microbiome characterization and identification

In spite of the variety of techniques used for assessing microbial presence in samples, alpha and beta diversity are commonly used. Alpha diversity is concerned with the number and distribution of operational taxonomic units (OTUs) in a sample. Shannon and Simpson indices are the usual alpha diversity metrics (Knight et al., 2018; Wagner et al., 2018). The difference between samples and a medium of distance is assessed by beta diversity (Ditz et al., 2020). It must be stated that both diversity metrics (alpha and beta) fail to adequately describe the details of the interactions among microbes as well as between microbes and the environment. 16S rRNA gene sequencing technique has been used for characterization of microbiome. It, however, cannot differentiate between dead and living microbes. Furthermore, sequences generated are short; hence, similar bacteria may have comparable sequences, hence creating a challenge in distinguishing between different microbial species. The specificity of 16S rRNA for bacteria and archea is another limitation (Ditz et al., 2020). Microbiome analysis can also provide information on the infectivity or otherwise an ailment. This is brought about by the analysis of bacteria diversity (Chiu and Miller, 2019).

Shotgun metagenomic sequencing is a newer technique that depends on sequencing unrestricted DNA in a sample. It is specific for DNAs in organisms, irrespective of the level of simplicity of the organism (Knight et al., 2018). It gives a nonbiased and detailed taxonomical analysis of microbiome relative to 16S rRNA (Ditz et al., 2020). It can also elucidate the metabolic ability of microbiome as well as probable interaction of microbiome and the host (Quince et al., 2017).

Metagenomics has been used in microbiome testing in both health and disease (Chiu and Miller, 2019). The role of microbiome in health and disease is being explored. There is, however, paucity of information on the clinical validation of microbiome testing for both diagnosis and disease treatment. There are hopes that microbiome analysis may be of importance in the treatment of Clostridium difficile and its related pathologies. The efficacy of fecal stool transplantation in treating the disease has been suggested (Shogbesan et al., 2018; van Nood et al., 2013).

Additionally, the utility of mNGS in characterizing microbiome is several studies has helped in developing probiotic mixtures, which can find an application in prophylactic and treatment of Clostridium difficile (Chiu and Miller, 2019).

Advanced gene sequencing techniques have been used in the identification and characterization of microbiome. This includes metatranscriptomics that focuses on RNA. This technique is directed at living microbiota. It gives insight into the functional ability of a microbe; therefore, it gives concurrent assessment of both the microbial and host transcriptome. This is advancement over metagenomics (Ditz et al., 2020).

Certain limitations have been identified in the use of metatranscriptomics. This includes its novelty, which presents with challenges associated with sample processing and analysis. Secondly, the vulnerability of the RNA with a short half-life is another limitation. This necessitates cautious sample collection and storage. This comes with extra cost, and finally, the outcome of the process may be biased toward microbial genes with elevated transcription rates. This also necessitates supplementation with DNA-dependent sequencing technique, to compare between RNA:DNA ratio and microbial transcription (Ditz et al., 2020).

Microbiome characterization using nucleic acid—based molecular techniques

The conventional method of microbiome characterization, which involves serial dilution, cultivation, subculturing, biochemical, and morphological characterization, is not only time-consuming, tedious, or rigorous. This often leads to misinterpretation or misjudgment due to several microorganisms having similar phenotypic properties. Recent advances in microbiome characterization have led to several time saving and more accurate methods of characterizing a specific microflora. The use of nucleic acid—based molecular techniques using the polymerase chain reaction to amplify either the deoxyribonucleic acid (DNA) or ribonucleic acid (RNA) sequence through 18S rRNA or 16S rRNA is extensively utilized for this purpose, and it is yielding more accurate result.

Technologies based on this research have increased the accuracy, effectiveness, and of great advantage over the traditional methods of microbiome analysis, which are widely capital intensive and time-consuming. The use of this method in identifying pathogenic microbes present in the microbiota of dairy has been very effective and increased the consumers' confidence in dairy industries. Fusco and Quero (2012) evaluated different nuclei acid–based methods for analyzing the microbiome of dairy for identifying and detecting pathogenic or disease-borne microbes in the biota. They discovered that though these techniques are time-consuming and effective, they still have their respective setbacks sometimes in reproducibility or sensitivity.

Microbiome characterization using omics approaches

The emergence of omics technologies has revealed and highlighted the importance of studying the microbiome. In the past decades, amazing discoveries providing a better understanding of the roles a pathogenic or nonpathogenic microorganism plays in microbiota, and its association to certain disease conditions has been achieved through the use of omics technologies to study the microbiome. Human diseases such as obesity, diabetes, cirrhosis, stroke, ulcer, mouth odor, cancer, metabolic syndrome, cardiovascular diseases, and atherosclerosis have been attributed to microbiome. This means the study of the microbiome of a particular environment is of great importance, and the emergence of the omics approach to the study of microbiota of a particular environment has been a game-changer. Omics approaches in the microbiome can easily be categorized based on whether they are targeted at molecular profiling (MPro) or molecular perturbation (MPer) (Sharma et al., 2020; Yao et al., 2015).

Omics approaches such as proteomics, transcriptomics/miRNAomics, genomics, genomics, inomics, and metabolomics have been very effective and useful in understanding the microbiome at various levels, from DNA (genomics), RNA (transcriptomics/miRNAomics), proteins (proteomics), metabolites (lipidomics, metabolome), phenotype (phenome, exposome), and inomics (element profiling, biochemical regulation).

Microbiome characterization: the landscape of genome and transcriptome

Characterizing the microbiome using the genome, DNA sequencing, genetic profiling, genetic mapping, recombinant DNA technology, and structural and functional analysis of the genome are important tools to consider. The advent of next-generation sequencing (NGS) to replace the old and less effective 16S rRNA sequencing has a better view of microbiomes. Transcriptome involves all the RNA transcripts of a cell; these can either be coding or noncoding RNAs, while coding RNA is responsible for proteins, noncoding RNA can either be short noncoding or long noncoding RNA. Short noncoding RNA transcripts include small nucleolar RNA (snoRNA), enhancer RNA (eRNA), Piwi-interacting RNA (piRNA), transfer RNA (tRNA), small interfering RNA (siRNA), and micro-RNA (miRNA); these RNA sequences are less than 200 nucleotides, while those that are more than 200 nucleotide sequences are termed the long noncoding RNA transcripts.

Microarray as a molecular biology tool has been involved in the study of thousands of gene expressions in a given environment. This method is, however, has lower throughput and sensitivity compared with the RNA-seq. RNA-seq is a molecular tool that enables the study and understanding of RNA-based regulation in a genome-wide manner (Li et al., 2017). To obtain a more accurate and comprehensive systematic transcriptome analysis, the use of a molecular tool that covers a larger dynamic range and has better single-nucleotide resolution and genome sequence independence makes RNA-seq more ideal than the conventional microarray (Wang et al., 2009). The measurement of the activity of thousands of genes in a microbiome through expression profiling is also an important factor during the molecular characterization of the microbiome.

Microbiome characterization: impact of sequencing and bioinformatics tools

Since accuracy, precision and reproducibility are desired in microbiological investigations; hence as the year succeeds years in scientific research and microbial evaluation, varieties of methods have popped up for the identification of microorganisms. Many of these approaches are aimed at quantification, identification, and characterization of the microbial communities. Before now primitive techniques such as Gram staining, microscopy, culture, and sensitivities studies have one time given an insight into the quantification and identification of microorganisms (Jarman et al., 2000). On that basis, more improved methods such as serology, flow cytometry, typing, and plasmid profiling have also at one time provided clues for microbial evaluations (Armstrong et al., 2001).

All the aforementioned techniques require preparations of microbial cultures having a long time for an appropriate analysis. The accuracy and precision of these techniques have many times been questioned as it is almost impossible or difficult to determine the exact microbial population (He et al., 2003). Bacterial fermentation assay, parasitic morphology, and cytopathology are forms of microbial phenotypic typing methods used in identification. No matter how long the assays are, they are not definite for a typical microbe because many microorganisms may have similar phenotypical and biochemical characteristics. The development of polymerase chain reaction (PCR) techniques that allow the amplification of desired genes has opened up reliable molecule methods based on DNA isolation and purification techniques. PCR has become more popular because it can be performed after culturing (Bansal, 2005).

Every microbe has its genome, which refers to its complete set of genes making the DNA contained in its chromosome. The bacterial genome is a circular supercoiled structure localized at the nucleoid of the cell. Most bacterial species have less than 5 MB in terms of size, but a few exceptions have been recorded from one of the members of the *Bacillus* spp. having more than 30 MB. However, the genome sizes of microbes can give many indications as to if the microbes are pathogenic or not since it has been reported that parasitic microbes have smaller genomic sizes although obligate pathogens tend to be smaller (Guzman et al., 2008). Since accuracy and precision were lacking in other microbial typing methods, recent methods involving DNA and protein sequencing started around the 1970s when Lamba viruses' genomes were sequenced by Sanger technology (Sanger et al., 1977).

Following the success of the first microbial genome, which was about 50,000 nucleotides, short nucleotides and smaller proteins were sequenced, but at that time sequencing of the whole bacterial genome, it was technically unachievable and economically impossible (Sanger et al., 1977). The technicality of microbial whole-genome sequencing was solved back then by breaking the genome into short fragments and then sequencing individually through a shotgun sequencing technique developed by Sanger. Through this technique, *Haemophilus influenzae* was the first bacterium to have it is whole-genome sequenced (Fleischmann et al., 1995). Shotgun sequencing only allows short strands of DNA fragments (about 100−1000 bp) to be sequenced. Longer nucleotides to be sequenced have to be cut into smaller fragments and then assembled thereafter their sequencing. In shotgun, DNA is intentionally broken into many segments, which are sequenced making use of the chain termination methods to obtain the read; after sequencing, computational programs and bioinformatics are used to assemble the broken fragment sequence into a long and continuous sequence (Sanger et al., 1977).

Drawbacks of microbial phenotypic typing

Microbial characterization can be achieved using two basic methods, which are phenotypic typing and genetic typing. Phenotypes to microbiologists are observable characteristics shown by the microorganism following their assays against some substrates where their responses can then be used for their characterization and evaluation. Consequently, phenotypes as the name implies also assesses observable features of microbial cells, morphology, cultural appearance, and their microscopic arrangements. Until recent, phenotypic typing such as biochemical tests was the most widely used method of microbial characterization due to its low cost of setups, reproducibility, and less labor-intensive.

In microbial phenotypic characterizations, cell size, shapes, antigenicity, and susceptibilities to antimicrobials are fundamentals for the decision taken. Such decisions are also influenced by some environmental factors such as pH, temperature, redox potentials, and water activities; in some cases, growth factors and growth cycles have all been known including the cultural media to influence results. Microbial phenotypic information has since been largely captured in free text, reviews, original papers, and microbiological compilation map known as *Bergey's Manual of Determinative Microbiology* (Holt, 1994). The database for this type of typing is also a limitation as most of the databases are oriented toward clinical applications, which may not completely give an account for microbes of industrial importance; aside from that, a relatively low number of organisms have been reported because only culture-dependent microorganisms can be assessed through microbial typing (Stager and Davis, 1992).

This old scheme of microbial characterization depends on the ability of the test microbe to grow on a culture media that selectively cultivates them. Growth factors may or may not be added to the media. In some cases, characterization may require the addition of inhibitors and degradable compounds in the culture medium, leading to observable metabolic activities. Hence the behaviors of already known cultures can then be matched with the metabolic responses from the unknown cultures. The drawbacks of microbial phenotypic characterization then include process tediousness; some may be labor-intensive and require a lot of material and time-consuming; and in addition, the technical skill of the analyst to interpret and arrive at a correct judgmental decision may often be biased and unreliable.

Understanding the microbial genome and sequencing

The bacterial genome is the complete and entire genes contained in their DNA. Their DNA is usually circular that is supercoiled and localized in their nucleoid. Microbial genomes, especially bacteria, are studied to be very small compared with other living identities (excluding viruses), in terms of size; they range from 100 kb to 15,000 kb nucleotides. As small as they are in genomic size, they are repositories for various genes coding for traits or enzymes, the genetic differences found among closely related species are fundamental for their biodiversity. The most bacterial cell has one chromosome, but exceptions have been reported, having two or more chromosomes. Examples of microbial cell with two or more chromosomes include *Vibrio* spp., *Burkholderia*, *Brucella* spp., and *Leptospira* spp.; these microbial cells all have linear DNA contrary to many others with circular DNA (Guzman et al., 2008).

Critical examination and study of microorganisms have been encountered many challenges; the most brainstorming of those challenges is the discovery of unknown microbes, which are not culture-dependent. Understanding how such unknown organism interacts and relates with their environments was a big milestone before the discovery of modern high-throughput sequencing (Nichols et al., 2011). Through microbial sequencing, the nucleotide composition of microorganisms is now known, for example, the guanosine—cytosine also known as G + C content has been used to differentiate between related species since G + C content varies from one organism to the other; however, it is relatively unvarying within a microbe of the same genus or species (Guzman et al., 2008).

Rather than wasting time on biochemical assays in other to determine the identity of a given microbial colony, microbial characterization through sequencing has made it possible to know all the genes that code for all biochemical functions that are necessary for survival. Furthermore, sequencing has shown that disease-causing bacteria may possess some genetic traits, which are fundamental for their virulency. However, there are noncoding regions located in the genome of prokaryotes. Typically, bacterial genes may be regulated by operons, where those genes are located adjacently to one another and are functionally related. One of the impacts of microbial characterization was to elucidate the regulative mechanisms of lactose operon as they are found in lactose-fermenting *Escherichia coli* (Allen et al., 2006).

The NGS technology focuses on many techniques such as sequencing by synthesis, ligation, and their subsequent annotations. Shotgun sequencing by Sanger has been replaced by NGS whose technique is based on sequencing by synthesis where a DNA molecule is attached to a primer on a slide and then amplified with DNA polymerase—an enzyme that synthesizes the polymerization of DNA. After sequencing, annotating the sequenced genes comes next. Annotation is the process of giving structural, functional, and other biological importance to the sequenced genes or proteins, and this is based on similarity to previously characterized sequences in public databases (Weinstock, 2000).

Information on many microbial cells and the functions those individual genes perform have been sequenced and annotated. The availability of sequenced genes of whole organisms allows the determination of microbial phylogenic relationships among individual organisms by many means of which the construction of a phylogenetic tree is the most common (Radhey and Gupta, 2016).

These advances have led to increasing acceptance among scientist knowing fully well that the problem of dependable and reproducible means classification of microorganisms and other prokaryotes have been solved to a greater extent. With few problems associated, however, the essential taxonomy is constructed to a large extent on the 16S rRNA gene sequences, which is generally accepted (Radhey and Gupta, 2016). Whole-genome sequencing has enabled new techniques for the determination of relatedness rather than 16S rRNA. These approaches include the determination of average nucleotide identity (ANI), and average amino acid identity (AAI) in shared genes and protein, respectively.

Other methods include the determination of conserved protein in the whole-genome sequences, maximum unique exact match indexed between two genomes and also the genome BLAST distance phylogeny. Average nucleotide identity (ANI) is a widely accepted useful tool for identifying new microbial species. "The ANI value of 95%–96% are found to be equivalent to 70% DDH or 98.65% 16S rRNA sequence similarity, and these values can be otherwise used as the benchmark for new prokaryotic species identification" (Radhey and Gupta, 2016).

Understanding microbial diversity

The study of microorganisms has received a lot of scientific attention; the early scientist solely depends on cultural morphologies, biochemical, and microscopy to assess microbial interaction. Staining techniques at that time significantly improved resolution focus and gave a fundamental answer to questions raised by improving resolution power. However, the number of microbial cells observed microscopically does not correlate to the number retrieved culturally (Staley and Konopka, 1985). The hypothesis was not clear enough, and the only explanation they could give was that those microbes needed a special grow factor. The science of metagenomics will make it possible to explore microorganisms in their natural habitat, which we know is a complex community where they reside. One of the impacts of microbial sequencing is to transform the field of biology, medicine, ecology, and biotechnology by circumventing the unculturable.

A large number of organisms have not been cultured; the cultivability of microbial cells is the major fundamental difficulty in assessing the microbial communities. Many techniques have been employed to give account and evaluation of microbial population from the natural environment; some of these techniques include microarray, DNA hybridization, cloning, and detection of specific genes using PCR (Mori and Kamagata, 2014). Shotgun sequencing has already popularized the use of housekeeping genes for microbial identification; the most profound means is the 16S rRNA whose amplicon can be sequencing to identify members of particular taxa after being compared with sequences from an online database.

The metagenomic project has now generated many microbial sequences that are presumed to be unculturable (Gilbert et al., 2014). Metagenomic approaches have opened up the biggest bottlenecks in clinical and environmental microbiology. Metagenomics is the most prominent approach recently used in evaluating microbial ecology; bioinformatic approaches make it possible to harness sufficient data from sequences generated. Several tools have been developed to analyze sequenced amplicons; such tools and pipelines include Web servers, QIIME, MEGAN, and Mothur among many others. To date, many analyses have been based on sequence comparisons and similarity searches against reference sequences available on the databases (Hiraoka et al., 2016).

Next-generation sequencing techniques for the microbiome

NGS allows for the metagenomics analysis of complex microbiomes and therefore delivers more information than just phylogenetic descriptions of the system. Today, due to its accuracy when it comes to large genomic data management, analysis interpretation, time efficiency, and cost sequencing, the NGS approach is fast become an interesting technology on the rise with wide application in the understanding and managing of infectious diseases (Deurenberg et al., 2017). It is involved in gene identification and molecular diagnosis of human diseases (Fang et al., 2016). In recent years, advances in this technique and interpretation of rigorous statistical and computational biology (Peter et al., 2019) have also found its usefulness in the food industries with several applications of NGS for improving food safety in our respective industries (Jagadeesan et al., 2019).

An important advantage of this type of sequencing technology is that, within a short period, a whole genome or thousands/hundreds of genes can be sequenced (hence, this type of sequencer is very cost-effective.

Clinically, this sequencing technology has provided a better understanding of the roles microbiome plays in disease conditions and also boosted the diagnosis and improved the quality of treatment options for several diseases ranging from genetic skin disease (Takeichi et al, 2013), neuroleptospirosis (Wilson et al., 2014), myelodysplastic syndromes (Duncavage and Tandon, 2015), rare pediatric or neonatal diseases (Bacchelli and Williams, 2016; Daoud et al., 2016; Lévesque et al., 2016; François-Heude et al., 2021), acute myeloid leukemia (Duncavage and Tandon, 2015), myeloid neoplasms (Carbonell et al., 2019), neurometabolic disorders (Tarailo-Graovac et al., 2017), epilepsy syndromes (Dunn et al., 2018), pneumonia in China (Xie et al., 2019), and management of cancer (Qin, 2019; Nagahashi et al., 2019).

NGS today is an interesting tool for a better understanding of the influence of the microbiome on antibiotic resistance, which has been a global concern over the years (Woollard et al., 2011; Grumaz et al., 2016; Zhong et al., 2021). In understanding the resistance gene responsible for antibiotic resistance in microbes, NGS has been rising to the global needs with reports on the use of this technique for *Neisseria gonorrhoeae* drug resistance (Graham et al., 2017), *Plasmodium falciparum* drug resistance markers (Kunasol et al., 2022), ciprofloxacin resistance markers (Stefan et al., 2016), antibiotic resistance in Cambodian *Helicobacter pylori* (Tuan et al., 2019), drug resistance associated with an acute lower respiratory infection (Chao et al., 2020), and drug resistance associated to *Plasmodium vivax* patient (Flannery et al., 2015).

NGS techniques of microbiomes

Phylogenetic and taxonomic classification of microbiota has improved over the past three decades with the innovations and advancements made in NGS. The 16S/18S/ITS rDNA amplicon sequencing and the metagenomic shotgun sequencing are the two most widely applied culture-independent NGS techniques for microbiome profiling.

16S/18S/ITS rDNA amplicon sequencing

The 16S/18S/ITS rDNA amplicon sequencing is a targeted sequencing technique that is used to sequence PCR products from specific regions on a functional gene and is commonly used to characterize microbiotas. In bacteria, the 16S rDNA sequence contains 9 hypervariable regions and 10 conserved regions. While the sequences in the conserved regions determine the similarities between species, the sequences in the hypervariable regions determine the differences between species and can be used to classify a bacterium (Cendron et al., 2020; Sirichoat et al., 2021). Primers that bind specifically to highly conserved regions and target one or more hypervariable regions in bacterial 16S rDNA are very critical for amplicon sequencing technique. During 16S/18S/ITS rDNA amplicon sequencing, the highly hypervariable loop region, is targeted for 16S primer binding. There are nine identified bacterial hypervariable regions (V1—V9) to which the 16S tag primers can bind to. However, the binding of 16S tag primers to the V1—V3 region is the most efficient for metataxonomic analysis with a higher level of data reproducibility (Meisel et al., 2016; Sirichoat et al., 2021).

Amplicon sequencing can also be used to sequence eukaryotic genomes and not only prokaryotic genomes. In eukaryotes such as fungi, conserved 18S rDNA sequences can be targeted using specific primers to amplify and sequence the microbe's genome. The most efficient conserved regions targeted for fungal amplicon sequencing are the 18S rDNA, D1/D2 domain of the 26S rDNA, and the internal transcribed spacers (ITS) (Kesmen et al., 2018; Tekpinar and Kalmer, 2019). The 18S rDNA contains eight hypervariable regions (V1—V5 and V7—V9), which are flanked on either side by conserved regions. The V4 region of 18S rDNA is the most widely applied for metataxonomic classifications because of its high evolution rate, high efficiency, high variability, and enormous database information (Han et al., 2022; Ki, 2012). ITS sequences can also be targeted for amplicon sequencing in eukaryotes. The most commonly used ITS sequences for microbial identification and classification are ITS1 and ITS2 because they exhibit some level of variability, which can be exploited in resolving the diversity between species and for phylogenetic analysis (Jillwin et al., 2020; Yang et al., 2018).

The amplicon sequencing technique commences with the collection of samples from an environment (usually the host) whose microbiota is under investigation, followed by microbial DNA isolation and annealing of specific primers to the conserved region in the genome of the microorganism. Next, PCR sequencing is used to amplify the target region (Dong et al., 2017). Roche 454 genome sequencer, Illumina NGS series, Ion Torrent sequencer, and PacBio SMRT sequencing are some common examples of available NGS platforms that can be used for sequencing.

The raw data generated after sequencing is subjected to stringent clean-up to exclude low-quality data while retaining high-quality metadata for downstream analysis. The metadata is analyzed using available analytical programming software such as Quantitative Insights Into Microbial Ecology (QIIME) (Kuczynski et al., 2011), Mothur (Schloss et al., 2009), Distance-based OTU and Richness (DOTUR) (Schloss and Handelsman, 2005), CD-HIT (Li and Godzik, 2006), Uni-Frac (Lozupone et al., 2006; Lozupone and Knight, 2005), BLAST and USEARCH (Edgar, 2010), and Comparative Microbiome Analysis (CoMA) (Hupfauf et al., 2020). The sequences from the analysis are assembled into operational taxonomic units (OTU), which are used to identify the microbial population in the microbiome and to build a taxonomic classification by comparing the OTUs to available information in the database.

Metagenomic shotgun sequencing

Metagenomic shotgun or metatranscriptomic sequencing is a more advanced sequencing technique with a wider range of species identification. Unlike the 16S/18S/ITS rDNA amplicon sequencing where a target region of the genomic material is first amplified using primers before sequencing, metagenomic sequencing can sequence the whole DNA/RNA or sections of the DNA/RNA in a nontargeted manner without prior amplification of a gene. Consequently, metagenomic sequencing provides more information about the structure and organization of the microbial genome, thus allowing for the discovery of low abundance microorganisms that may be difficult to culture (Ferravante et al., 2021; Pérez-Cobas et al., 2020). Similar to the 16S/18S/ITS rDNA amplicon sequencing, metagenomic sequencing begins with the collection of samples and DNA isolation.

To improve the quality of the metadata generated and the accuracy of microbes identified, it is important to apply a more targeted DNA isolation technique to minimize contamination of microbial DNA samples with host DNA samples. The DNA is processed by fragmenting the DNA and amplifying each fragment independently to produce contigs (Meisel et al., 2016). The contigs are assembled into their respective genomes, and the source of the DNA can be determined by comparing the assembled contigs to online references available in databases such as SILVA (Quast et al., 2013). Once the source of the DNA is identified, the taxonomic classification of the gene sequences at the species or strain level is executed.

Today, there are many sequencing platforms available in the market for microbiome profiling. These platforms can be classified as first, second, third, or fourth generation sequencers depending on their operational principle. The first-generation sequencing platform worked on the principle of DNA chain termination by ddNTPs. The second-generation or next-generation sequencing platforms apply different operational principles such as pyrosequencing technology, reversible terminator technology, and sequencing by ligation. The third-generation sequencing platforms use single-molecule sequencing of ion semiconductor sequencing. The fourth-generation sequencing platforms use a variety of technologies such as DNA nanoball sequencing, nanopore sequencing, and polony-based sequencing by ligation (Gupta and Gupta, 2020). Herein, we describe some commonly applied sequencing platforms for microbiome profiling.

Roche 454 genome sequencer

Roche 454 genome sequencing machine was introduced to the market in 2005 by 454 Life Sciences (which was later acquired by Roche Holding AG in 2007). One of the intriguing features of this NGS platform is the long reads and fast run time, which make the sequencing machine very accurate in its sequence production in a short time. Roche 454 genome sequencer utilizes clonal amplification by emulsion PCR and pyrosequencing technology to sequence a DNA sample. Roche 454 genome sequencer processes amplicons emanating from fragmented double-stranded DNA, which have been broken down to between 400 and 600 base pairs using restriction enzymes. Generic adapters (short sequences of DNA) are attached to the end of each DNA fragment for binding to tiny resin beads suspended in a water-in-oil emulsion.

The complementarity of DNA sequences on the resin beads to sequences on the adapter ensures the binding of each DNA fragment to one resin bead. Many copies of the DNA fragments are made using emulsion PCR to create millions of DNA copies that are identical to one another. Resin beads that are not attached to any DNA fragment or those that are attached to more than one DNA fragment are filtered off. Only DNA fragments attached to a single resin bead are collected into wells of a sequencing plate. Enzymes (such as DNA polymerase, ATP sulfurylase, luciferase), primers, and sequencing buffers are added to the sequencing wells. As the sequencing commences, a particular nucleotide base is added to the wells. For instance, a wave of adenine is added into the wells followed by a wave of cytosine, guanine, and thymine. Each nucleotide base binds complementarily to the DNA fragments with the release of a pyrophosphate molecule. The pyrophosphate molecule is enzymatically converted to ATP, which is required by luciferase to convert luciferin into oxyluciferin (a luminescent product of luciferin). The flashes of light that are emitted are recorded by a camera. The amount of light intensity recorded by the camera indicates the number of the same nucleotide base that is bound to DNA fragments in each well. By analyzing the pattern of the light produced, the nucleotide sequence of the parent DNA can be predicted.

Illumina sequencer

The Illumina sequencing technology was introduced to the market in 2006 by Illumina Innovative Technologies. The Illumina sequencer is available in different sequencing platforms including the HiSeq, MiniSeq, MiSeq, NextSeq series, and NovaSeq platforms. Illumina sequencer utilizes clonal amplification and sequencing-by-synthesis to generate millions of reads in a relatively shorter time and at a lesser cost. The initial preparation of DNA samples for this technology is similar to that of the Roche 454 genome sequencer. Illumina sequencer processes amplicons from fragmented double-stranded DNA that have been broken down to between 200 and 600 base pairs. Again, generic adapters are attached to the ends of each fragmented DNA. The adapter-bound fragmented double-stranded DNA are incubated in sodium hydroxide to unwind the DNA fragments into single strands.

The single-stranded fragments are washed through a flow cell carrying attached oligoprimers. Only single-stranded fragments that are complementary to the oligoprimers bind to the oligoprimers, while unbound single-stranded fragments are washed away.

The bound DNA fragments are replicated to produce several copies of identical double-stranded DNA fragments, which are broken down to single strands using heat. The enzyme DNA polymerase and primers are introduced into the flow cell, and DNA replication is initiated to produce multiple identical copies of the DNA fragments. Fluorescently labeled terminators (nucleotide bases that can terminate DNA synthesis) are added to the flow cell and washed away after a short time. Once a fluorescently labeled terminator is added to the growing DNA strand, the activity of the DNA polymerase is terminated until the terminator is cut off. The process is repeated several times with the addition of nucleotide bases and subsequent termination of the process by a fluorescently labeled terminator. During the DNA synthesis and termination process, lasers that activate the fluorescently labeled terminators are passed over the flow cell to give off a fluorescence that can be recorded on a computer. The recorded fluorescence is analyzed and used to identify the sequence of the original DNA sample.

Ion Torrent sequencer

The Ion Torrent sequencing technology was introduced to the market in 2010 by Ion Torrent Systems Inc. which was later acquired by Life Technologies. The release and detection of protons from growing DNA strands in microwells of the Ion Personal Genome Machine (PGM) is the principle that drives the functioning of the Ion Torrent sequencer. The processing of DNA sample for Ion Torrent sequencing technique is similar to DNA processing for Roche 454 genome sequencer except that the read length of the fragmented double-stranded DNA for Ion Torrent sequencing is about 200–600 base pairs. Fragmented DNA with a reading length higher than the specified length for the sequencer is removed during the size-selection process.

The fragmented double-stranded DNAs are bound to beads in microwells of a plate, and nucleotide incorporation is carried out in the presence of DNA polymerase. Upon addition of nucleotide bases (A, C, G or T) into the microwells containing different fragmented DNA, protons are released from the growing DNA strands, and the pH of the microenvironment is slightly reduced. The Ion Torrent sequencer has ion semiconductor sequencing chips that can detect the change in pH within the microenvironment. As the process is repeated several times, the change in pH is recorded and the change in pH is the basis upon which the sequence of the original DNA is determined.

PacBio sequencer

The PacBio sequencing technology was introduced to the market in 2011 by Pacific Biosciences. The PacBio sequencing technology is a single-molecule real-time (SMRT) sequencing technology that provides real-time monitoring of DNA synthesis. The template DNA used in this technology is purified double-stranded DNA with ligated hairpins at the ends of the template and is called a SMRTbell. The SMRTbell is loaded into a specialized SMRT cell where sequencing is performed. Each SMRT cell has at least

150,000 zero mode waveguides (ZMW) attached to its bottom surface (Rhoads and Au, 2015). The ZMW (specially designed nanoholes) are used to detect and monitor single-molecule sequencing reactions in an SMRT cell.

DNA polymerase is immobilized at the bottom surface of the ZMW; then the SMRTbell is introduced into the SMRT cell where it binds to the DNA polymerase via its ligated hairpins. Fluorescently labeled nucleotides (blue, yellow, red, and green for A, C, G, and T, respectively) are added to the bottom of the SMRT cell where they bind to the active site of the DNA polymerase and are sequentially incorporated into the growing DNA strand. As laser lights are aimed at the SMRT cells and through the ZMW nanoholes, the fluorescently labeled nucleotides are incorporated into the growing DNA strand fluoresce. Since the fluorescently labeled nucleotides have varying emission spectra, the amount of fluorescence detected and recorded indicates the nucleotide base added to the growing DNA strand. Consequently, DNA synthesis is monitored in real time.

Conclusion

Advances in metagenomics have been driven by the development of second- and third-generation sequencing technologies, which have been developed to produce hundreds of giga-bases of DNA sequences. Microbial sequencing has offered possibilities for identifying the least abundant microbes in a given environment. Current sequencing machinery, in combination with bioinformatics, has enabled metagenomic studies friendly, relatively inexpensive, and fast technique for many laboratories. Microbial sequencing can also provide insight into newer enzymes and other biocatalysts with major importance in biotechnology, energy, drug delivery, and developments.

References

Abu-Ali, G.S., Mehta, R.S., Lloyd-Price, J., Mallick, H., Branck, T., Ivey, K.L., et al., 2018. Metatranscriptome of human faecal microbial communities in a cohort of adult men. Nat. Microbiol. 3, 356–366.

Adetunji, C.O., Olaniyan, O.T., Adeyomoye, O., Dare, A., Adeniyi, M.J., Enoch, A., Rebezov, M., Garipova, L., Ali Shariati, M., 2022a. eHealth, mHealth, and telemedicine for COVID-19 pandemic. In: Pani, S.K., Dash, S., dos Santos, W.P., Chan Bukhari, S.A., Flammini, F. (Eds.), Assessing COVID-19 and Other Pandemics and Epidemics Using Computational Modelling and Data Analysis. Springer, Cham. https://doi.org/10.1007/978-3-030-79753-9_10.

Adetunji, C.O., Olaniyan, O.T., Adeyomoye, O., Dare, A., Adeniyi, M.J., Enoch, A., Rebezov, M., Petukhova, E., Ali Shariati, M., 2022b. Machine learning approaches for COVID-19 pandemic. In: Pani, S.K., Dash, S., dos Santos, W.P., Chan Bukhari, S.A., Flammini, F. (Eds.), Assessing COVID-19 and Other Pandemics and Epidemics Using Computational Modelling and Data Analysis. Springer, Cham. https://doi.org/10.1007/978-3-030-79753-9_8.

Adetunji, C.O., Olaniyan, O.T., Adeyomoye, O., Dare, A., Adeniyi, M.J., Enoch, A., Rebezov, M., Isabekova, O., Ali Shariati, M., 2022c. Smart sensing for COVID-19 pandemic. In: Pani, S.K., Dash, S., dos Santos, W.P., Chan Bukhari, S.A., Flammini, F. (Eds.), Assessing COVID-19 and Other Pandemics and Epidemics Using Computational Modelling and Data Analysis. Springer, Cham. https://doi.org/10.1007/978-3-030-79753-9_9.

Adetunji, C.O., Olaniyan, O.T., Adeyomoye, O., Dare, A., Adeniyi, M.J., Enoch, A., Rebezov, M., Petukhova, E., Ali Shariati, M., 2022d. Internet of health things (IoHT) for COVID-19. In: Pani, S.K., Dash, S., dos Santos, W.P., Chan Bukhari, S.A., Flammini, F. (Eds.), Assessing COVID-19 and Other Pandemics and Epidemics Using Computational Modelling and Data Analysis. Springer, Cham. https://doi.org/10.1007/978-3-030-79753-9_5.

Adetunji, C.O., Olaniyan, O.T., Adeyomoye, O., Dare, A., Adeniyi, M.J., Enoch, A., Rebezov, M., Koriagina, N., Ali Shariati, M., 2022e. Diverse techniques applied for effective diagnosis of COVID-19. In: Pani, S.K., Dash, S., dos Santos, W.P., Chan Bukhari, S.A., Flammini, F. (Eds.), Assessing COVID-19 and Other Pandemics and Epidemics Using Computational Modelling and Data Analysis. Springer, Cham. https://doi.org/10.1007/978-3-030-79753-9_3.

Adetunji, C.O., Samuel, M.O., Adetunji, J.B., Oluranti, O.I., 2022f. Corn silk and health benefits. In: Medical Biotechnology, Biopharmaceutics, Forensic Science and Bioinformatics, first ed. Imprint CRC Press, ISBN 9781003178903, p. 12. https://doi.org/10.1201/9781003178903-11. First Published 2022.

Adetunji, C.O., Wilson, N., Olayinka, A.S., Olugbemi, O.T., Akram, M., Laila, U., Samuel, M.O., Oshinjo, A.M., Adetunji, J.B., Okotie, G.E., Esiobu, N.D., 2022g. Computational intelligence techniques for combating COVID-19. In: Medical Biotechnology, Biopharmaceutics, Forensic Science and Bioinformatics, first ed. Imprint CRC Press, ISBN 9781003178903, p. 12. https://doi.org/10.1201/9781003178903-16. First Published 2022.

Adetunji, C.O., Olugbemi, O.T., Akram, M., Laila, U., Samuel, M.O., Michael Oshinjo, A., Adetunji, J.B., Okotie, G.E., Esiobu, N.D., Oyedara, O.O., Adeyemi, F.M., 2022h. Application of computational and bioinformatics techniques in drug repurposing for effective development of potential drug candidate for the management of COVID-19. In: Medical Biotechnology, Biopharmaceutics, Forensic Science and Bioinformatics, first ed. Imprint CRC Press, ISBN 9781003178903, p. 14. https://doi.org/10.1201/9781003178903-15. First Published 2022.

Adetunji, C.O., Wilson, N., Olayinka, A.S., Olugbemi, O.T., Akram, M., Laila, U., Samuel Olugbenga, M., Michael Oshinjo, A., Adetunji, J.B., Okotie, G.E., Esiobu, N.D., 2022i. Machine learning and behaviour modification for COVID-19. In: Medical Biotechnology, Biopharmaceutics, Forensic Science and Bioinformatics, first ed. Imprint CRC Press, ISBN 9781003178903, p. 17. https://doi.org/10.1201/9781003178903-17. First Published 2022.

Allen, T.E., Price, N.D., Joyce, A.R., Palsson, B.Ø., 2006. Long-range periodic patterns in microbial genomes indicate significant multi-scale chromosomal organization. PLoS Comput. Biol. 2, e2. https://doi.org/10.1371/journal.pcbi.0020002.

Armstrong, D.W., Schneiderheinze, J.M., Kullman, J.P., He, L., 2001. Rapid CE microbial assays for consumer products that contain active bacteria. FEMS Microbiol. Lett. 194, 33–37.

Bacchelli, C., Williams, H.J., 2016. Opportunities and technical challenges in next-generation sequencing for the diagnosis of rare pediatric diseases. Expert Rev. Mol. Diagn. 16 (10), 1073–1082.

Bansal, A.K., 2005. Bioinformatics in microbial biotechnology—a mini-review. Microb. Cell Factories 4, 19. https://doi.org/10.1186/1475-2859-4-19.

Carbonell, D., Suárez-González, J., Chicano, M., Andrés-Zayas, C., Triviño, J.C., Rodríguez-Macías, G., Bastos-Oreiro, M., Font, P., Ballesteros, M., Muñiz, P., Balsalobre, P., 2019. Next-generation sequencing improves diagnosis, prognosis and clinical management of myeloid neoplasms. Cancers 11 (9), 1364.

Cendron, F., Niero, G., Carlino, G., Penasa, M., Cassandro, M., 2020. Characterizing the fecal bacteria and archaea community of heifers and lactating cows through 16S rRNA next-generation sequencing. J. Appl. Genet. 61 (4), 593–605. https://doi.org/10.1007/s13353-020-00575-3.

Chao, L., Li, J., Zhang, Y.N., Pu, H., Yan, X., 2020. Application of next generation sequencing-based rapid detection platform for microbiological diagnosis and drug resistance prediction in acute lower respiratory infection. Ann. Transl. Med. 8 (24).

Chiu, C.Y., Miller, S.A., 2019. Clinical metagenomics. Microbial genomics. Nat. Rev. Genet. 20, 341–355.

Cho, I., Blaser, M.J., 2012. The human microbiome: at the interface of health and disease. Nat. Rev. Genet. 13, 260–270.

Clemente, J.C., Ursell, L.K., Parfrey, L.W., Knight, R., 2012. The impact of the gut microbiota on human health: an integrative view. Cell 148, 1258–1270.

Daoud, H., Luco, S.M., Li, R., Bareke, E., Beaulieu, C., Jarinova, O., Carson, N., Nikkel, S.M., Graham, G.E., Richer, J., Armour, C., 2016. Next-generation sequencing for the diagnosis of rare diseases in the neonatal intensive care unit. Can. Med. Assoc. J. 188 (11), 254—260.

Deurenberg, R.H., Bathoorn, E., Chlebowicz, M.A., Couto, N., Ferdous, M., García-Cobos, S., Kooistra-Smid, A.M., Raangs, E.C., Rosema, S., Veloo, A.C., Zhou, K., 2017. Application of next-generation sequencing in clinical microbiology and infection prevention. J. Biotechnol. 243, 16—24.

Ditz, B., Christenson, S., Rossen, J., et al., 2020. Thorax 75, 338—344.

Dong, X., Kleiner, M., Sharp, C.E., Thorson, E., Li, C., Liu, D., Strous, M., 2017. Fast and simple analysis of MiSeq amplicon sequencing data with MetaAmp. Front. Microbiol. 8 (AUG), 1—8. https://doi.org/10.3389/fmicb.2017.01461.

Duncavage, E.J., Tandon, B., 2015. The utility of next-generation sequencing in diagnosis and monitoring of acute myeloid leukemia and myelodysplastic syndromes. Int. J. Lit. Humanit. 37, 115—121.

Dunn, P., Albury, C.L., Maksemous, N., Benton, M.C., Sutherland, H.G., Smith, R.A., Haupt, L.M., Griffiths, L.R., 2018. Next-generation sequencing methods for diagnosis of epilepsy syndromes. Front. Genet. 9, 20. https://doi.org/10.3389/fgene.2018.00020.

Edgar, R.C., 2010. Search and clustering orders of magnitude faster than BLAST. Bioinformatics 26 (19), 2460—2461. https://doi.org/10.1093/bioinformatics/btq461.

Emwas, A.H., 2015. The strengths and weaknesses of NMR spectroscopy and mass spectrometry with particular focus on metabolomics research. Methods Mol. Biol. 1277, 161—193.

Esiobu, N.D., Ogbonna, J.C., Adetunji, C.O., Obembe, O.O., Ezeonu, I.M., Ibrahim, A.B., Ubi, B.E., 2022. Microbiomes and Emerging Applications. First Published 2022. eBook Published 11 May 2022, first ed. Imprint CRC Press, Boca Raton, ISBN 9781003180241, p. 186. https://doi.org/10.1201/9781003180241. eBook ISBN 9781003180241. Subjects Bioscience, Engineering and Technology.

Fang, M., Abolhassani, H., Lim, C.K., Zhang, J., Hammarström, L., 2016. Next-generation sequencing data analysis in primary immunodeficiency disorders—future directions. J. Clin. Immunol. 36 (1), 68—75.

Ferravante, C., Memoli, D., Palumbo, D., Ciaramella, P., Di Loria, A., D'Agostino, Y., Giurato, G., 2021. HOME-BIO (sHOtgun MEtagenomic analysis of BIOlogical entities): a specific and comprehensive pipeline for metagenomic shotgun sequencing data analysis. BMC Bioinf. 22 (7), 1—10. https://doi.org/10.1186/s12859-021-04004-y.

Flannery, E.L., Wang, T., Akbari, A., Corey, V.C., Gunawan, F., Bright, A.T., Abraham, M., Sanchez, J.F., Santolalla, M.L., Baldeviano, G.C., Edgel, K.A., 2015. Next-generation sequencing of *Plasmodium vivax* patient samples shows evidence of direct evolution in drug-resistance genes. ACS Infect. Dis. 1 (8), 367—379.

Fleischmann, R., Adams, M., White, O., Clayton, R.A., Kirkness, E.F., Kerlavage, A.R., Bult, C.J., Tomb, J.F., Dougherty, B.A., Merrick, J.M., 1995. Whole-genome random sequencing and assembly of *Haemophilus influenza*. Science 269, 496—512.

François-Heude, M.C., Walther-Louvier, U., Espil-Taris, C., Beze-Beyrie, P., Rivier, F., Baudou, E., Uro-Coste, E., Rigau, V., Negrier, M.L.M., Rendu, J., Morales, R.J., 2021. Evaluating next-generation sequencing in neuromuscular diseases with neonatal respiratory distress. Eur. J. Paediatr. Neurol. 31, 78—87.

Fusco, V., Quero, G.M., 2012. Nucleic acid-based methods to identify, detect and type pathogenic bacteria occurring in milk and dairy products. Struct. Funct. Food Eng. 371—404.

Gilbert, J.A., Jansson, J.K., Knight, R., 2014. The Earth Microbiome project: successes and aspirations. BMC Biol. 12, 69.

Graham, R.M.A., Doyle, C.J., Jennison, A.V., 2017. Epidemiological typing of *Neisseria gonorrhoeae* and detection of markers associated with antimicrobial resistance directly from urine samples using next-generation sequencing. Sex. Transm. Infect. 93 (1), 65—67.

Grumaz, S., Stevens, P., Grumaz, C., Decker, S.O., Weigand, M.A., Hofer, S., Brenner, T., von Haeseler, A., Sohn, K., 2016. Next-generation sequencing diagnostics of bacteremia in septic patients. Genome Med. 8 (1), 1—13.

Gupta, A.K., Gupta, U.D., 2020. Next-generation sequencing and its applications. In: Verma, A.S., Singh, A. (Eds.), Animal Biotechnology: Models in Discovery and Translation, second ed. Elsevier Inc, pp. 395—421. https://doi.org/10.1016/B978-0-12-811710-1.00018-5.

Guzman, P.E., Romeu, A., Garcia-Vallve, S., 2008. Completely sequenced genomes of pathogenic bacteria: a review. Enferm. Infecc. Microbiol. Clín. 26, 88—89. https://doi.org/10.1157/13115544.

Han, Y., Guo, C., Guan, X., McMinn, A., Liu, L., Zheng, G., ... Wang, M., 2022. Comparison of deep-sea picoeukaryotic composition estimated from the V4 and V9 regions of 18S rRNA gene with a focus on the Hadal zone of the Mariana trench. Microb. Ecol. 83 (1), 34—47. https://doi.org/10.1007/s00248-021-01747-2.

He, L., Jepsen, R.J., Evans, L.E., Armstrong, D.W., 2003. Electrophoretic behaviour and potency assessment of boar sperm using a capillary electrophoresis-laser induced fluorescence system. Anal. Chem. 75, 825—834.

Hiraoka, S., Yang, C.C., Iwasaki, W., 2016. Metagenomics and bioinformatics in microbial ecology: current status and beyond. Microb. Environ. 31 (3), 204—212. https://doi.org/10.1264/jsme2.ME16024.

Holt, J.G., 1994. Bergey's Manual of Determinative Microbiology. Lippincott Williams and Wilkins, Hagerstown, MD, p. 9.

Hupfauf, S., Etemadi, M., Juárez, M.F.D., Gómez-Brandón, M., Insam, H., Podmirseg, S.M., 2020. CoMA—an intuitive and user-friendly pipeline for amplicon-sequencing data analysis. PLoS One 15 (12), 1—28. https://doi.org/10.1371/journal.pone.0243241.

Jagadeesan, B., Gerner-Smidt, P., Allard, M.W., Leuillet, S., Winkler, A., Xiao, Y., Chaffron, S., Van Der Vossen, J., Tang, S., Katase, M., McClure, P., Kimura, B., Chai, L.C., Chapman, J., Grant, K., 2019. The use of next-generation sequencing for improving food safety: translation into practice. Food Microbiol. 79, 96—115.

Jarman, K.H., Cebula, S.T., Saenz, A., Peterson, C.E., Valentine, N.B., Kingsley, M.T., Wahl, K.L., 2000. An algorithm for automated bacterial identification using matrix-assisted laser desorption/ionization mass spectrometry. Anal. Chem. 72, 1217—1223.

Jillwin, J., Rudramurthy, S.M., Singh, S., Bal, A., Das, A., Radotra, B., ... Chakrabarti, A., 2020. Molecular identification of pathogenic fungi in formalin-fixed and paraffin-embedded tissues. J. Med. Microbiol. 70 (2), 1—8. https://doi.org/10.1099/JMM.0.001282.

Kesmen, Z., Büyükkiraz, M.E., Özbekar, E., Çelik, M., Özkök, F.Ö., Kılıç, Ö., ... Yetim, H., 2018. Assessment of multi fragment melting analysis system (MFMAS) for the identification of food-borne yeasts. Curr. Microbiol. 75 (6), 716—725. https://doi.org/10.1007/s00284-018-1437-9.

Ki, J.S., 2012. Hypervariable regions (V1-V9) of the dinoflagellate 18S rRNA using a large dataset for marker considerations. J. Appl. Phycol. 24 (5), 1035—1043. https://doi.org/10.1007/s10811-011-9730-z.

Knight, R., Vrbanac, A., Taylor, B.C., et al., 2018. Best practices for analysing microbiomes. Nat. Rev. Microbiol. 16, 410—422.

Kuczynski, J., Stombaugh, J., Walters, W.A., González, A., Caporaso, J.G., Knight, R., 2011. Using QIIME to analyze 16S rRNA gene sequences from microbial communities. Curr. Protoc. Bioinformatics (Suppl. 36), 1—28. https://doi.org/10.1002/0471250953.bi1007s36.

Kunasol, C., Dondorp, A.M., Batty, E.M., Nakhonsri, V., Sinjanakhom, P., Day, N.P., Imwong, M., 2022. Comparative analysis of targeted next-generation sequencing for *Plasmodium falciparum* drug resistance markers. Sci. Rep. 12 (1), 1—10.

Lévesque, S., Auray-Blais, C., Gravel, E., Boutin, M., Dempsey-Nunez, L., Jacques, P.E., Chenier, S., Larue, S., Rioux, M.F., Al-Hertani, W., Nadeau, A., 2016. Diagnosis of late-onset Pompe disease and other muscle disorders by next-generation sequencing. Orphanet J. Rare Dis. 11 (1), 1—10.

Li, W., Godzik, A., 2006. Cd-hit: a fast program for clustering and comparing large sets of protein or nucleotide sequences. Bioinformatics 22 (13), 1658—1659. https://doi.org/10.1093/bioinformatics/btl158.

Li, X., Mei, H., Chen, F., Tang, Q., Yu, Z., Cao, X., He, J., 2017. Transcriptome landscape of Mycobacterium smegmatis. Front. Microbiol. 8, 2505. https://doi.org/10.3389/fmicb.2017.02505.

Lozupone, C., Knight, R., 2005. UniFrac: a new phylogenetic method for comparing microbial communities. Appl. Environ. Microbiol. 71 (12), 8228—8235. https://doi.org/10.1128/AEM.71.12.8228-8235.2005.

Lozupone, C., Hamady, M., Knight, R., 2006. UniFrac—an online tool for comparing microbial community diversity in a phylogenetic context. BMC Bioinf. 7, 1—14. https://doi.org/10.1186/1471-2105-7-371.

Marchesi, J.R., Ravel, J., 2015. The vocabulary of microbiome research: a proposal. Microbiome 3, 31.

Meisel, J.S., Hannigan, G.D., Tyldsley, A.S., SanMiguel, A.J., Hodkinson, B.P., Zheng, Q., Grice, E.A., 2016. Skin microbiome surveys are strongly influenced by experimental design. J. Invest. Dermatol. 136 (5), 947–956. https://doi.org/10.1016/j.jid.2016.01.016.

Michael, O.S., Oluranti, O.I., Oshinjo, A.M., Adetunji, C.O., Olaniyi, K.S., Adetunji, J.B., 2022a. Microbiome Reshaping and Epigenetic Regulation, first ed. Imprint CRC Press, ISBN 9781003180241, p. 22. https://doi.org/10.1201/9781003180241-6. First Published 2022.

Michael, O.S., Oluranti, O.I., Oshinjo, A.M., Adetunji, C.O., Adetunji, J.B., Esiobu, N.D., 2022b. Microbiota transplantation, health implications, and the way forward. In: Microbiomes and Emerging Applications, first ed. Imprint CRC Press, ISBN 9781003180241, p. 19. https://doi.org/10.1201/9781003180241-5. First Published 2022.

Mori, K., Kamagata, Y., 2014. The challenges of studying the anaerobic microbial world. Microb. Environ. 29, 335–337, 2014.

Nagahashi, M., Shimada, Y., Ichikawa, H., Kameyama, H., Takabe, K., Okuda, S., Wakai, T., 2019. Next-generation sequencing-based gene panel tests for the management of solid tumors. Cancer Sci. 110 (1), 6–15. https://doi.org/10.1111/cas.13837.

Nichols, R.J., Sen, S., Choo, Y.J., Beltrao, P., Zietek, M., Chaba, R., Lee, S., Kazmierczak, K.M., Lee, K.J., Wong, A., Shales, M., Lovett, S., Winkler, M.E., Krogan, N.J., Typas, A., Gross, C.A., 2011. Phenotypic landscape of a bacterial cell. Cell 144 (1), 143–156. https://doi.org/10.1016/j.cell.2010.11.052.

Olaniyan, O.T., Adetunji, C.O., Adeniyi, M.J., Hefft, D.I., 2022a. Machine learning techniques for high-performance computing for IoT applications in healthcare. In: Deep Learning, Machine Learning and IoT in Biomedical and Health Informatics, first ed. Imprint CRC Press, ISBN 9780367548445, p. 13. https://doi.org/10.1201/9780367548445-20. First Published 2022.

Olaniyan, O.T., Adetunji, C.O., Adeniyi, M.J., Hefft, D.I., 2022b. Computational intelligence in IoT healthcare, 2022m. In: Deep Learning, Machine Learning and IoT in Biomedical and Health Informatics, first ed. Imprint CRC Press, ISBN 9780367548445, p. 13. https://doi.org/10.1201/9780367548445-19. First Published 2022.

Oyedara, O.O., Adeyemi, F.M., Adetunji, C.O., Elufisan, T.O., 2022. Repositioning antiviral drugs as a rapid and cost-effective approach to discover treatment against SARS-CoV-2 infection. In: Medical Biotechnology, Biopharmaceutics, Forensic Science and Bioinformatics, first ed. Imprint CRC Press, ISBN 9781003178903, p. 12. https://doi.org/10.1201/9781003178903-10. First Published 2022.

Pérez-Cobas, A.E., Gomez-Valero, L., Buchrieser, C., 2020. Metagenomic approaches in microbial ecology: an update on whole-genome and marker gene sequencing analyses. Microb. Genom. 6 (8), 1–22. https://doi.org/10.1099/mgen.0.000409.

Peter, S.C., Dhanjal, J.K., Malik, V., Radhakrishnan, N., Jayakanthan, M., Sundar, D., Sundar, D., Jayakanthan, M., 2019. In: Ranganathan, S., Gribskov, M., Nakai, K., Schönbach, C. (Eds.), Encyclopedia of Bioinformatics and Computational Biology, pp. 661–676.

Qin, D., 2019. Next-generation sequencing and its clinical application. Cancer Biol. Med. 16 (1), 4–10. https://doi.org/10.20892/j.issn.2095-3941.2018.0055.

Quast, C., Pruesse, E., Yilmaz, P., Gerken, J., Schweer, T., Yarza, P., … Glöckner, F.O., 2013. The SILVA ribosomal RNA gene database project: improved data processing and web-based tools. Nucleic Acids Res. 41 (1), 590–596. https://doi.org/10.1093/nar/gks1219.

Quince, C., Walker, A.W., Simpson, J.T., et al., 2017. Shotgun metagenomics, from sampling to analysis. Nat. Biotechnol. 35, 833–844.

Radhey, Gupta, S., July 2016. Impact of genomics on the understanding of microbial evolution and classification: the importance of Darwin's views on classification. FEMS Microbiol. Rev. 40 (Issue 4), 520–553. https://doi.org/10.1093/femsre/fuw011.

Rhoads, A., Au, K.F., 2015. PacBio sequencing and its applications. Dev. Reprod. Biol. 13 (5), 278–289. https://doi.org/10.1016/j.gpb.2015.08.002.

Sanger, F., Nicklen, S., Coulson, A.R., 1977. DNA sequencing with chain-terminating inhibitors. Proc. Natl. Acad. Sci. USA 74, 5463–5467. https://doi.org/10.1073/pnas.74.12.5463.

Schloss, P.D., Handelsman, J., 2005. Introducing DOTUR, a computer program for defining operational taxonomic units and estimating species richness. Appl. Environ. Microbiol. 71 (3), 1501–1506. https://doi.org/10.1128/AEM.71.3.1501-1506.2005.

Schloss, P.D., Westcott, S.L., Ryabin, T., Hall, J.R., Hartmann, M., Hollister, E.B., Weber, C.F., 2009. Introducing mothur: open-source, platform-independent, community-supported software for describing and comparing microbial communities. Appl. Environ. Microbiol. 75 (23), 7537—7541. https://doi.org/10.1128/AEM.01541-09.

Sharma, J., Balakrishnan, L., Kaushik, S., Kashyap, M.K., 2020. Multi-omics approaches to study signaling pathways. Front. Bioeng. Biotechnol. 8.

Shogbesan, O., et al., 2018. A Systematic review of the efficacy and safety of fecal microbiota transplant for *Clostridium difficile* infection in immunocompromised patients. Chin. J. Gastroenterol. Hepatol. 1394379, 2018.

Sirichoat, A., Sankuntaw, N., Engchanil, C., Buppasiri, P., Faksri, K., Namwat, W., Lulitanond, V., 2021. Comparison of different hypervariable regions of 16S rRNA for taxonomic profiling of vaginal microbiota using next-generation sequencing. Arch. Microbiol. 203 (3), 1159—1166. https://doi.org/10.1007/s00203-020-02114-4.

Stager, C.E., Davis, J.R., July 1992. Automated systems for identification of microorganisms. Clin. Microbiol. Rev. 5 (3), 302—327. https://doi.org/10.1128/CMR.5.3.302. PMID: 1498768; PMCID: PMC358246.

Staley, J.T., Konopka, A., 1985. Measurement of *in situ* activities of nonphotosynthetic microorganisms in aquatic and terrestrial habitats. Annu. Rev. Microbiol. 39, 321—346. https://doi.org/10.1146/annurev.mi.39.100185.001541.

Stefan, C.P., Koehler, J.W., Minogue, T.D., 2016. Targeted next-generation sequencing for the detection of ciprofloxacin resistance markers using molecular inversion probes. Sci. Rep. 6 (1), 1—12.

Stewart, C.J., Ajami, N.J., O'Brien, J.L., Hutchinson, D.S., Smith, D.P., Wong, M.C., et al., 2018. Temporal development of the gut microbiome in early childhood from the TEDDY study. Nature 562, 583—588.

Takeichi, T., Nanda, A., Liu, L., Salam, A., Campbell, P., Fong, K., Akiyama, M., Ozoemena, L., Stone, K.L., Al-Ajmi, H., Simpson, M.A., 2013. Impact of next-generation sequencing on diagnostics in a genetic skin disease clinic. Exp. Dermatol. 22 (12), 825—831.

Tarailo-Graovac, M., Wasserman, W.W., Van Karnebeek, C.D., 2017. Impact of next-generation sequencing on diagnosis and management of neurometabolic disorders: current advances and future perspectives. Expert Rev. Mol. Diagn. 17 (4), 307—309.

Tekpinar, A.D., Kalmer, A., 2019. Utility of various molecular markers in fungal identification and phylogeny. Nova Hedwigia 109 (1), 187—224. https://doi.org/10.1127/nova_hedwigia/2019/0528.

Tuan, V.P., Narith, D., Tshibangu-Kabamba, E., Dung, H.D.Q., Viet, P.T., Sokomoth, S., Binh, T.T., Sokhem, S., Tri, T.D., Ngov, S., Tung, P.H., 2019. A next-generation sequencing-based approach to identify genetic determinants of antibiotic resistance in Cambodian *Helicobacter pylori* clinical isolates. J. Clin. Med. 8 (6), 858.

van Nood, E., et al., 2013. Duodenal infusion of donor feces for recurrent *Clostridium difficile*. N. Engl. J. Med. 368, 407—415.

Vatanen, T., Franzosa, E.A., Schwager, R., Tripathi, S., Arthur, T.D., Vehik, K., et al., 2018. The human gut microbiome in early-onset type 1 diabetes from the TEDDY study. Nature 562, 589—594.

Wagner, B.D., Grunwald, G.K., Zerbe, G.O., et al., 2018. On the use of diversity measures in longitudinal sequencing studies of microbial communities. Front. Microbiol. 9.

Wang, Z., Gerstein, M., Snyder, M., 2009. RNA-Seq: a revolutionary tool for transcriptomics. Nat. Rev. Genet. 10, 57—63. https://doi.org/10.1038/nrg2484.

Weinstock, G.M., 2000. Genomics and bacterial pathogenesis. Emerg. Infect. Dis. 6, 496—504. https://doi.org/10.3201/eid0605.000509.

Wilson, M.R., Naccache, S.N., Samayoa, E., Biagtan, M., Bashir, H., Yu, G., Salamat, S.M., Somasekar, S., Federman, S., Miller, S., Sokolic, R., 2014. Actionable diagnosis of neuroleptospirosis by next-generation sequencing. N. Engl. J. Med. 370 (25), 2408—2417.

Woollard, P.M., Mehta, N.A., Vamathevan, J.J., Van Horn, S., Bonde, B.K., Dow, D.J., 2011. The application of next-generation sequencing technologies to drug discovery and development. Drug Discov. Today 16 (11—12), 512—519.

Xie, Y., Du, J., Jin, W., Teng, X., Cheng, R., Huang, P., Xie, H., Zhou, Z., Tian, R., Wang, R., Feng, T., 2019. Next-generation sequencing for the diagnosis of severe pneumonia: China, 2010—2018. J. Infect. 78 (2), 158—169.

Yang, R.H., Su, J.H., Shang, J.J., Wu, Y.Y., Li, Y., Bao, D.P., Yao, Y.J., 2018. Evaluation of the ribosomal DNA internal transcribed spacer (ITS), specifically ITS1 and ITS2, for the analysis of fungal diversity by deep sequencing. PLoS One 13 (10), 1—17. https://doi.org/10.1371/journal.pone.0206428.

Yao, Z., Petschnigg, J., Ketteler, R., Stagljar, I., 2015. Application guide for omics approaches to cell signalling. Nat. Chem. Biol. 11, 387—397. https://doi.org/10.1038/nchembio.1809.

Zhang, X., Li, L., Butcher, J., Stintzi, A., Figeys, D., 2019. Advancing functional and translational microbiome research using meta-omics approaches. Microbiome 7, 154. https://doi.org/10.1186/s40168-019-0767-6.

Zhong, Y., Xu, F., Wu, J., Schubert, J., Li, M.M., 2021. Application of next-generation sequencing in laboratory medicine. Ann. Lab. Med. 41 (1), 25—43.

Zhou, W., Sailani, M.R., Contrepois, K., Zhou, Y., Ahadi, S., Leopold, S.R., et al., 2019. Longitudinal multi-omics of host—microbe dynamics in prediabetes. Nature 569, 663—671.

Zierer, J., Jackson, M.A., Kastenmuller, G., Mangino, M., Long, T., Telenti, A., et al., 2018. The fecal metabolome as a functional readout of the gut microbiome. Nat. Genet. 50, 790—795.

Further reading

Liu, Z., Zhu, L., Roberts, R., Tong, W., 2019. Toward clinical implementation of next-generation sequencing-based genetic testing in rare diseases: where are we? Trends Genet. 35 (11), 852—867.

CHAPTER 5

COVID-19 and microbiome

Olugbenga Samuel Michael[1,2,7], Juliana Bunmi Adetunji[3],
Olufunto Olayinka Badmus[4], Emmanuel Damilare Areola[5],
Ayomide Michael Oshinjo[1], Charles Oluwaseun Adetunji[6] and
Oluwafemi Adebayo Oyewole[8]

[1]Cardiometabolic, Microbiome and Applied Physiology Laboratory, Department of Physiology, College of Health Sciences, Bowen University, Iwo, Osun State, Nigeria; [2]Department of Physiology, University of Tennessee Health Science Center, Memphis, TN, United States; [3]Nutritional and Toxicological Research Laboratory, Department of Biochemistry Sciences, Osun State University, Osogbo, Osun State, Nigeria; [4]Department of Physiology and Biophysics, Cardiorenal, and Metabolic Diseases Research Center, University of Mississippi Medical Center, Jackson, MS, United States; [5]Department of Physiology, College of Health Sciences, University of Ilorin, Ilorin, Kwara, Nigeria; [6]Applied Microbiology, Biotechnology and Nanotechnology Laboratory, Department of Microbiology, and Directorate of Research and Innovation, Edo State University Uzairue, Iyamho, Auchi, Edo State, Nigeria; [7]Department of Medical Pharmacology and Physiology, University of Missouri, Columbia, MO, United States; [8]Department of Microbiology, Federal University of Technology, Minna, Nigeria

Introduction

Partial to total global lockdown that crippled economies, education, and social activities because of severe acute respiratory syndrome coronavirus 2 (SARS-CoV-2) would be unthinkable in the 21st century due to advancement of medicine, science, and technology. However, the world could not withstand the deleterious effects and mortality rates of the deadly virus, which has been recorded as a global pandemic that have killed approximately 6 millions of people across the globe (Kolahchi et al., 2021; Naseer et al., 2023).

There were about 3.3 million COVID-19 infected people globally as of May 2020. Interestingly, most individuals who have COVID-19 present with mild symptoms, but the disease can also be aggressively dangerous causing increased hospitalizations, respiratory failure, intensive care unit admissions, or death. This disease has stretched and broke the most sophisticated health systems in the world within few months (Onder et al., 2020).

There are interesting association between respiratory viral infection and impaired microbiome dynamics resulting in severe gastrointestinal (GI) dysbiosis (Yildiz et al., 2018). Approximately 2%–20% COVID-19 patients had altered gut functions with symptoms such as diarrhea. The GI system has been suggested to be the site of action and replication of SARS-CoV-2 outside respiratory system because more than 50% COVID-19 patients have SARS-CoV-2 detected in their stool samples and anal swabs (Cheung et al., 2020; Wölfel et al., 2020). Hence, COVID-19 severity is connected to increase SARS-CoV-2 levels in the feces due to altered intestinal and fecal microbiome composition.

Severity of COVID-19 differs among populations from mild symptoms to severe multiorgan damage especially among individuals with preexisting conditions such as metabolic syndrome, obesity, hypertension, and diabetes known to be linked with gut

An Introduction to the Microbiome in Health and Diseases
ISBN 978-0-323-91190-0,
https://doi.org/10.1016/B978-0-323-91190-0.00005-9

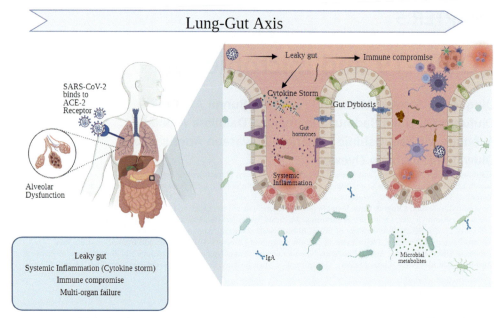

Figure 5.1 COVID-19 and microbiome interaction. COVID-19 infection causes gut microbiome dysbiosis altering the composition of the microbiome resulting in immune compromise, cytokine storm, systemic inflammation, and multiorgan failure. Cross-talks exist between the gut and lung microbiome mediating the COVID-19 disease progression and severity. *(Created with Biorender.com.)*

microbiome dysbiosis. GI dysfunctions such as abdominal pain, inflammation, vomiting, and diarrhea are initial presentations associated with SARS-CoV-2 infection. Therefore, existing dysbiosis in individuals with metabolic disorders, diabetes, and inflammatory bowel disease is connected to severe COVID-19 infection in this population (Liu and Lou, 2020). Furthermore, COVID-19 infection is much more in the elderly because microbial diversity is lower with advancing age.

Microbiome involvement in COVID-19 infection and severity cannot be overemphasized Fig. 5.1 in transmission of the virus among humans (Xiao et al., 2020). SARS-CoV-2 is transmitted among humans through airborne respiratory droplets and fecal—oral route (Gu et al., 2020a; D'amico et al., 2020). Angiotensin-converting enzyme 2 (ACE2) receptor is used by SARS-CoV-2 to enter host. ACE2 is localized within GI and respiratory system. Microbial ecology of the gut and intestinal inflammation are regulated by ACE2 (Hashimoto et al., 2012). In the elderly, there is increased expression of ACE2 receptor in duodenum suggesting involvement of gut microbiome in COVID-19 infection severity and its association with aging (Yang et al., 2020).

Covid-19, ACE2, and microbiome

Many body organs, from the respiratory system to the cardiovascular and brain systems, are affected by SARS-CoV-2 (Noris et al., 2020). To facilitate cellular entrance, the coronavirus' S protein attaches to the ACE2 during infection (Singh et al., 2020). This interaction causes the viral RNA to be released into respiratory epithelial cells, where it may replicate and spread to other cells. ACE2 regulates blood pressure and fluid balance (Cole-Jeffrey et al., 2015; Patel et al., 2016). ACE2 is found in a variety of tissues, such as colon, liver, kidneys, heart, and the lungs (Hamming et al., 2004). In recent years, research has indicated high ACE2 expression in the gut, implying its important function in the digestive system (Harmer et al., 2002; Ziegler et al., 2020). By regulating amino acid balance, antimicrobial protein production, and interacting with the gut microbiome, ACE2 is now recognized to modulate intestinal immunity (Vuille-dit-Bille et al., 2014, 2020). The gut is regarded as a major site of coronavirus infection because of the bonding of ACE2 for entry, replication, and spread. The subsequent impaired ACE2 activity is linked to perturbed microbiome homeostasis triggering events that present as GI symptoms of COVID-19 (Perlot and Penninger, 2013; Kotfis and Skonieczna-Zydecka, 2020). This aligns with the prevailing theory that ACE2 expression is downregulated in the infection, triggering microbiome alteration and affecting homeostatic balance with a predisposition toward pathogenic microbes and abnormal levels of gut metabolites such as amino acids, bile acids, and SCFAs (He et al., 2020; Yu et al., 2021). For example, Zuo and colleagues observed the growth of harmful fungal infections in the digestive tracts of infected participants resulting in a prolonged clinical course beyond negative nasopharyngeal results (Zuo et al., 2020a,b). Similarly, other investigators have reported the proliferation of pathogenic bacteria and reduced beneficial bacteria (*Lactobacillus and Bifidobacterium*) as well as antiinflammatory bacteria (*Faecalibacterium prausnitzii*) (Xu et al., 2020; Zuo et al., 2021). COVID-19 has also lowered butyrate concentrations, a critical gut metabolite with immunomodulatory properties in another investigation (Liu et al., 2018; Briguglio et al., 2020). COVID-19-induced gut dysbiosis influences pathogenicity and inflammatory cytokine concentrations via reducing the immunomodulatory capability of gut microbiota (Yeoh et al., 2021). Furthermore, bacteria from the Bacteriodetes family, negatively associated with fecal SARS-CoV-2 load, were demonstrated to suppress ACE2 production (Zuo et al., 2021). In summary, these findings suggest that COVID-19 patients lose their ACE2-protective properties, resulting in gut dysbiosis, defective gut barrier permeability, and hence altered immunological response (inflammation) along the gut—lung axis.

COVID-19 and fecal microbiome transplantation

This condition is largely considered a respiratory infection due to the predominance of indicators of pulmonary dysfunction (Lauer et al., 2020; Rodriguez-Morales et al., 2020). GI issues occurred in patients with COVID-19 (Mao et al., 2020). Approximately 39% of COVID-19 patients face GI complications (Jin et al., 2020). Numerous investigations discovered viral RNA shedding in stool samples of infected patients (Parasa et al., 2020). Digestive issues and fecal viral particles suggest that the coronavirus interacts with the GI system, which is home to the gut microbiota that has immunomodulatory and antiinflammatory effects (Adak and Khan, 2018). Interestingly, infected patients have shown considerable alteration in the respiratory and GI microbiome (Sulaiman et al., 2021; Xu et al., 2021). Several researchers have considered this dysbiosis as a contributor to the GI dysfunction observed in affected patients (Yeoh et al., 2021). Thus, fecal microbiota transplantation (FMT), a gut microbiota-targeted procedure, is thought to hold a plausible therapeutic effect for COVID-19 patients. The procedure involves the transfer of gut microflora sourced from fecal material of normal donor to the affected intestinal tract (Wang et al., 2019; Michael et al., 2022). This transfer has been demonstrated to restore altered microbial communities and treat dysbiotic disease conditions such as recurrent *Clostridioides difficile* infection, diabetes, and colitis (Paramsothy et al., 2017; Michael et al., 2022). Bidirectional communication connecting lungs and microbiota is thought to be involved in FMT's therapeutic impact. Activities along this gut—lung axis can directly (via ACE2 and gut metabolites) and indirectly (via immunomodulation) impact the GI tract and the lungs. In an investigation by Zuo et al., a significant positive association was noted between 23 bacterial taxa and the progression of the COVID-19 infection (Zuo et al., 2020a,b). The scientists discovered a link connecting virus concentration and specific *Bacteroides* species linked to lower GI ACE2 expression (Ng and Tilg, 2020). Furthermore, COVID-19 patients were shown to have abundant harmful bacteria and low beneficial bacteria (Yeoh et al., 2021). One of the advantageous bacteria, *F. prausnitzii*, is a key generator of SCFAs with robust antiinflammatory capacity, which may protect against COVID-19-induced inflammation and progression of the infection in patients (Włodarczyk et al., 2022). Furthermore, FMT also indirectly impacts the immune system, which is responsible for handling the wave of released cytokines in affected patients (Huang et al., 2020). Despite multiple supporting calls and hypotheses around FMT as a potential emerging COVID-19 therapy, there are limited investigations around FMT as treatment (Apartsin and Smirnova, 2020; Nejadghaderi et al., 2020; Wang et al., 2022). An early investigation was conducted at Nanjing Medical University, China, to determine the efficacy of washed microbiota transplantation combined with conventional treatment on the disease course and severity. However, the study trial was withdrawn to adhere to new government policies on disease control (Zhang, 2020). Nonetheless, several ongoing clinical trials in China, France, and Switzerland are looking at the link between the intestinal microbiota's makeup (and transplantation) and the intensity, mortality, and well-being in infected patients (Buehler, 2020; Nseir, 2020; Wu et al., 2021). Interestingly, few investigations hint at the potential of FMT for COVID-19 treatment. For example, Liu and others performed FMT on 11 convalescent COVID-19 patients to observe how it affected gut microbiota—immune system interactions (Liu et al., 2021). Forty-five percent (five participants) of the study group experienced an improvement in GI issues after the FMT procedure (Liu et al., 2021). Although overall diversity remained constant, increased microbial diversity was observed (increased abundance of *Bifidobacterium* and *Faecalibacterium*). Hence, FMT could be useful in patients experiencing GI complications even after negative nasopharyngeal clearance. The findings are also consistent with a recent case report by Biliński and colleagues (Biliński et al., 2021) that demonstrate the influence of FMT on disease trajectory in infected patients. They observed reduced severity of COVID-19 symptoms in two FMT-treated patients despite risk factors for adverse symptoms. The researchers have progressed further into initiating clinical investigations around FMT's immunomodulatory action (Biliński et al., 2021).

Microbiome and severity of COVID-19

Most nations have had to spend fortunes on the advent of fatal illnesses due to COVID-19, leading to a huge burden on the economy and healthcare systems. The disease is highly infectious and can survive on steel and plastic residing in aerosols (van Doremalen et al., 2020). Main cause of death after COVID-19 infection is attendant wounds in the lungs and development of acute respiratory illness (Huang et al., 2020; Qiu et al., 2019; Zhu et al. 2020). Patients infected with COVID-19 who were put in intensive care unit due to severe symptoms such as ARDS had a larger amount of proinflammatory cytokines, compared with patients without severe symptoms (Del Valle et al., 2020). Interestingly, gut inflammation and dysfunction seen in COVID-19 patients may be due to alteration of microbiome balance in gut (de Oliveira et al., 2021; Chakraborty et al., 2022; Zhang et al., 2023).

Deleterious outcomes of COVID-19 are mostly caused by abnormal immune responses and other separate inflammatory conditions such as metabolic syndrome. It is known that the severity of COVID-19 is related to elevated plasma proinflammatory cytokines, which point to dysregulated immune response and overt organ damage (Tay et al., 2020; Vabret et al., 2020). Some inflammatory conditions do not resolve after recovery of COVID-19 patients (Cheung et al., 2020; Galeotti and Bayry, 2020) showing the disease does not only progress via inflammatory responses but leaves the victims in a malignant state characterized by inflammation after recovery, which makes them susceptible to complications from other inflammation–related diseases such as hypertension and metabolic syndrome. In line with these findings, studies have elucidated the severity of COVID-19 infection based on extent of inflammatory responses judged by the occurrence of inflammatory mediators and markers in the blood stream (Chakraborty et al., 2022). It is also worthy of note that COVID-19 infection is still being studied and the processes that mediates the aggressive inflammatory responses associated with the disease is still unclear (Tay et al., 2020). Nevertheless, recent investigation of the human microbiomes along with COVID-19 disease received the attention of research works and posits the human microbiome dysconfiguration as a potential mechanism for the characteristic inflammatory response to COVID-19 infection (Yeoh et al., 2021; de Nies et al., 2023) (Fig. 5.1).

Invasion of the lungs by SARS-CoV-2 is what leads to overt COVID-19, which may be followed by complications that are related to generation of immune responses and lung infection. Since the coronavirus becomes part of the lung microbiota, human microbiome might play a significant function in advancement of COVID-19 disease (Yuki et al., 2020). The microbiota in humans varies from individual to individual and from one race to another (Brooks et al., 2018). Several studies have associated human microbiota with a number of disorders despite the fact that its role in these disorders has not yet been fully clarified. Some of the disorders that have strong association with human microbiota include obesity-related diseases, metabolic disorder, and Parkinson's disease (Lynch and Pedersen, 2016; Lee et al., 2020) among others.

Some scientific remarks emanating from many investigations have been made suggestions of significant involvement of the gut to COVID-19 infection. Observations such as detection of virus RNA in feces (Wölfel et al., 2020; Xu et al., 2020), infection and replication of SARS-CoV-2 in small intestine cells of human (Lamers et al., 2020), and changed gut microbiota in subjects infected with SARS-CoV-2 are striking (Zuo et al., 2020a,b).

Microbiome and hypertension in COVID-19

Hypertension is sustained elevated blood pressure above systolic of 140 mmHg and diastolic of 100 mmHg has many risk factors and is the basis for the complications associated with several communicable and noncommunicable diseases (Mills et al., 2020). With the deleterious microvascular and macrovascular complications associated with hypertension, several end-organ damage can develop within days to weeks of sustained uncontrolled elevated blood pressure (Schmieder, 2010; Stewart, 2023). With the current advent of pandemic, it is worthwhile to elucidate the association of hypertension with diseases such as COVID-19, which become a global health and economic burden (Schiffrin et al., 2020; Peng et al., 2021; Gallo et al., 2022). This will help to discover the contributions of hypertension to the severity and consequently mortality of COVID-19 and other infectious diseases (Peng et al., 2021; Shibata et al., 2023). Nevertheless, it is also expedient to understand the position of microbiomes in the association of hypertension with COVID-19 (Avery et al., 2021; Matsumoto et al., 2023).

Hypertension has been shown to increase the possibility of deleterious illnesses associated with COVID-19. Lippi et al. showed that hypertension increases the mortality of COVID-19 by approximately two and a half folds and up to nearly 30% of COVID-19 in-patients are diagnosed with hypertension (Lippi et al., 2019). Therefore, there appears to be a correlation between the incidence of hypertension and a worse case of COVID-19. Despite being ignored, intestinal dysbiosis is an underlying cause of hypertension (Yang et al., 2015; Santisteban et al., 2017). The findings of research works show that hypertensive individuals have a quite different microbiome compared with nonhypertensive individuals. A study observed that there was decrease in genus *Bifidobacterium* in subjects with hypertension (Santisteban et al., 2017). Durgan et al. proved that the transplant of dysbiotic fecal material from hypertensive rats to normotensive rats caused increased BP in the normal rats (Durgan et al., 2016, 2017). A similar study confirmed this finding, also noting that transplant of healthy fecal material from normotensive rats to hypertensive rats led to a reduction in BP of the hypertensive rats (Robles-Vera et al., 2017, 2020). Just as evidence accumulate that associates intestinal dysbiosis with incidence of hypertension, there are also evidence that GI symptoms may contribute to worsened cases of COVID-19. It is therefore evident that a healthy gut microbiome may reduce the risk for complications associated with hypertension and COVID-19 (Segal et al., 2020; Rocchi et al., 2022; Seong et al., 2023). It is plausible that the influence of the gut microbiota on immune responses and inflammation may be the mechanism by which it affects hypertensive traits and COVID-19 severity (Segal et al., 2020; Rocchi et al., 2022; Seong et al., 2023; Martín Giménez et al., 2023).

Gut microbiota, immune response, and COVID-19

The GI tract, also known as digestive system, plays vital roles in fighting various pathogenic infections (Zuo et al., 2020a,b). There are several microorganisms living inside the digestive tracts of an individual, and these include virus, fungi, bacteria, archeae, and other forms of life cooperatively known as gut flora, gut microbiome, or gut microbiota (Shreiner et al., 2015). The gut microbiota is a multifaceted ecological unit where the host obtains a beneficial relationship (Ferreira et al., 2020). The microbiome community influences several physiological functions in the body including preservation of the intestinal integrity, energy production, defense against disease causing organisms, and maintenance of immunological responses (Thursby and Juge, 2017; Wang et al., 2017a; Hasan et al., 2019). Nevertheless, when there is an alteration in the gut microbiota, and there is an imbalance in intestinal homeostasis due to factors such as dietary patterns, medication, aging, stress, environmental factors, and diseases, a condition known as dysbiosis arises (Hasan et al., 2019; Moran-Ramos et al., 2020).

There are existing reports revealing microbiome—immune interactions (Wu and Wu, 2012; Pagliari et al., 2018; Yoo et al., 2020). A balance between removal of disease-causing pathogens and the preservation of healthy self-tissue is maintained by immune system (Wu and Wu, 2012). Interestingly, microbiome is essential for the maintenance of immune system, and it influences adaptive and innate immunity (Smith et al., 2007). However, gut dysbiosis results in dysregulation of immune responses causing diseases such as metabolic disorder, inflammation, oxidative stress, insulin resistance, cardiovascular disease, autoimmune disease, diabetes, and a range of several cancer diseases (Yoo et al., 2020). Gut microbiota that communicates with the immune system to influence immune responses produces metabolites such as the short-chain fatty acids (SCFAs) (Belkaid and Hand, 2014). The SCFAs are produced in the gut lumen when bacteria digest complex carbohydrates by means of fermentation, and the different types of SCFAs include acetate, butyrate, and propionate (Kau et al., 2011).

Almost 20% of patients diagnosed with COVID-19 disease end up displaying severe complications involving an excessively active immune response similar to cytokine storm syndrome, which often results in a serious acute breathing disorder (Moore and June 2020). Even though this disease primarily infects the respiratory tract, accumulating evidence is suggesting that the disease has a connection with the GI tract in human. This is evident by GI symptoms, ranging from nausea, diarrhea, vomiting, abdominal pain, and intestinal inflammation, observed in patients diagnosed with COVID-19 disease (Ferreira et al., 2020; Cheung et al., 2020; Zuo et al., 2021), and these GI symptoms are often indicators of altered microbiome homeostasis in COVID-19.

Several studies recorded that structural makeup of gut microbiome can influence lung's vulnerability to viral infections such as COVID-19 infection, and likewise, virus infecting the lungs in cases of COVID-19 infection can broadly alter gut microbiome composition (van der Lelie and Taghavi, 2020; Zuo et al., 2021). Scientists have observed that impaired microbiome homeostasis is mostly found in people highly vulnerable to severe COVID-19, such as patients suffering from cardiometabolic disorders, aging-related, and autoimmune diseases (Li et al., 2019). COVID-19 patients showed few and suppressed symbiotic bacteria and beneficial bacteriophages compared with healthy controls, whereas there they have an abundance of opportunistic pathogens including fungi, bacteria, and eukaryotic viruses, usually connected with worsening of diseases. Unfortunately, this condition extends even after the disease resolved, and this can exert a harmful effect on the host immune system (Zuo et al., 2020a,b). In this same study, there was great quantity of opportunistic bacteria such as *Coprobacillus*, and *Clostridium hathewayi, Clostridium ramosum*, and reduced antiinflammatory bacterium such as *F. prausnitzii* in the feces of hospitalized COVID-19 patients with severe infection (Zuo et al., 2020a,b). Likewise, in an ecological network analysis of COVID-19 patients, it was found that there were positive correlations of opportunistic bacteria dominating the ecosystem of the gut flora as a result of SARS-CoV-2 infection (Gu et al., 2020b).

Two-hospital cohort study revealed that 100 confirmed COVID-19 patients exhibited significantly altered gut microbiome composition compared with individuals without COVID-19 individuals, regardless of whether COVID-19 patients have recovered or received medication. In this study, commensal bacteria in the gut that act as immunomodulators, such as *bifidobacteria, Eubacterium rectal*, and *F. prausnitzii* were diminished in patients diagnosed with COVID-9 disease and remained low in stool samples collected up to 30 days after resolution of the disease. Also, the disturbed gut microbiota is associated with increased levels of inflammatory biomarkers and cytokines (Yeoh et al., 2021). Likewise, Ferreira et al. documented that elevated vulnerability to poor outcome in patients diagnosed with COVID-19 can be notably predicted by an alteration of the gut microbiome (Ferreira et al., 2020).

Therefore, an obvious connection exists between gut flora and COVID-19 infection, ultimately affecting immune responses. Gut microbiome dysbiosis can contribute to the likelihood of having COVID-19 as well as determining the severity of COVID-19 infection; likewise, COVID-19 infection can disrupt intestinal homeostasis and influence gut microbiota (Ferreira et al., 2020; Yeoh et al., 2021). This understanding will be helpful in discovering biomarkers and producing microbiome-based therapy for COVID-19 treatment and prevention.

Neonatal microbiome and COVID-19

The most important source of gut organisms in a newborn is from the maternal gut microbiota, to which the newborn is exposed (Hartz et al., 2015). The structural makeup and variety of microbiome in early life of neonate influence acquired and innate immune systems with subsequent health consequences in adulthood (Ebrahimi et al., 2021). The intrauterine environment such as amniotic fluid, fetal meconium, and placenta possess their microbiome (Dalby and Hall, 2020; Xiao and Zhao, 2023). Hence, the gut microbiome begins to develop during fetal life. In about 1 week after birth, babies would have developed gut microbiome, suggesting that babies are exposed to these bacteria from the placenta before birth, rather than after birth (Hartz et al., 2015). Neonatal microbiome is further determined by other important factors aside maternal microbiome, and these include mode of infant delivery, feeding choice, lifestyle, living environment, and use of antibiotics (Greenwood et al., 2014; Hartz et al., 2015). Breast milk transmits gut microbiome to an infant, and breastfeeding improves gut microbiome compared with babies fed with formula (Penders et al., 2013; Jost et al., 2014). Breastfeeding can directly influence early-life gut microbiome by exposing the neonate to the breast milk microbiota. Also, breastfeeding can in some other ways affect the gut microbiota of neonates through factors such as secretory IgA, human milk oligosaccharides, and antimicrobial factors since they possess the ability to affect bacterial growth and metabolism (van den Elsen et al., 2019). Oligosaccharide, a vital prebiotic found in breast milk of human being, aids the development of a beneficial gut microbe such as *Bifidus*-predominant flora (Stiverson et al., 2014). *Bifidobacterium* represented the most predominant genus that dominates the microbiota of infant gut that were vaginally delivered and breastfed (Turroni et al., 2018), and it is of importance to note that *Bifidobacteria* present in the infant gut possess immunomodulatory function enhancing childhood defense against infections (Hidalgo-Cantabrana et al., 2017; Saturio et al., 2021; Stuivenberg et al., 2022; Gavzy et al., 2023). Also, from earlier studies, there are reports that the route of delivery, either through vagina or caesarean section, affects the colonization of bacteria in the gut of the neonates. Babies born through the vagina have different gut bacteria microbiome when compared with babies delivered by caesarean section (Dominguez-Bello et al., 2010; Dunn et al., 2017). Shao et al. documented that babies born via caesarean section were deprived of beneficial bacteria transmitted from the maternal, but they had more opportunistic bacteria that are common in hospital environments, and were more likely to have antimicrobial resistance (Shao et al., 2019). Also, Kim et al. documented that caesarean section babies were deprived of vaginal microbiome, an essential factor in developing neonatal microbiome (Kim et al., 2020). Also, it has been observed that because babies born through caesarean section did not pass through the birth passageway that has been occupied with maternal microbiome, they have delay in the development of their gut microbiome (Dominguez-Bello et al., 2010).

COVID-19 infection has been connected with lower severity, morbidity, and death among infants and children; however, the complications of reduced microbiome in the neonates remain a challenge secondary to the pandemic (Romano-Keeler et al., 2021). Earlier study reported that women infected with COVID-19 have higher occurrence of cesarean section than women without COVID-19 (Giaxi et al., 2020); this can impact the gut diversity of the neonates. Some of the reasons for increased cesarean section in women infected with COVID-19 include fetal distress, deterioration of the mother's condition, previous cesarean delivery, and severe preeclampsia (Chen et al., 2020; Ashokka et al., 2020). Unfortunately, cesarean section does not prevent COVID-19 transmission from mother-to-child better than vaginal delivery will do (Cai et al., 2021). Additionally, babies delivered prematurely have altered gut microbiome unlike the termed babies because of the babies' exposure to a different condition (Henderickx et al., 2019). Moreover, the bacterial composition and the frequency of bacteria in breast milk of mothers with premature delivery were lower and not like milk of mothers that gave birth to mature babies (Moles et al., 2015). Unfortunately, earlier studies revealed that about 47% of hospitalized pregnant women diagnosed with COVID-19 give birth to their babies prematurely (Galang et al., 2020; Mullins et al., 2020; Dashraath et al., 2021)

Furthermore, route, delivery time, and choice of feeding an infant also influence microbiome development in the early stages of life. It has been recommended that COVID-19 mothers continue to breastfeed their infants since there has been no evidence of the virus in the breast milk, neither is there any evidence of transmission between mother and child (Heydarpoor and Valizadeh, 2020; Salvatori et al., 2020; WHO, 2020a,b). Breast milk contains probiotics such as *Bifidobacterium* and *Lactobacillus* that contribute to the gut microbiota of breastfed infant (Fernández et al., 2013). However, so many factors can limit a COVID-19-infected mother from breastfeeding the neonate, and these factors include the severity of the mother's illness, misconceptions about safety of breastfeeding, hospital policies, and whether the mother is quarantined after discharge (WHO, 2020a,b). Unfortunately, pumped breast milk can be an option in a situation where the mother cannot directly breastfeed the baby; however, the severity of the mother's sickness has been associated with reduced milk richness, reduced bacterial variety, and altered microbiota composition of breast milk especially bifidobacteria (Moossavi et al., 2019; Moossavi and Azad, 2020).

Previous studies documented that several COVID-19 pregnant patients used antibiotics without bacterial infection (Hantoushzadeh et al., 2020; Wang et al., 2020). Antibiotics have been linked with decreased amount of commensal bacteria in gut microbiome of infants (Yassour et al., 2016; Ebrahimi et al., 2021). From earlier studies, maternal use of antibiotics during and even before pregnancy has destructive effects of altering microbiome composition in infants with resulting effects on the immune system (Greenwood et al., 2014).

Neonates and children are less vulnerable to COVID-19 like adults, but secondary effects of disrupted gut microbiota are demonstrated in neonates with mothers infected with COVID-19. Factors contributing to reduced microbial diversity in the early years of life of neonates born from COVID-19-infected mothers include increased incidence of cesarean delivery, limited direct breastfeeding, and increased use of antibiotics, and this eventually lowers immune response.

Probiotics and COVID-19

In 2020, Pourhossein and Moravejolahkami documented the efficacy of probiotics against all kinds of viral infections through stimulation of antibodies and antiinflammatory interleukins production (Pourhossein and Moravejolahkami, 2020). Consequently, upon probiotics supplementation, there is a decrease in the viral load, which could be deduced from the immune system modulation by the available microbiomes against COVID-19 virus directly or via depreciation in secondary risk infections involving multiple COVID-19 experiments by extended antibiotic exposure (Adnan and Dewi, 2020). During the pandemic, the National Health Commission China declares that probiotics can be used by COVID-19 threatened patients for protection against secondary infections and for intestinal microbiome balance restoration. Moreover, COVID-19 fatality in aged individuals has been linked to the decrease in the available microbiome in the guts. Nagpal et al. (2018) established that changes in taxonomic and aging are related to the reduction in the microbiome of metabolic benefit, which includes SCFAs down generation. Therefore, ingested probiotic supplements could be enhanced by gut microbiota, hence resulting in an improved immunity for the immunocompromised individuals.

SARS-CoV-2 causes lung function compromised through its attachment to ACE2 receptors present on the epithelial cells alveolar. Tian et al. (2020) reported that viral RNA was revealed to be present in stool of COVID-19 individuals, while Gou et al. (2020) documented that predisposition of several groups to COVID-19 could be connected to the microbiota composition.

Consequently, Kang et al. (2013) reported that *Bifidobacterium* and *Lactobacillus* found in probiotics were capable of inhibiting respiratory infection by common cold and SARS-CoV-2. Supplementation containing probiotics was suggested to reduce severity or/and duration of COVID -19 in individuals. Therefore, cytokines stimulation due to aggressive inflammatory response in COVID-19 patients could be responsible for the high mortality rate. Hence, the ability to inhibit the production of the cytokines would serve a beneficial potential in managing COVID-19. Cytokines inhibition in COVID-19 threatened individuals using available therapeutics will be of immense benefit in hyperinflammatory conditions that might be revealed in severe COVID-19 casualty individuals (Montazersaheb et al., 2022).

Preclinical studies on pneumonia and influenza using probiotics administered through nasal or oral route were found to be capable of minimizing bronchial epithelial alterations, reducing lung viral loads and weight loss, and extending survival rate (Kumova et al., 2019; Shahbazi et al., 2020). Interestingly, probiotics protection was linked to timely innate immune system recruitment via NK lymphocytes and proinflammatory cytokines such as TNF-α and IL-6, etc., and the alveolar macrophages. The process is connected to the decline stimulated by the antiinflammatory mediators such as Tregs cells (regulatory T cells) and interleukin 10 in the lungs, to prevent autoimmune diseases and tissue injuries. Consequently, vitamin D and vitamin D receptor (VDR) modulation by probiotics and the ability to balance the microbiome composition and growth in the gut via the GI, which suggest its immune modulatory potential on cytokine storm. Hence, prevention of secondary infection in COVID-19 with the use of probiotics having antiinflammatory potentials might be beneficial in maintaining the equilibrium of intestinal microecology (Kurian et al., 2021; Elekhnawy and Negm, 2022).

Furthermore, Minato et al. (2020) established that carboxypeptidase gotten from *Paenibacillus* sp. B38 was able to decrease hypertension-mediated cardiac hypertrophy and fibrosis caused by angiotensin II in mice. Meanwhile, COVID-19 is connected to thrombotic and coagulopathy complications, which serve as a critical pointer to infections caused by SARS-CoV-2 (Becker, 2020). It was documented that lactic acid bacteria (LAB) probiotic is vital in controlling the host immunocoagulative response during the infections caused by a respiratory syncytial virus, IFV, and *Streptococcus pneumonia* (Zelaya et al., 2014).

In 2019, Yu and coresearchers evaluated the therapeutic efficacy of a probiotic derived from *Weissella cibaria* possesses potential to repress NF-κB as a result of its biological activity, which contributed to the antiinflammatory potentials (Yu et al., 2019). In another experiment, Schmitter et al. (2016) demonstrated availability of *Lactobacillus paracasei* and *Lactobacillus plantarum* strains could possess a potent attribute in decreasing response to inflammation by the immune system. Meanwhile, *Lactobacillus reuteri* and some *Lactobacillus* strains probiotics could be a biofilm that can produce certain biological factors with antiinflammatory features (Ayyanna et al., 2018).

Conte and Toraldo (2020) described that an alternative approach might be of benefit in curbing the COVID-19 infection inflammation through immune system enhancement using pre- or probiotics therapy especially in the event of a failure in vaccine production. Also, Bottari et al. (2020) documented that mucosal immunity development via IgA secretion stimulation, phagocytosis function, and regulatory cell adjustment stimulation can be the source of probiotic immune system benefit. More so, it was established that lots of scientific reports are available to support the efficacy of some nutrients and probiotics through the enhancement of immune function, which could manage and obstruct COVID-19 caused infections (Jayawardena et al., 2020).

In all, Jamal and Khaled in 2021 reported that the immune, functional, and health advantages of both prebiotic and probiotic have now been researched into because it could be a potent biological agent against COVID-19 infection (Jamal and Khaled, 2021).

Butyrate and COVID-19 management

Numerous probiotic and micronutrients containing supplements have been tested to inhibit inflammation and restore the microbiota of the gut. Though, butyric acid, an anaerobic bacteria fermentation product of nondigestible carbohydrates in the gut, could also be used to attenuate the deadly action of the immune response in aggravating inflammation. Reduction of the deleterious immune response and hyperinflammation is caused from SCFAs synthesized by the gut microbiota. Butyrate, the most often studied SCFA, has been documented for possessing antiinflammatory properties through the depletion of the cytokine storm. Additionally, butyrate was reported to ameliorate hyperinflammation and lower oxidative damage (Parada Venegas et al., 2019; Nithin et al., 2021; Jardou et al., 2021).

COVID-19 risk has been documented to be on the increase in hypertensive patients and also exhibit some GI symptoms that could affect the communications between the gut—lung and hence might result in multiorgan pathologies with diverse cardiovascular manifestations (Sharma et al., 2020). It was established by Li et al. (2020) that there is an elevation in key molecules of SARS-Cov-2 infection, which are ACE2 expression and transmembrane protease serine-2 (TMPRSS2) as observed in spontaneously hypertensive rats gut epithelium. Furthermore, butyrate (histone deacetylase), an SCFA to the gut microbiome, keeps changing in hypertensive and COVID-19-infected individuals (Sharma et al., 2020).

In 2021, Li and coresearcher demonstrated the downregulation of essential genes in SARS-CoV-2 by butyrate. The authors report elevated COVID-19 infection risk in hypertensive patient is linked to butyric acid—producing microbiota cumulative depletion in the gut (Li et al., 2021).

However, continuous synthesis of SCFAs such as butyrate as a metabolite of the microbiome can improve health conditions (Tayyeb et al., 2021). The mechanism of its action could be via histone deacetylase (HDAC) inhibition or activated G protein—coupled receptors (GPCRs). HDAC inhibition by SCFAs has a great impact on several immunoregulatory cells (He et al., 2020). The HDAC inhibitor sodium butyric acid can reduce GI symptoms linked with CoVID-19 by suppressing gut ACE2 expression epithelial cells (Takahashi et al., 2021). Some anaerobic bacteria in the colon, which include *F. prausnitzii, Roseburia intestinalis, Mitsuokella multiacida, Megasphaera elsdenii*, and *Clostridium butyricum*, are responsible for the synthesis of butyrate as a source of fuel (Pituch et al., 2013). Supplementing diet in obese mice with sodium butyrate boosts the number of beneficial bacteria such as Bifidobacteriaceae, *Coprococcus, Ruminococcus, Actinobacteria*, and Lachnospiraceae enhancing the intestinal barrier integrity (Fang et al., 2019).

In mice, pretreatment with butyrate can reduce oxidative stress, inflammation, and cardiac depression merged with septic shock (Wang et al., 2017b). Hence, suggestion has been made to the use of prophylactic medication such as sodium butyric acid lowering the activity of myeloperoxidase and infiltration of inflammatory cell in the lungs, which is linked to the reduction of HMGB1 expression, proinflammatory cytokines, and NF-κB expression.

The potential of butyrate in inflammation and tissue-protective effects in the lungs and gut and its modulatory potentials on gut microbiome diversity leading to more synthesis of butyrate is a pointer to the beneficial effect on COVID-19.

Gut microbiome and COVID-19-associated gastrointestinal dysfunctions

The GI symptoms arising from COVID-19 are abdominal pain, diarrhea, and/or vomiting (Villapol, 2020). Gut microbial changes accompanied by increased inflammatory cytokines are results of dysfunctions intestine. Also, patients with symptoms show changes in gut microbiome composition and function that could be involved in intestinal barrier dysfunction and immune activation (Marasco et al., 2021). Villapol reported that early detection and treatment of COVID-19 could be recognized by GI symptoms that ensue in respiratory difficulties. Metabolic microbiome composition and its products are useful tools to the identification of biomarkers of novel disease and its therapeutic targets (Villapol, 2020). Authors also documented that diverse diet strengthen intestinal barrier with reduced proinflammatory markers (Olendzki et al., 2022; Prame Kumar et al., 2023; Liang et al., 2023). It was documented that frequent dysfunction of the respiratory system is linked with decreased biodiversity and changes in gut microbiomes after prolonged COVID-19 infection (Vestad et al., 2022).

Summary and conclusion

It has been documented that COVID-19 has affected people around the globe. Moreover, a lot of effort are in place to develop a mRNA vaccines that could help in mitigating against the spread of the ravaging deadly virus that keeps mutating and spreading across the countries of the world. Interestingly, some microbiome-derived metabolites (acetate and butyrate) have been shown to be promising in boosting immune response to the virus. However, the efficacy of microbiome-based therapeutics against COVID-19 is yet to be clinically validated. Hence, future investigations may provide information on the variation in microbiome among individuals, their differences in susceptibility to COVID-19, variation in development of severity to the viral infection, and microbiome composition. Since microbiome has immunomodulatory effects that have been suggested to offer some protection against COVID-19, further clarification into the immunoregulatory activities of the microbiome especially in individuals with low susceptibility or asymptomatic or mild symptoms to SARS-CoV-2 is required. Microbiome-based therapeutics could also be a complementary intervention alongside COVID-19 vaccines to improve their efficacy and prolong the defense against COVID-19 infection.

References

Adak, A., Khan, M.R., 2018. An insight into gut microbiota and its functionalities. Cell. Mol. Life Sci. 76 (3), 473−493. https://doi.org/10.1007/s00018-018-2943-4.

Adnan, M.L., Dewi, M.D., 2020. Potential effects immunomodulators on probiotics in COVID-19 preventing infection in the future. A narrative review. Int. J. Med. Stud. https://doi.org/10.5195/ijms.2020.486.

Apartsin, K., Smirnova, V., 2020. Convalescent faecal microbiota transplantation as a possible treatment for COVID-19. Clin. Res. Hepatol. Gastroenterol. 44 (5). https://doi.org/10.1016/j.clinre.2020.08.003. Article e113-e114.

Ashokka, B., Loh, M.H., Tan, C.H., Su, L.L., Young, B.E., Lye, D.C., et al., 2020. Care of the pregnant woman with coronavirus disease 2019 in labor and delivery: anesthesia, emergency cesarean delivery, differential diagnosis in the acutely ill parturient, care of the newborn, and protection of the healthcare personnel. Am. J. Obstet. Gynecol. 223 (1), 66−74.e3.

Avery, E.G., Bartolomaeus, H., Maifeld, A., Marko, L., Wiig, H., Wilck, N., Rosshart, S.P., Forslund, S.K., Müller, D.N., April 2, 2021. The gut microbiome in hypertension: recent advances and future perspectives. Circ. Res. 128 (7), 934−950. https://doi.org/10.1161/CIRCRESAHA.121.318065.

Ayyanna, R., Ankaiah, D., Arul, V., 2018. Anti-inflammatory and antioxidant properties of probiotic bacterium Lactobacillus mucosae AN1 and Lactobacillus fermentum SNR1 in wistar albino rats. Front. Microbiol. 9, 3063. https://doi.org/10.3389/fmicb.2018.03063.

Becker, R.C., 2020. COVID-19 update: covid-19-associated coagulopathy. J. Thromb. Thrombolysis 50, 54−67.

Belkaid, Y., Hand, T.W., 2014. Role of the microbiota in immunity and inflammation. Cell 157, 121−141.

Biliński, J., Winter, K., Jasiński, M., Szcześ, A., Bilinska, N., Mullish, B.H., Małecka-Panas, E., Basak, G.W., 2021. Rapid resolution of COVID-19 after faecal microbiota transplantation. Gut. gutjnl−2021−325010. https://doi.org/10.1136/gutjnl-2021-325010.

Bottari, B., Castellone, V., Neviani, E., 2020. Probiotics and covid-19. Int. J. Food Sci. Nutr. https://doi.org/10.1080/09637486.2020.1807475.

Briguglio, M., Pregliasco, F.E., Lombardi, G., Perazzo, P., Banfi, G., 2020. The malnutritional status of the host as a virulence factor for new coronavirus sars-cov-2. Front. Med. 7. https://doi.org/10.3389/fmed.2020.00146.

Brooks, A.W., Priya, S., Blekhman, R., Bordenstein, S.R., 2018. Gut microbiota diversity across ethnicities in the United States. PLoS Biol. 16, e2006842. https://doi.org/10.1371/journal.pbio.2006842. PMID: 30513082.

Buehler, S., 2020. Microbiota in COVID-19 Patients for Future Therapeutic and Preventive Approaches (MICRO-COV). https://clinicaltrials.gov/ct2/show/NCT04410263.

Cai, J., Tang, M., Gao, Y., Zhang, H., Yang, Y., Zhang, D., Wang, H., Liang, H., Zhang, R., Wu, B., February 17, 2021. Cesarean section or vaginal delivery to prevent possible vertical transmission from a pregnant mother confirmed with COVID-19 to a neonate: a systematic review. Front. Med. 8, 634949.

Chakraborty, C., Sharma, A.R., Bhattacharya, M., Dhama, K., Lee, S.S., July 7, 2022. Altered gut microbiota patterns in COVID-19: markers for inflammation and disease severity. World J. Gastroenterol. 28 (25), 2802—2822. https://doi.org/10.3748/wjg.v28.i25.2802.

Chen, H., Guo, J., Wang, C., Luo, F., Yu, X., Zhang, W., et al., March 7, 2020. Clinical characteristics and intrauterine vertical transmission potential of COVID-19 infection in nine pregnant women: a retrospective review of medical records. Lancet 395 (10226), 809—815.

Cheung, K.S., Hung, I.F.N., Chan, P.P.Y., et al., 2020. Gastrointestinal manifestations of SARS-CoV-2 infection and virus load in fecal samples from the Hong Kong cohort and systematic review and meta-analysis. Gastroenterology 159, 81—95.

Cheung, E.W., Zachariah, P., Gorelik, M., et al., 2020. Multisystem inflammatory syndrome related to COVID-19 in previously healthy children and adolescents in New York City. JAMA 324, 294.

Cole-Jeffrey, C.T., Liu, M., Katovich, M.J., Raizada, M.K., Shenoy, V., 2015. ACE2 and microbiota. J. Cardiovasc. Pharmacol. 66 (6), 540—550. https://doi.org/10.1097/fjc.0000000000000307.

Conte, L., Toraldo, D.M., 2020. Targeting the gut-lung microbiota axis by means of a high-fibre diet and probiotics may have anti-inflammatory effects in COVID-19 infection. Ther. Adv. Respir. Dis. 14. 1753466620937170. https://doi.org/10.1177/1753466620937170.

Dalby, M.J., Hall, L.J., May 22, 2020. Recent advances in understanding the neonatal microbiome. F1000Res 9, F1000. https://doi.org/10.12688/f1000research.22355.1. Faculty Rev-422.

Dashraath, P., Wong, J.L., Lim, M.X., Lim, L.M., Li, S., Biswas, A., et al., 2021. Coronavirus disease 2019 (COVID-19) pandemic and pregnancy. Obstet. Anesth. Digest 41 (1), 7.

de Nies, L., Galata, V., Martin-Gallausiaux, C., Despotovic, M., Busi, S.B., Snoeck, C.J., Delacour, L., Budagavi, D.P., Laczny, C.C., Habier, J., Lupu, P.C., Halder, R., Fritz, J.V., Marques, T., Sandt, E., O'Sullivan, M.P., Ghosh, S., Satagopam, V., Consortium, C.-V., Krüger, R., Fagherazzi, G., Ollert, M., Hefeng, F.Q., May, P., Wilmes, P., March 9, 2023. Altered infective competence of the human gut microbiome in COVID-19Microbiome, 11 (1), 46. https://doi.org/10.1186/s40168-023-01472-7.

de Oliveira, G.L.V., Oliveira, C.N.S., Pinzan, C.F., de Salis, L.V.V., Cardoso, C.R.B., February 24, 2021. Microbiota modulation of the gut-lung Axis in COVID-19. Front. Immunol. 12, 635471. https://doi.org/10.3389/fimmu.2021.635471.

Del Valle, D.M., Kim-Schulze, S., Huang, H.H., Beckmann, N.D., Nirenberg, S., Wang, B., Lavin, Y., Swartz, T.H., Madduri, D., Stock, A., Marron, T.U., Xie, H., Patel, M., Tuballes, K., Van Oekelen, O., Rahman, A., Kovatch, P., Aberg, J.A., Schadt, E., Jagannath, S., Mazumdar, M., Charney, A.W., Firpo-Betancourt, A., Mendu, D.R., Jhang, J., Reich, D., Sigel, K., Cordon-Cardo, C., Feldmann, M., Parekh, S., Merad, M., Gnjatic, S., October 2020. An inflammatory cytokine signature predicts COVID-19 severity and survival. Nat. Med. 26 (10), 1636—1643. https://doi.org/10.1038/s41591-020-1051-9.

Dominguez-Bello, M.G., Costello, E.K., Contreras, M., Magris, M., Hidalgo, G., Fierer, N., et al., June 29, 2010. Delivery mode shapes the acquisition and structure of the initial microbiota across multiple body habitats in newborns. Proc. Natl. Acad. Sci. U. S. A. 107 (26), 11971—11975.

Dunn, A.B., Jordan, S., Baker, B.J., Carlson, N.S., 2017. The maternal infant microbiome: considerations for labor and birth. MCN Am. J. Matern./Child Nurs. 42 (6), 318—325.

Durgan, D., 2017. Obstructive sleep apnea-induced hypertension: role of the gut microbiota. Curr. Hypertens. Rep. 19 (4), 35.

Durgan, D., Ganesh, B., Cope, J., et al., 2016. Role of the gut microbiome in obstructive sleep apnea-induced hypertension. Hypertension 67 (2), 469—474.

D'amico, F., Baumgart, D.C., Danese, S., Peyrin-Biroulet, L., 2020. Diarrhea during COVID-19 infection: pathogenesis, epidemiology, prevention, and management. Clin. Gastroenterol. Hepatol. 18, 1663—1672. https://doi.org/10.1016/j.cgh.2020.04.001.

Ebrahimi, S., Khatami, S., Mesdaghi, M., December 2021. The effect of COVID-19 pandemic on the infants' microbiota and the probability of development of allergic and autoimmune diseases. Int. Arch. Allergy Immunol. 1—8.

Elekhnawy, E., Negm, W.A., 2022. The potential application of probiotics for the prevention and treatment of COVID-19. Egypt J. Med. Hum. Genet. 23 (1), 36. https://doi.org/10.1186/s43042-022-00252-6.

Fang, W., Xue, H., Chen, X., et al., 2019. Supplementation with sodium butyrate modulates the composition of the gut microbiota and ameliorates high- fat diet-induced obesity in mice. J. Nutr. 149 (5), 747—754.

Fernández, L., Langa, S., Martín, V., Jiménez, E., Martín, R., Rodríguez, J.M., 2013. The microbiota of human milk in healthy women. Cell. Mol. Biol. 59 (1), 31—42.

Ferreira, C., Viana, S.D., Reis, F., 2020. Is gut microbiota dysbiosis a predictor of increased susceptibility to poor outcome of COVID-19 patients? An update. Microorganisms 9 (1), 53.

Galang, R.R., Chang, K., Strid, P., Snead, M.C., Woodworth, K.R., House, L.D., et al., 2020. Severe coronavirus infections in pregnancy: a systematic review. Obstet. Gynecol. 136 (2), 262—272.

Galeotti, C., Bayry, J., 2020. Autoimmune and inflammatory diseases following COVID-19. Nat. Rev. Rheumatol. 16, 413—414.

Gallo, G., Calvez, V., Savoia, C., March 2022. Hypertension and COVID-19: current evidence and perspectives. High Blood Pres. Cardiovasc. Prev. 29 (2), 115—123. https://doi.org/10.1007/s40292-022-00506-9.

Gavzy, S.J., Kensiski, A., Lee, Z.L., Mongodin, E.F., Ma, B., Bromberg, J.S., December 2023. Bifidobacterium mechanisms of immune modulation and tolerance. Gut Microb. 15 (2), 2291164. https://doi.org/10.1080/19490976.2023.2291164.

Giaxi, P., Maniatelli, E., Vivilaki, V., July 2, 2020. Evaluation of mode of delivery in pregnant women infected with COVID-19. Eur. J. Midwifery 4, 28.

Gou, W., Fu, Y., Yue, L., et al., 2020. Gut Microbiota May Underlie the Predisposition of Healthy Individuals to COVID-19 medRxiv 2020.04.22.20076091; Available from: https://www.medrxiv.org/content/10.1101/2020.04.22.20076091v1. (Accessed 11 August 2020).

Greenwood, C., Morrow, A.L., Lagomarcino, A.J., Altaye, M., Taft, D.H., Yu, Z., , … Schibler, K.R., 2014. Early empiric antibiotic use in preterm infants is associated with lower bacterial diversity and higher relative abundance of enterobacter. J. Pediatr. 165 (1), 23—29.

Gu, S., Chen, Y., Wu, Z., Chen, Y., Gao, H., Lv, L., et al., 2020a. Alterations of the Gut microbiota in patients with coronavirus disease 2019 or H1N1 Influenza. Clin. Infect. Dis.

Gu, J., Han, B., Wang, J., 2020b. COVID-19: gastrointestinal manifestations and potential fecal-oral transmission. Gastroenterology 158, 1518—1519. https://doi.org/10.1053/j.gastro.2020.02.054.

Hamming, I., Timens, W., Bulthuis, M.L., Lely, A.T., Navis, G., van Goor, H., June 2004. Tissue distribution of ACE2 protein, the functional receptor for SARS coronavirus. A first step in understanding SARS pathogenesis. J. Pathol. 203 (2), 631—637. https://doi.org/10.1002/path.1570.

Hantoushzadeh, S., Shamshirsaz, A.A., Aleyasin, A., Seferovic, M.D., Aski, S.K., Arian, S.E., et al., 2020. Maternal death due to COVID-19. Am. J. Obstet. Gynecol. 223 (1), 109.e1—109.e16.

Hartz, L.E., Bradshaw, W., Brandon, D.H., 2015. Potential NICU environmental influences on the neonate's microbiome: a systematic review. Adv. Neonatal Care 15 (5), 324—335.

Hasan, N., Yang, H., 2019. Factors affecting the composition of the gut microbiota, and its modulation. Peer J. 7, e7502.

Hashimoto, T., Perlot, T., Rehman, A., et al., 2012. ACE2 links amino acid malnutrition to microbial ecology and intestinal inflammation. Nature 487, 477—481.

Harmer, D., Gilbert, M., Borman, R., Clark, K.L., 2002. Quantitative mRNA expression profiling of ACE 2, a novel homologue of angiotensin converting enzyme. FEBS (Fed. Eur. Biochem. Soc.) Lett. 532 (1—2), 107—110. https://doi.org/10.1016/s0014-5793(02)03640-2.

He, Y., Wang, J., Li, F., Shi, Y., 2020. Main clinical features of COVID-19 and potential prognostic and therapeutic value of the microbiota in SARS-CoV-2 infections. Front. Microbiol. 11, 1302. https://doi.org/10.3389/fmicb.2020.01302.

Henderickx, J.G.E., Zwittink, R.D., van Lingen, R.A., Knol, J., Belzer, C., April 2, 2019. The preterm gut microbiota: an inconspicuous challenge in nutritional neonatal care. Front. Cell. Infect. Microbiol. 9, 85.

Heydarpoor, Z., Valizadeh, L., December 1, 2020. Breastfeeding and COVID-19. Bull. Acad. Natl. Med. 23 (5), 646—653.

Hidalgo-Cantabrana, C., Delgado, S., Ruiz, L., Ruas-Madiedo, P., Sánchez, B., Margolles, A., June 2017. Bifidobacteria and their health-promoting effects. Microbiol. Spectr. 5 (3). https://doi.org/10.1128/microbiolspec.BAD-0010-2016.

Huang, C., Wang, Y., Li, X., Ren, L., Zhao, J., Hu, Y., Zhang, L., Fan, G., Xu, J., Gu, X., Cheng, Z., Yu, T., Xia, J., Wei, Y., Wu, W., Xie, X., Yin, W., Li, H., Liu, M., , … Cao, B., 2020. Clinical features of patients infected with 2019 novel coronavirus in Wuhan, China. Lancet 395 (10223), 497–506. https://doi.org/10.1016/s0140-6736(20)30183-5.

Jardou, M., Lawson, R., September 2021. Supportive therapy during COVID-19: the proposed mechanism of short-chain fatty acids to prevent cytokine storm and multi-organ failure. Med. Hypotheses 154, 110661. https://doi.org/10.1016/j.mehy.2021.110661.

Jayawardena, R., Sooriyaarachchi, P., Chourdakis, M., Jeewandara, C., Ranasinghe, P., 2020. Enhancing immunity in viral infections, with special emphasis on COVID-19: a review. Diabetes Metabol. Syndr. 14 (4), 367–382. https://doi.org/10.1016/j.dsx.2020.04.015.

Jin, X., Lian, J.-S., Hu, J.-H., Gao, J., Zheng, L., Zhang, Y.-M., Hao, S.-R., Jia, H.-Y., Cai, H., Zhang, X.-L., Yu, G.-D., Xu, K.-J., Wang, X.-Y., Gu, J.-Q., Zhang, S.-Y., Ye, C.-Y., Jin, C.-L., Lu, Y.-F., Yu, X., , … Yang, Y., 2020. Epidemiological, clinical and virological characteristics of 74 cases of coronavirus-infected disease 2019 (COVID-19) with gastrointestinal symptoms. Gut 69 (6), 1002–1009. https://doi.org/10.1136/gutjnl-2020-320926.

Jost, T., Lacroix, C., Braegger, C.P., Rochat, F., Chassard, C., 2014. Vertical mother-neonate transfer of maternal gut bacteria via breastfeeding. Environ. Microbiol. 16 (9), 2891–2904.

Nithin, K.K., Patil, P., Bhandary, S.K., Haridas, V., Suchetha Kumari, N., Sarathkumar, E., Shetty, P., April 6, 2021. Is butyrate a natural alternative to dexamethasone in the management of CoVID-19? F1000Res 10, 273. https://doi.org/10.12688/f1000research.51786.1.

Kang, E.J., Kim, S.Y., Hwang, I.H., 2013. The effect of probiotics on prevention of common cold: a meta-analysis of randomized controlled trial studies. Korean J. Fam. Med. 34, 2–10.

Kau, A.L., Ahern, P.P., Griffin, N.W., Goodman, A.L., Gordon, J.I., 2011. Human nutrition, the gut microbiome and the immune system. Nature 474, 327–336.

Kim, G., Bae, J., Kim, M.J., et al., 2020. Delayed establishment of gut microbiota in infants delivered by cesarean section. Front. Microbiol. 11, 2099.

Kolahchi, Z., De Domenico, M., Uddin, L.Q., Cauda, V., Grossmann, I., Lacasa, L., Grancini, G., Mahmoudi, M., Rezaei, N., 2021. COVID-19 and its global economic impact. Adv. Exp. Med. Biol. 1318, 825–837. https://doi.org/10.1007/978-3-030-63761-3_46.

Kotfis, K., Skonieczna-Żydecka, K., 2020. COVID-19: gastrointestinal symptoms and potential sources of SARS-CoV-2 transmission. Anaesthesiol. Intensive Ther. 52 (2), 171–172. https://doi.org/10.5114/ait.2020.93867.

Kumova, O.K., Fike, A.J., Thayer, J.L., 2019. Lung transcriptional unresponsiveness and loss of early influenza virus control in infected neonates is prevented by intranasal Lactobacillus rhamnosus GG. PLoS Pathog. 15.

Kurian, S.J., Unnikrishnan, M.K., Miraj, S.S., Bagchi, D., Banerjee, M., Reddy, B.S., Rodrigues, G.S., Manu, M.K., Saravu, K., Mukhopadhyay, C., Rao, M., August 2021. Probiotics in prevention and treatment of COVID-19: current perspective and future prospects. Arch. Med. Res. 52 (6), 582–594. https://doi.org/10.1016/j.arcmed.2021.03.002.

Lamers, M.M., Beumer, J., van der Vaart, J., et al., 2020. SARS-CoV-2 productively infects human gut enterocytes. Science 369, eabc1669.

Lauer, S.A., Grantz, K.H., Bi, Q., Jones, F.K., Zheng, Q., Meredith, H.R., Azman, A.S., Reich, N.G., Lessler, J., 2020. The incubation period of coronavirus disease 2019 (COVID-19) from publicly reported confirmed cases: estimation and application. Ann. Intern. Med. 172 (9), 577–582. https://doi.org/10.7326/m20-0504.

Lee, H.L., Lobbestael, E., Vermeire, S., Sabino, J., Cleynen, I., 2020. Inflammatory bowel disease and Parkinson's disease: common pathophysiological links. Gut. https://doi.org/10.1136/gutjnl-2020-322429. PMID: 33067333 5. Wang T, Cai G, Qiu Y.

Li, N., Ma, W.-T., Pang, M., Fan, Q.-L., Hua, J.-L., 2019. The commensal microbiota and viral infection: a comprehensive review. Front. Immunol. 10, e1551.

Li, J., Stevens, B.R., Richards, E.M., Raizada, M.K., 2020. SARS-CoV-2 receptor ACE2 (Angiotensin-Converting Enzyme 2) is upregulated in colonic organoids from hypertensive rats. Hypertension 76, e26—e28. https://doi.org/10.1161/HYPERTENSIONAHA.120.15725.

Li, J., Richards, E.M., Handberg, E.M., Pepine, C.J., Raizada, M.K., 2021. Butyrate regulates COVID-19—relevant genes in gut epithelial organoids from normotensive rats. Hypertension 77, e13—e16. https://doi.org/10.1161/HYPERTENSIONAHA.120.16647.

Liang, L., Saunders, C., Sanossian, N., March 7, 2023. Food, gut barrier dysfunction, and related diseases: a new target for future individualized disease prevention and management. Food Sci. Nutr. 11 (4), 1671—1704. https://doi.org/10.1002/fsn3.3229.

Lippi, G., Wong, J., Henry, B., 2019. Hypertension and its severity or mortality in Coronavirus Disease 2019 (COVID-19): a pooled analysis [published online ahead of print March 31, 2020]. Pol. Arch. Intern. Med. https://doi.org/10.20452/pamw.15272.

Liu, Y., Lou, X., 2020. Type 2 diabetes mellitus-related environmental factors and the gut microbiota: emerging evidence and challenges. Clinics 75, e1277. https://doi.org/10.6061/clinics/2020/e1277.

Liu, H., Wang, J., He, T., Becker, S., Zhang, G., Li, D., Ma, X., 2018. Butyrate: a double-edged sword for health? Adv. Nutr. 9 (1), 21—29. https://doi.org/10.1093/advances/nmx009.

Liu, F., Ye, S., Zhu, X., He, X., Wang, S., Li, Y., Lin, J., Wang, J., Lin, Y., Ren, X., Li, Y., Deng, Z., 2021. Gastrointestinal disturbance and effect of fecal microbiota transplantation in discharged COVID-19 patients. J. Med. Case Rep. 15 (1). https://doi.org/10.1186/s13256-020-02583-7.

Lynch, S.V., Pedersen, O., December 15, 2016. The human intestinal microbiome in health and disease. N. Engl. J. Med. 375 (24), 2369—2379. https://doi.org/10.1056/NEJMra1600266.

Mao, R., Qiu, Y., He, J.-S., Tan, J.-Y., Li, X.-H., Liang, J., Shen, J., Zhu, L.-R., Chen, Y., Iacucci, M., Ng, S.C., Ghosh, S., Chen, M.-H., 2020. Manifestations and prognosis of gastrointestinal and liver involvement in patients with COVID-19: a systematic review and meta-analysis. The Lancet Gastroenterology & Hepatology 5 (7), 667—678. https://doi.org/10.1016/s2468-1253(20)30126-6.

Marasco, G., Lenti, M.V., Cremon, C., Barbaro, M.R., Stanghellini, V., Di Sabatino, A., Barbara, G., 2021. Implications of SARS-CoV-2 infection for neurogastroenterology. In: Neurogastroenterology and Motility. https://doi.org/10.1111/nmo.14104.

Martín Giménez, V.M., Modrego, J., Gómez-Garre, D., Manucha, W., de Las Heras, N., July 31, 2023. Gut microbiota dysbiosis in COVID-19: modulation and approaches for prevention and therapy. Int. J. Mol. Sci. 24 (15), 12249. https://doi.org/10.3390/ijms241512249.

Matsumoto, C., Shibata, S., Kishi, T., Morimoto, S., Mogi, M., Yamamoto, K., Kobayashi, K., Tanaka, M., Asayama, K., Yamamoto, E., Nakagami, H., Hoshide, S., Mukoyama, M., Kario, K., Node, K., Rakugi, H., March 2023. Long COVID and hypertension-related disorders: a report from the Japanese society of hypertension project team on COVID-19. Hypertens. Res. 46 (3), 601—619. https://doi.org/10.1038/s41440-022-01145-2.

Michael, O.S., Oluranti, O.I., Oshinjo, A.M., Adetunji, C.O., Adetunji, J.B., Esiobu, N.D., 2022. Microbiota transplantation, health implications, and the way forward. In: Book: Microbiomes and Emerging Applications, first ed.st Edition. Imprint CRC Press, ISBN 9781003180241, p. 19. https://doi.org/10.1201/9781003180241-5. First Published 2022.

Mills, K.T., Stefanescu, A., He, J., April 2020. The global epidemiology of hypertension. Nat. Rev. Nephrol. 16 (4), 223—237. https://doi.org/10.1038/s41581-019-0244-2.

Minato, T., Nirasawa, S., Sato, T., 2020. B38-CAP is a bacteria-derived ACE2-like enzyme that suppresses hypertension and cardiac dysfunction. Nat. Commun. 11, 1058, 2020.

Moles, L., Manzano, S., Fernández, L., Montilla, A., Corzo, N., Ares, S., et al., 2015. Bacteriological, biochemical, and immunological properties of colostrum and mature milk from mothers of extremely preterm infants. J. Pediatr. Gastroenterol. Nutr. 60 (1), 120—126.

Montazersaheb, S., Hosseiniyan Khatibi, S.M., Hejazi, M.S., Tarhriz, V., Farjami, A., Ghasemian Sorbeni, F., Farahzadi, R., Ghasemnejad, T., May 26, 2022. COVID-19 infection: an overview on cytokine storm and related interventions. Virol. J. 19 (1), 92. https://doi.org/10.1186/s12985-022-01814-1.

Moore, J.B., June, C.H., 2020. Cytokine release syndrome in severe COVID-19. Lessons from arthritis and cell therapy in cancer patients point totherapy for severe disease. Science 368, 473—474.

Moossavi, S., Azad, M.B., 2020. Origins of human milk microbiota: new evidence and arising questions. Gut Microb. 12 (1), 1667722.

Moossavi, S., Sepehri, S., Robertson, B., Bode, L., Goruk, S., Field, C.J., et al., February 13, 2019. Composition and variation of the human milk microbiota are influenced by maternal and early-life factors. Cell Host Microb. 25 (2), 324–335. e4.

Moran-Ramos, S., Lopez-Contreras, B.E., Villarruel-Vazquez, R., Ocampo-Medina, E., Macias-Kauffer, L., Martinez-Medina, J.N., Villamil-Ramirez, H., León-Mimila, P., Del Rio-Navarro, B.E., Ibarra-Gonzalez, I., et al., 2020. Environmental and intrinsic factors shaping gut microbiota composition and diversity and its relation to metabolic health in children and early adolescents: a population-based study. Gut Microb. 11, 900–917.

Mullins, E., Evans, D., Viner, R.M., O'Brien, P., Morris, E., 2020. Coronavirus in pregnancy and delivery: rapid review. Ultrasound Obstet. Gynecol. 55 (5), 586–592.

Nagpal, R., Mainali, R., Ahmadi, S., 2018. Gut microbiome, and aging: physiological and mechanistic insights. Nutr. Healthy Aging 4, 267–285.

Naseer, S., Khalid, S., Parveen, S., Abbass, K., Song, H., Achim, M.V., January 30, 2023. COVID-19 outbreak: impact on global economy. Front. Public Health 10, 1009393. https://doi.org/10.3389/fpubh.2022.1009393.

Nejadghaderi, S.A., Nazemalhosseini-Mojarad, E., Aghdaei, H.A., 2020. Fecal microbiota transplantation for COVID-19; a potential emerging treatment strategy. Med. Hypotheses 110476. https://doi.org/10.1016/j.mehy.2020.110476.

Ng, S.C., Tilg, H., 2020. COVID-19 and the gastrointestinal tract: more than meets the eye. Gut 69 (6), 973–974. https://doi.org/10.1136/gutjnl-2020-321195.

Noris, M., Benigni, A., Remuzzi, G., 2020. The case of complement activation in COVID-19 multiorgan impact. Kidney Int. 98 (2), 314–322. https://doi.org/10.1016/j.kint.2020.05.013.

Nseir, S., 2020. Bacterial and Fungal Microbiota of Patients with Severe Viral Pneumonia with COVID-19 (MICROVID) (n.d.). https://clinicaltrials.gov/ct2/show/NCT04359706.

Olendzki, B., Bucci, V., Cawley, C., Maserati, R., McManus, M., Olednzki, E., Madziar, C., Chiang, D., Ward, D.V., Pellish, R., Foley, C., Bhattarai, S., McCormick, B.A., Maldonado-Contreras, A., 2022. Dietary manipulation of the gut microbiome in inflammatory bowel disease patients: pilot study. Gut Microb. 14 (1), 2046244. https://doi.org/10.1080/19490976.2022.2046244.

Onder, G., Rezza, G., Brusaferro, S., 2020. Case-fatality rate and characteristics of patients dying in relation to COVID-19 in Italy. JAMA. https://doi.org/10.1001/jama.2020.4683.

Pagliari, D., Saviano, A., Newton, E.E., Serricchio, M.L., Dal Lago, A.A., Gasbarrini, A., Cianci, R., February 1, 2018. Gut microbiota-immune system crosstalk and pancreatic disorders. Mediat. Inflamm. 7946431, 2018.

Parada Venegas, D., De la Fuente, M.K., Landskron, G., González, M.J., Quera, R., Dijkstra, G., Harmsen, H.J.M., Faber, K.N., Hermoso, M.A., March 11, 2019. Short chain fatty acids (SCFAs)-Mediated gut epithelial and immune regulation and its relevance for inflammatory bowel diseases. Front. Immunol. 10, 277. https://doi.org/10.3389/fimmu.2019.00277.

Paramsothy, S., Kamm, M.A., Kaakoush, N.O., Walsh, A.J., van den Bogaerde, J., Samuel, D., Leong, R.W.L., Connor, S., Ng, W., Paramsothy, R., Xuan, W., Lin, E., Mitchell, H.M., Borody, T.J., 2017. Multidonor intensive faecal microbiota transplantation for active ulcerative colitis: a randomised placebo-controlled trial. Lancet 389 (10075), 1218–1228. https://doi.org/10.1016/s0140-6736(17)30182-4.

Parasa, S., Desai, M., Thoguluva Chandrasekar, V., Patel, H.K., Kennedy, K.F., Roesch, T., Spadaccini, M., Colombo, M., Gabbiadini, R., Artifon, E.L.A., Repici, A., Sharma, P., 2020. Prevalence of gastrointestinal symptoms and fecal viral shedding in patients with coronavirus disease 2019. JAMA Netw. Open 3 (6). https://doi.org/10.1001/jamanetworkopen.2020.11335. Article e2011335.

Patel, V.B., Zhong, J.-C., Grant, M.B., Oudit, G.Y., 2016. Role of the ace2/angiotensin 1–7 axis of the renin–angiotensin system in heart failure. Circ. Res. 118 (8), 1313–1326. https://doi.org/10.1161/circresaha.116.307708.

Penders, J., Gerhold, K., Stobberingh, E.E., Thijs, C., Zimmermann, K., Lau, S., et al., September 2013. Establishment of the intestinal microbiota and its role for atopic dermatitis in early childhood. J. Allergy Clin. Immunol. 132 (3), 601–607.e8.

Peng, M., He, J., Xue, Y., Yang, X., Liu, S., Gong, Z., November 1, 2021. Role of hypertension on the severity of COVID-19: a review. J. Cardiovasc. Pharmacol. 78 (5), e648–e655. https://doi.org/10.1097/FJC.0000000000001116.

Perlot, T., Penninger, J.M., 2013. ACE2 – from the renin–angiotensin system to gut microbiota and malnutrition. Microb. Infect. 15 (13), 866–873. https://doi.org/10.1016/j.micinf.2013.08.003.

Pituch, A., Walkowiak, J., Banaszkiewicz, A., 2013. Butyric acid in functional constipation. Przegląd Gastroenterol. 8 (5), 295–298.

Pourhossein, M., Moravejolahkami, A.R., 2020. Probiotics in viral infections, with a focus on COVID-19: a systematic review. Prepints. https://doi.org/10.22541/au.158938616.61042433.

Prame Kumar, K., Ooi, J.D., Goldberg, R., December 1, 2023. The interplay between the microbiota, diet and T regulatory cells in the preservation of the gut barrier in inflammatory bowel disease. Front. Microbiol. 14, 1291724. https://doi.org/10.3389/fmicb.2023.1291724.

Qiu, H., Wu, J., Hong, L., Luo, Y., Song, Q., Chen, D., 2019. Clinical and epidemiological features of 36 children with coronavirus disease 2019 (COVID-19) in Zhejiang, China: an observational cohort study. Lancet Infect. Dis.

Robles-Vera, I., Toral, M., Romero, M., Jiménez, R., Sánchez, M., Pérez-Vizcaíno, F., Duarte, J., April 2017. Antihypertensive effects of probiotics. Curr. Hypertens. Rep. 19 (4), 26. https://doi.org/10.1007/s11906-017-0723-4.

Robles-Vera, I., de la Visitación, N., Toral, M., Sánchez, M., Romero, M., Gómez-Guzmán, M., Yang, T., Izquierdo-García, J.L., Guerra-Hernández, E., Ruiz-Cabello, J., Raizada, M.K., Pérez-Vizcaíno, F., Jiménez, R., Duarte, J., October 2020. Probiotic Bifidobacterium breve prevents DOCA-salt hypertension. Faseb. J. 34 (10), 13626–13640. https://doi.org/10.1096/fj.202001532R.

Rocchi, G., Giovanetti, M., Benedetti, F., Borsetti, A., Ceccarelli, G., Zella, D., Altomare, A., Ciccozzi, M., Guarino, M.P.L., September 15, 2022. Gut microbiota and COVID-19: potential implications for disease severity. Pathogens 11 (9), 1050. https://doi.org/10.3390/pathogens11091050.

Rodriguez-Morales, A.J., Cardona-Ospina, J.A., Gutiérrez-Ocampo, E., Villamizar-Peña, R., Holguin-Rivera, Y., Escalera-Antezana, J.P., Alvarado-Arnez, L.E., Bonilla-Aldana, D.K., Franco-Paredes, C., Henao-Martinez, A.F., Paniz-Mondolfi, A., Lagos-Grisales, G.J., Ramírez-Vallejo, E., Suárez, J.A., Zambrano, L.I., Villamil-Gómez, W.E., Balbin-Ramon, G.J., Rabaan, A.A., Harapan, H., , … Sah, R., 2020. Clinical, laboratory and imaging features of COVID-19: a systematic review and meta-analysis. Trav. Med. Infect. Dis. 34, 101623. https://doi.org/10.1016/j.tmaid.2020.101623.

Romano-Keeler, J., Zhang, J., Sun, J., 2021. COVID-19 and the neonatal microbiome: will the pandemic cost infants their microbes? Gut Microb. 13 (1), 1–7.

Salvatori, G., De Rose, D.U., Concato, C., Alario, D., Olivini, N., Dotta, A., et al., 2020. Managing COVID- 19-positive maternal-infant dyads: an Italian experience. Breastfeed. Med. 15 (5), 347–348.

Santisteban, M., Qi, Y., Zubcevic, J., et al., 2017. Hypertension-linked pathophysiological alterations in the gut. Circ. Res. 120 (2), 312–323.

Saturio, S., Nogacka, A.M., Alvarado-Jasso, G.M., Salazar, N., de Los Reyes-Gavilán, C.G., Gueimonde, M., Arboleya, S., November 23, 2021. Role of bifidobacteria on infant health. Microorganisms 9 (12), 2415. https://doi.org/10.3390/microorganisms9122415.

Schiffrin, E.L., Flack, J.M., Ito, S., Muntner, P., Webb, R.C., April 29, 2020. Hypertension and COVID-19. Am. J. Hypertens. 33 (5), 373–374. https://doi.org/10.1093/ajh/hpaa057.

Schirmer, M., Smeekens, S.P., Vlamakis, H., Jaeger, M., Oosting, M., Franzosa, E.A., et al., 2016. Linking the human gut microbiome to inflammatory cytokine production capacity. Cell 167, 1125–1136. https://doi.org/10.1016/j.cell.2016.10.020. PMID: 27814509.

Schmieder, R.E., December 2010. End organ damage in hypertension. Dtsch. Arztebl. Int. 107 (49), 866–873. https://doi.org/10.3238/arztebl.2010.0866.

Segal, J.P., Mak, J.W.Y., Mullish, B.H., Alexander, J.L., Ng, S.C., Marchesi, J.R., November 24, 2020. The gut microbiome: an under-recognised contributor to the COVID-19 pandemic? Therap. Adv. Gastroenterol. 13. https://doi.org/10.1177/1756284820974914, 1756284820974914.

Seong, H., Kim, J.H., Han, Y.H., Seo, H.S., Hyun, H.J., Yoon, J.G., Nham, E., Noh, J.Y., Cheong, H.J., Kim, W.J., Lim, S., Song, J.Y., March 24, 2023. Clinical implications of gut microbiota and cytokine responses in coronavirus disease prognosis. Front. Immunol. 14, 1079277. https://doi.org/10.3389/fimmu.2023.1079277.

Shahbazi, R., Yasavoli-Sharahi, H., Alsadi, N., Ismail, N., Matar, C., October 22, 2020. Probiotics in treatment of viral respiratory infections and neuroinflammatory disorders. Molecules 25 (21), 4891. https://doi.org/10.3390/molecules25214891.

Shao, Y., Forster, S.C., Tsaliki, E., et al., 2019. Stunted microbiota and opportunistic pathogen colonization in caesarean-section birth. Nature 574 (7776), 117–121. https://doi.org/10.1038/s41586-019-1560-1.

Sharma, R.K., Stevens, B.R., Obukhov, A.G., Grant, M.B., Oudit, G.Y., Li, Q., Richards, E.M., Pepine, C.J., Raizada, M.K., 2020. ACE2 (Angiotensin-Converting Enzyme 2) in cardiopulmonary diseases: ramifications for the control of SARS-CoV-2. Hypertension 76, 651–661. https://doi.org/10.1161/HYPERTENSIONAHA.120.15595.

Shibata, S., Kobayashi, K., Tanaka, M., Asayama, K., Yamamoto, E., Nakagami, H., Hoshide, S., Kishi, T., Matsumoto, C., Mogi, M., Morimoto, S., Yamamoto, K., Mukoyama, M., Kario, K., Node, K., Rakugi, H., March 2023. COVID-19 pandemic and hypertension: an updated report from the Japanese Society of Hypertension project team on COVID-19. Hypertens. Res. 46 (3), 589–600. https://doi.org/10.1038/s41440-022-01134-5.

Shreiner, A.B., Kao, J.Y., Young, V.B., 2015. The gut microbiome in health and in disease. Curr. Opin. Gastroenterol. 31 (1), 69–75.

Singh, S.P., Pritam, M., Pandey, B., Yadav, T.P., 2020. Microstructure, pathophysiology, and potential therapeutics of COVID-19: a comprehensive review. J. Med. Virol. 93 (1), 275–299. https://doi.org/10.1002/jmv.26254.

Smith, K., McCoy, K.D., Macpherson, A.J., April 2007. Use of axenic animals in studying the adaptation of mammals to their commensal intestinal microbiota. Semin. Immunol. 19 (2), 59–69.

Stewart, M.H., July 1, 2023. Hypertensive crisis: diagnosis, presentation, and treatment. Curr. Opin. Cardiol. 38 (4), 311–317. https://doi.org/10.1097/HCO.0000000000001049.

Stiverson, J., Williams, T., Chen, J., Adams, S., Hustead, D., Price, P., , … Yu, Z., 2014. Prebiotic oligosaccharides: comparative evaluation using in vitro cultures of infants' fecal microbiomes. Appl. Environ. Microbiol. 80 (23), 7388–7397.

Stuivenberg, G.A., Burton, J.P., Bron, P.A., Reid, G., January 25, 2022. Why are bifidobacteria important for infants? Microorganisms 10 (2), 278. https://doi.org/10.3390/microorganisms10020278.

Sulaiman, I., Chung, M., Angel, L., Tsay, J.-C.J., Wu, B.G., Yeung, S.T., Krolikowski, K., Li, Y., Duerr, R., Schluger, R., Thannickal, S.A., Koide, A., Rafeq, S., Barnett, C., Postelnicu, R., Wang, C., Banakis, S., Pérez-Pérez, L., Shen, G., , … Segal, L.N., 2021. Microbial signatures in the lower airways of mechanically ventilated COVID-19 patients associated with poor clinical outcome. Nat. Microbiol. https://doi.org/10.1038/s41564-021-00961-5.

Takahashi, Y., Hayakawa, A., Sano, R., et al., 2021. Histone deacetylase inhibitors suppress ACE2 and ABO simultaneously, suggesting a preventive potential against COVID-19. Sci. Rep. 11 (1), 3379.

Tay, M.Z., Poh, C.M., Rénia, L., MacAry, P.A., Ng, L.F.P., June 2020. The trinity of COVID-19: immunity, inflammation and intervention. Nat. Rev. Immunol. 20 (6), 363–374. https://doi.org/10.1038/s41577-020-0311-8.

Tayyeb, J.Z., Popeijus, H.E., Mensink, R.P., et al., 2021. Butyric acid added apically to intestinal caco-2 cells elevates hepatic ApoA-I transcription and rescues lower ApoA-I expression in inflamed HepG2 cells Co-cultured in the basolateral compartment. Biomolecules 11 (1), 71.

Thursby, E., Juge, N., May 16, 2017. Introduction to the human gut microbiota. Biochem. J. 474 (11), 1823–1836.

Tian, Y., Rong, L., Nian, W., 2020. Review article: gastrointestinal features in COVID-19 and the possibility of faecal transmission. Aliment. Pharmacol. Ther. 51, 843–851, 2020.

Turroni, F., Milani, C., Duranti, S., et al., 2018. Bifidobacteria and the infant gut: an example of co-evolution and natural selection. Cell. Mol. Life Sci. 75 (1), 103–118.

Vabret, N., Britton, G.J., Gruber, C., et al., 2020. Immunology of COVID-19: current state of the science. Immunity 52, 910–941.

van den Elsen, L.W.J., Garssen, J., Burcelin, R., Verhasselt, V., February 27, 2019. Shaping the gut microbiota by breastfeeding: the gateway to allergy prevention? Front. Pediatr. 7, 47.

van der Lelie, D., Taghavi, S., July 21, 2020. COVID-19 and the gut microbiome: more than a. Gut Feeling. mSystems. 5 (4), 004533–e520.

van Doremalen, N., Bushmaker, T., Morris, D.H., Holbrook, M.G., Gamble, A., Williamson, B.N., et al., 2020. Aerosol and surface stability of SARS-CoV-2 as compared with SARS-CoV-1. N. Engl. J. Med. 382 (16), 1564—1567.

Vestad, B., Ueland, T., Lerum, T.V., Dahl, T.B., Holm, K., Barratt-Due, A., Kåsine, T., Dyrhol-Riise, A.M., Stiksrud, B., Tonby, K., Hoel, H., Olsen, I.C., Henriksen, K.N., Tveita, A., Manotheepan, R., Haugli, M., Eiken, R., Berg, Å., Halvorsen, B., Lekva, T., Ranheim, T., Michelsen, A.E., Kildal, A.B., Johannessen, A., Thoresen, L., Skudal, H., Kittang, B.R., Olsen, R.B., Ystrøm, C.M., Skei, N.V., Hannula, R., Aballi, S., Kvåle, R., Skjønsberg, O.H., Aukrust, P., Hov, J.R., Trøseid, M., NOR-Solidarity study group, June 2022. Respiratory dysfunction three months after severe COVID-19 is associated with gut microbiota alterations. J. Intern. Med. 291 (6), 801—812. https://doi.org/10.1111/joim.13458.

Villapol, S., 2020. Gastrointestinal symptoms associated with COVID-19: impact on the gut microbiome. In: Translational Research. https://doi.org/10.1016/j.trsl.2020.08.004.

Vuille-dit-Bille, R.N., Camargo, S.M., Emmenegger, L., Sasse, T., Kummer, E., Jando, J., Hamie, Q.M., Meier, C.F., Hunziker, S., Forras-Kaufmann, Z., Kuyumcu, S., Fox, M., Schwizer, W., Fried, M., Lindenmeyer, M., Götze, O., Verrey, F., 2014. Human intestine luminal ACE2 and amino acid transporter expression increased by ACE-inhibitors. Amino Acids 47 (4), 693—705. https://doi.org/10.1007/s00726-014-1889-6.

Vuille-dit-Bille, R.N., Liechty, K.W., Verrey, F., Guglielmetti, L.C., 2020. SARS-CoV-2 receptor ACE2 gene expression in small intestine correlates with age. Amino Acids 52 (6—7), 1063—1065.

Wang, B., Yao, M., Lv, L., Ling, Z., Li, L., 2017a. The human microbiota in health and disease. Engineering 3, 71—82.

Wang, F., Jin, Z., Shen, K., et al., 2017b. Butyrate pretreatment attenuates heart depression in a mice model of endotoxin-induced sepsis via antiinflammation and anti-oxidation. Am. J. Emerg. Med. 35 (3), 402—409.

Wang, J.-W., Kuo, C.-H., Kuo, F.-C., Wang, Y.-K., Hsu, W.-H., Yu, F.-J., Hu, H.-M., Hsu, P.-I., Wang, J.-Y., Wu, D.-C., 2019. Fecal microbiota transplantation: review and update. J. Formos. Med. Assoc. 118, S23—S31. https://doi.org/10.1016/j.jfma.2018.08.011.

Wang, X., Zhou, Z., Zhang, J., Zhu, F., Tang, Y., Shen, X., 2020. A case of 2019 novel coronavirus in a pregnant woman with preterm delivery. Clin. Infect. Dis. 71 (15), 844—846.

Wang, B., Zhang, L., Wang, Y., Dai, T., Qin, Z., Zhou, F., Zhang, L., 2022. Alterations in microbiota of patients with COVID-19: potential mechanisms and therapeutic interventions. Signal Transduct. Targeted Ther. 7 (1). https://doi.org/10.1038/s41392-022-00986-0.

WHO, 2020a. WHO frequently asked questions : breastfeeding and COVID-19 for health care workers. J. Hum. Lactation 36 (3), 392—396.

WHO, 2020b. Coronavirus Disease (COVID-2019) Situation Reports [Available from: https://www.who.int/emergencies/diseases/novel-coronavirus-2019/situation-reports.

Włodarczyk, J., Czerwiński, B., Fichna, J., December 2022. Short-chain fatty acids-microbiota crosstalk in the coronavirus disease (COVID-19). Pharmacol. Rep. 74 (6), 1198—1207. https://doi.org/10.1007/s43440-022-00415-7.

Wölfel, R., Corman, V.M., Guggemos, W., et al., 2020. Virological assessment of hospitalized patients with COVID-2019. Nature 581, 465—469.

Wu, H.J., Wu, E., 2012. The role of gut microbiota in immune homeostasis and autoimmunity. Gut Microb. 3 (1), 4—14.

Wu, L.-h., Ye, Z.-n., Peng, P., Xie, W.-r., Xu, J.-t., Zhang, X.-y., Xia, H.H.-x., He, X.-x., 2021. Efficacy and safety of washed microbiota transplantation to treat patients with mild-to-severe COVID-19 and suspected of having gut microbiota dysbiosis: study protocol for a randomized controlled trial. Curr. Med. Sci. 41 (6), 1087—1095. https://doi.org/10.1007/s11596-021-2475-2.

Xiao, L., Zhao, F., April 2023. Microbial transmission, colonisation and succession: from pregnancy to infancy. Gut 72 (4), 772—786. https://doi.org/10.1136/gutjnl-2022-328970.

Xiao, F., Tang, M., Zheng, X., et al., 2020. Evidence for gastrointestinal infection of SARS-CoV-2. Gastroenterology 158, 1831—1833.e3.

Xu, Y., Li, X., Zhu, B., Liang, H., Fang, C., Gong, Y., Guo, Q., Sun, X., Zhao, D., Shen, J., Zhang, H., Liu, H., Xia, H., Tang, J., Zhang, K., Gong, S., 2020. Characteristics of pediatric SARS-CoV-2 infection and potential evidence for persistent fecal viral shedding. Nat. Med. 26 (4), 502—505. https://doi.org/10.1038/s41591-020-0817-4.

Xu, R., Lu, R., Zhang, T., Wu, Q., Cai, W., Han, X., Wan, Z., Jin, X., Zhang, Z., Zhang, C., 2021. Temporal association between human upper respiratory and gut bacterial microbiomes during the course of COVID-19 in adults. Commun. Biol. 4 (1). https://doi.org/10.1038/s42003-021-01796-w.

Yang, T., Santisteban, M.M., Rodriguez, V., Li, E., Ahmari, N., Carvajal, J.M., Zadeh, M., Gong, M., Qi, Y., Zubcevic, J., Sahay, B., Pepine, C.J., Raizada, M.K., Mohamadzadeh, M., June 2015. Gut dysbiosis is linked to hypertension. Hypertension 65 (6), 1331—1340. https://doi.org/10.1161/HYPERTENSIONAHA.115.05315.

Yang, T., Chakraborty, S., Saha, P., Mell, B., Cheng, X., Yeo, J.Y., et al., 2020. Gnotobiotic rats reveal that gut microbiota regulates colonic mRNA of Ace2 the receptor for SARS-CoV-2 infectivity. Hypertension 76, E1—E3.

Yassour, M., Vatanen, T., Siljander, H., Hämäläinen, A.M., Härkönen, T., Ryhänen, S.J., et al., June 15, 2016. Natural history of the infant gut microbiome and impact of antibiotic treatment on bacterial strain diversity and stability. Sci. Transl. Med. 8 (343), 343ra81.

Yeoh, Y.K., Zuo, T., Lui, G.C., Zhang, F., Liu, Q., Li, A.Y., Chung, A.C., Cheung, C.P., Tso, E.Y., Fung, K.S., Chan, V., Ling, L., Joynt, G., Hui, D.S., Chow, K.M., Ng, S.S.S., Li, T.C., Ng, R.W., Yip, T.C., Wong, G.L., Chan, F.K., Wong, C.K., Chan, P.K., Ng, S.C., April 2021. Gut microbiota composition reflects disease severity and dysfunctional immune responses in patients with COVID-19. Gut 70 (4), 698—706. https://doi.org/10.1136/gutjnl-2020-323020.

Yildiz, S., Mazel-Sanchez, B., Kandasamy, M., et al., 2018. Influenza A virus infection impacts systemic microbiota dynamics and causes quantitative enteric dysbiosis. Microbiome 6, 9.

Yoo, J.Y., Groer, M., Dutra, S.V.O., Sarkar, A., McSkimming, D.I., October 15, 2020. Gut microbiota and immune system interactions. Microorganisms 8 (10), 1587.

Yu, H.S., Lee, N.K., Choi, A.J., Choe, J.S., Bae, C.H., Paik, H.D., July 28, 2019. Anti-inflammatory potential of probiotic strain Weissella cibaria JW15 isolated from kimchi through regulation of NF-κB and MAPKs pathways in LPS-induced RAW 264.7 cells. J. Microbiol. Biotechnol. 29 (7), 1022—1032. https://doi.org/10.4014/jmb.1903.03014.

Yu, Z., Yang, Z., Wang, Y., Zhou, F., Li, S., Li, C., Li, L., Zhang, W., Li, X., July 10, 2021. Recent advance of ACE2 and microbiota dysfunction in COVID-19 pathogenesis. Heliyon 7 (7), e07548. https://doi.org/10.1016/j.heliyon.2021.e07548.

Yuki, K., Fujiogi, M., Koutsogiannaki, S., 2020. COVID-19 pathophysiology: a review. Clin. Immunol. 215, 108427.

Zelaya, H., Villena, J., Lopez, A.G., 2014. Modulation of the inflammation-coagulation interaction during pneumococcal pneumonia by immunobiotic *Lactobacillus rhamnosus* CRL1505: role of Toll-like receptor 2. Microbiol. Immunol. 58, 416—426.

Zhang, H., Li, H.-B., Lyu, J.-R., Lei, X.-M., Li, W., Wu, G., Lyu, J., Dai, Z.-M., 2020. Specific ACE2 expression in small intestinal enterocytes may cause gastrointestinal symptoms and injury after 2019-nCoV infection. Int. J. Infect. Dis. 96, 19—24. https://doi.org/10.1016/j.ijid.2020.04.027.

Zhang, F., Lau, R.I., Liu, Q., Su, Q., Chan, F.K.L., Ng, S.C., May 2023. Gut microbiota in COVID-19: key microbial changes, potential mechanisms and clinical applications. Nat. Rev. Gastroenterol. Hepatol. 20 (5), 323—337. https://doi.org/10.1038/s41575-022-00698-4.

Zhu, N., Zhang, D., Wang, W., Li, X., Yang, B., Song, J., Zhao, X., Huang, B., Shi, W., Lu, R., Niu, P., Zhan, F., Ma, X., Wang, D., Xu, W., Wu, G., Gao, G.F., Tan, W., China Novel Coronavirus Investigating and Research Team, February 20, 2020. A novel coronavirus from patients with pneumonia in China, 2019. N. Engl. J. Med. 382 (8), 727—733. https://doi.org/10.1056/NEJMoa2001017.

Ziegler, C.G.K., Allon, S.J., Nyquist, S.K., Mbano, I.M., Miao, V.N., Tzouanas, C.N., Cao, Y., Yousif, A.S., Bals, J., Hauser, B.M., Feldman, J., Muus, C., Wadsworth, M.H., Kazer, S.W., Hughes, T.K., Doran, B., Gatter, G.J., Vukovic, M., Taliaferro, F., , ... Zhang, K., 2020. SARS-CoV-2 receptor ACE2 is an interferon-stimulated gene in human airway epithelial cells and is detected in specific cell subsets across tissues. Cell 181 (5), 1016—1035.e19. https://doi.org/10.1016/j.cell.2020.04.035.

Zuo, T., Liu, Q., Zhang, F., Lui, G.C.-Y., Tso, E.Y., Yeoh, Y.K., Chen, Z., Boon, S.S., Chan, F.K., Chan, P.K., Ng, S.C., 2020a. Depicting SARS-CoV-2 faecal viral activity in association with gut microbiota composition in patients with COVID-19. Gut. https://doi.org/10.1136/gutjnl-2020-322294.

Zuo, T., Zhang, F., Lui, G.C.Y., Yeoh, Y.K., Li, A.Y.L., Zhan, H., Wan, Y., Chung, A.C.K., Cheung, C.P., Chen, N., Lai, C.K.C., Chen, Z., Tso, E.Y.K., Fung, K.S.C., Chan, V., Ling, L., Joynt, G., Hui, D.S.C., Chan, F.K.L., Ng, S.C., 2020b. Alterations in gut microbiota of patients with COVID-19 during time of hospitalization. Gastroenterology 159 (3), 944–955.e8. https://doi.org/10.1053/j.gastro.2020.05.048.

Zuo, T., Wu, X., Wen, W., Lan, P., September 21, 2021. Gut microbiome alterations in COVID-19. Dev. Reprod. Biol. S1672-0229(21)00206-0.

Further reading

Khaled, J.M.A., 2021. Probiotics, prebiotics, and COVID-19 infection: a review article. Saudi J. Biol. Sci. 28 (1), 865–869. https://doi.org/10.1016/j.sjbs.2020.11.025.

Trompette, A., Gollwitzer, E.S., Yadava, K., Sichelstiel, A.K., Sprenger, N., Ngom-Bru, C., Blanchard, C., Junt, T., Nicod, L.P., Harris, N.L., Marsland, B.J., 2014. Gut microbiota metabolism of dietary fiber influences allergic airway disease and hematopoiesis. Nat. Med. 20 (2), 159–166. https://doi.org/10.1038/nm.3444.

CHAPTER 6

Introduction to plant microbiome

Olulope Olufemi Ajayi[1], Charles Oluwaseun Adetunji[2], Olugbenga Samuel Michael[3,4,8], Frank Abimbola Ogundolie[5], Juliana Bunmi Adetunji[6] and Oluwafemi Adebayo Oyewole[7]

[1]Department of Biochemistry, Edo State University Uzairue, Edo State, Nigeria; [2]Applied Microbiology, Biotechnology and Nanotechnology Laboratory, Department of Microbiology, and Directorate of Research and Innovation, Edo State University Uzairue, Iyamho, Auchi, Edo State, Nigeria; [3]Cardiometabolic, Microbiome and Applied Physiology Laboratory, Department of Physiology, College of Health Sciences, Bowen University, Iwo, Osun State, Nigeria; [4]Department of Physiology, University of Tennessee Health Science Center, Memphis, TN, United States; [5]Department of Biotechnology, Baze University, Abuja, Nigeria; [6]Nutritional and Toxicological Research Laboratory, Department of Biochemistry Sciences, Osun State University, Osogbo, Osun State, Nigeria; [7]Department of Microbiology, Federal University of Technology, Minna, Nigeria; [8]Department of Medical Pharmacology and Physiology, University of Missouri, Columbia, MO, United States

Introduction

In plants, the role of the microbiome in their development has been discovered to be of great importance. The microorganism present in a specific habitat at a given time is the microbiome. These organisms can be pathogenic, indigenous flora, beneficial microbes that can either directly or indirectly enhance the plant's growth and/or health. Plant growth rate and protection can be influenced by the presence of the microbiome in the root environment (Lundberg et al., 2012). The presence of the microbiome at the root of plants has been reported to help break down nutrients resulting in easy uptake of nutrients in plants, which translates to a healthy and faster growth rate in plants (Wei et al., 2019; Trivedi et al., 2020). This action makes available important nutrients at the roots of the plants and thus initiates easy uptake of the nutrients. The plant-mediated microbiome has been reported to confer different forms of resistance against pathogens, thereby ensuring that the plant is healthy. This resistance can be through commensal–pathogen interaction or it can be just by modulating the plant defense mechanism directly (Berg and Koskella, 2018).

A key mechanism through which plants and soil flora interact is referred to as root exudation. The root exudation metabolites produced by the coexistence of their interactions are potent and sharpen the composition of the plant-root flora, thereby conferring some level of specific functions on the plants such as growth factor, disease resistance, enduring abiotic and biotic stress, salt tolerance among others (Bais et al., 2006; Rudrappa et al., 2008; Beyer et al., 2012; Rajniak et al., 2018; Stringlis et al., 2018; Korenblum et al., 2020).

The presence of microbes at the root of plants has also been linked to improving plant resistance to stress (Ansari, 2018; Liu et al., 2020). Several reports have established that organisms present in plant root microbiota through their plant–root–metabolite and

An Introduction to the Microbiome in Health and Diseases
ISBN 978-0-323-91190-0,
https://doi.org/10.1016/B978-0-323-91190-0.00006-0

microbes interaction cause balance in the microbial population with beneficial microbes creating the much-needed protection against the pathogenic ones (Chiu et al., 2017). Protective pro- or antiinflammatory immune activities of microorganisms can also be induced by altering the balance between host T cell subsets (Round and Mazmanian, 2009; Pang et al., 2012; Michael et al., 2022a,b; Esiobu et al., 2022; Adetunji et al., 2022a,b,c,d,e,f,g,h,i; Olaniyan et al., 2022a,b; Oyedara et al., 2022). By tilting the balance toward proinflammatory pathways, commensals help generate host-mediated attacks against pathogens (Chiu et al., 2017).

Microorganisms constitute a significant component of life. Microbiomes are key in host interaction and may be seen as an extension of host genome (Turner et al., 2013). Their interaction with the environment has been well documented. Some of them are beneficial to plant and animal health while some are pathogenic. Extensive studies have been carried out on microbiome—animal host relationship, much is yet to be known of plant microbiome.

There are evidence that plant microbiome determines plant health and yield (Turner et al., 2013). Plant microbiome aids modulation of gene expression and absorption of nutrients, serves as antipathogens, and ensures plant health (Orozco-Mosqueda et al., 2018). A slight alteration in host genome can affect plant microbiome. The interaction of microbe with plant has been reported in mycorrhizal fungi. Certain other microbes also enhance root nutrient uptake (Turner et al., 2013).

The relationship between microbes and plants occurs in three stages: firstly, metabolic gating, in which access to the plant by the microbe is hampered by definite plant metabolites. Plant nutrients that are beneficial to certain microbes or certain toxic compounds are synthesized by the plant in this stage. Mechanisms involving vesicle transport and diffusion are used by plant roots to release these compounds. Specifically, flavonoids and lipids are released by ABC transporters, while carbohydrates are released by anion channels (Glick and Gamalero, 2021). Secondly, twin receptor recognition entails the binding of signaling molecules of the microbe and plant to their respective receptors, hence either activating interaction or immunity and thirdly the incorporation of environmental signals with immune balance (Thoms et al., 2021). Microbes enhance plant development via mechanisms involving maximization of nutrients, production of phytohormones, and stimulation of immune system in the plants and hampering the growth of pathogenic plants (Song et al., 2020).

Microbiomes are associated with different parts of the plant. Definite microbiomes are found in the phyllosphere, rhizosphere, and endosphere of the same plant (Turner et al., 2013). Hence, microbiomes in each part of the plant have varying response to environmental factors. The link between above- and belowground tissues of a plant has been reported (Turner et al., 2013).

The survival of plant microbiome is both within and outside of a plant. Plant microbiomes also exert advantageous effects such as enhancement of plant growth and development as well as protection against plant pathogens (Orozco-Mosqueda et al., 2018).

Therefore, exploitation of bacterial microbiome could help in regulating plant development as well as protecting plants against disease-causing microbes; this limits the use of synthetic chemicals with potential adverse effects on the plant and the soil. A typical example is the plant growth—promoting bacteria (PGPB) that aid nutrient uptake and modify hormone level in plants (Orozco-Mosqueda et al., 2018). Iron and phosphorus are some of the nutrients, while hormones include auxins and cytokinins.

Another regulatory mechanism of PGPB is its ability of synthesizing 1-aminocyclopropane-1-carboxylate deaminase, an enzyme that reduces the production of ethylene. The protective mechanism exerted by PGPB against pathogens is via the production of degradative enzymes including chitinases and proteases. 2,4-Diacetylphloroglucinol, bacillomycin, mycosubtilin, iturin, and bacteriocins are other compounds with antibiotic potentials produced by PGPB (Orozco-Mosqueda et al., 2018).

Quite a number of plant growth—promoting rhizobacteria are antipathogenic. Hence, they produce antimicrobial compounds against plant pathogens (Rezzonico et al., 2005). Fluorescens produces diacetylphloroglucinol (DAPG), an antifungal agent. Diacetylphloroglucinol is seen to be toxic to some microbes but enhance the viability of others (Meyer et al., 2009).

There are reports that plants secrete phytotoxic compounds in the roots sequel to aerobic respiration in a bid to avert damage occasioned by flooding. Some of the compounds are ethanol, lactic acid, and alanine (Kozlowski, 1984; Badri and Vivanco, 2009; Hartman and Tringe, 2019).

It was indicated that a conglomerate of microorganisms rather than lone microorganism is helpful to plants (Song et al., 2020). This was observed in a study involving *Xanthomonas*, *Stenotrophomonas*, and *Microbacterium* in which the three bacterial genera were enhanced in the rhizosphere of *Arabidopsis thaliana* in *Hyaloperonospora arabidopsidis*—initiated defense, which led to plant protection and growth (Berendsen et al., 2018). In another study, endosphere and rhizosphere microbiomes were found to inhibit plant diseases (Duran et al., 2017, 2018). In separate studies, synergistic association between endophytes and certain bacteria boosted stress tolerance to drought in some plants (Khan et al., 2016). There are indications that plant diseases were hampered by microbiome of endosphere and rhizosphere (Song et al., 2020). In another study, wilt disease caused by *Ralstonia solanacearum* was suppressed by certain microbes in tomato plant rhizosphere (Kwak et al., 2018).

Phytohormones play significant roles in plant immunity. There is therefore an association between the composition of plant microbiome and phytohormones defense signaling. Lebeis et al. (2015) indicated that salicylate and its associated activities influenced the composition of root microbiome. In a study on the isolation of plant microbiota using apple carposphere and lettuce root microbiome, significant variation was observed in the efficiency of washing protocols. It was therefore concluded that four consecutive washes elevated the alpha diversity in apple relative to a lone wash (Sare et al., 2020). This enabled the quantity of culturable microbes with uneven yield (Sare et al., 2020).

Determination of plant microbiomes

The detection of microbes by plants is via pattern recognition receptors that get bound to microbe-associated molecular patterns. This elicits necessary defense that limits the growth of a number of pathogenic microorganisms (Finkel et al., 2017). The -omics enables the determination of the complex factors that influence plant microbiome. These methods include metaproteomics, metagenomics, metabolomics, etc. A higher efficiency output is achievable when any two of these methods are coupled together. For instance, the coupling of metaproteomic with metagenomic methods shows the proteins associated with the production of methane in rice rhizosphere (Knief et al., 2012).

A report indicated the use of metatranscriptomics in the determination of alteration in structure of plant rhizosphere microbiomes (Turner et al., 2013). Another study highlighted the comparable distribution of microbial phyla in studies on rhizosphere microbiomes (Lundberg et al., 2012; Bulgarelli et al., 2012; Turner et al., 2013). Complex techniques including iTAG and HTS-dependent marker gene tags enhance the determination of the constitution, arrangement, and spatial spread of microbes. This has been used on some crops to enhance their health. These crops include citrus, sugarcane, and some cereals (Srivastava et al., 2022). Whole-genome shotgun sequencing as well as amplicon sequencing elucidates comprehensive genomic information. A study on microbiome of citrus rhizosphere from about six continents using this technique showed a variety of microbes including *Mesorhizobium, Burkholderia, Cupriavidus* spp., etc. (Srivastava et al., 2022).

Reports on the determination of plant-dependent microbiome are sparse and could also be complex. This is contingent on the interaction of plant with soil metabolites, which is complex, heterogenous, and dynamic; therefore, sensitive methods such as liquid chromatography—mass spectrometry and proton nuclear magnetic resonance spectroscopy are required. Mass spectrometry imaging is a sensitive method that enables the determination of real-time spatial arrangement of metabolites. It has also been used in determining the interactions between microbes and plant—microbe interactions (Song et al., 2020). In a study on plant genotypes and bacterial microbiome in olive plants, it was reported that microbiome varied among the various parts of the olive plant. It was also reported that the microbiome composition depends on plant genotype (Malacrino et al., 2022). In their study, fruits were observed to have a higher percentage of microbiome in comparison with the leaves and soil. Furthermore, a higher amount of amplicon sequence variants was found in fruits than other plant parts.

Environmental factors and microbiomes

Stress has considerable effects on metabolism in plants. The stress could be abiotic or biotic. Extreme temperatures, flooding, high salinity, and presence of toxicants are examples of abiotic factors. Others including soil organic content, soil moisture, soil texture, soil pH, temperature, and root exudates are examples of these factors (Fierer, 2017). Furthermore, other factors include pathogens' presence, human activities, and climatic conditions that could affect the epiphytes and endophytes (Babalola et al., 2020). Disease-causing microbes constitute the biotic factors. There are indications that these environmental stresses usually overwhelm plants (Glick and Gamalero, 2021). Microbiota present in the rhizosphere providentially protects plants against these stresses. This could be a reason while plants opt for PGPB.

Almost all environmental stresses activate more production of ethylene (Abeles et al., 1992). Low-level stress results in the low production of ethylene, which activates defensive genes that protect the plant. This could be achieved using PGPB capable of synthesizing ACC deaminase. Reports showed alteration in microbial population occasioned by biotic stress. For instance, a report showed the association of soil microbiome with strawberry resistant to disease-causing microbes, in which higher valuable bacteria were found in the microbiome of the resistant plants (Lazcano et al., 2021). In another study on pepper plant, there was a change in root exudation contingent on aphid infestation. This resulted in the recruitment of rhizobacteria, which hampered the plant's resistance aphid infestation (Kim et al., 2016).

Root microbiome was reported to play significant roles in eliciting stress tolerance in plant hosts. This was reported in a study aimed at determining the responses of plants to abiotic stress including flooding, drought, etc. (Hartman and Tringe, 2019). Root exudates play a significant role in the recruitment of plant microbiomes. The relationship between root-dependent microbiomes and root exudates has also been reported (Babalola et al., 2020). The pathways associated with the metabolism of terpenoid were enhanced by rhizospheric bacteria related to wheat plants not treated with inorganic fertilizer (Kavamura et al., 2018).

Of all abiotic stresses, attention has focused more on salt stress probably due to relatively water shortage in over 50% of global arable land. High soil salinity and drought have been reported to inhibit crop productivity and adversely affect the plants' microbiomes. In a study, drought delayed the growth of sorghum root microbiome (Xu et al., 2018). Soil hydration improved rhizosphere microbiome (Mavrodi et al., 2018). Extreme temperatures due to climatic alteration have been reported to adversely affect microbiomes of the rhizosphere and phyllosphere microbiomes of some plants (Hough et al., 2020). Studies have also shown the impact of soil nitrogen and carbon on microbiome. Minimal carbon and nitrogen levels adversely affected soil microbiome.

Anthropogenic factors and plant microbiomes

Anthropocene is defined as the effect of anthropogenic activities on the ecosystem. Anthropogenic activities have significant impact on the planet. This invariably affects microbiomes. Evidence abound to the triggers of anthropocene, which includes excessive land cultivation, high population density in particular areas, and increase in global human being population. The impact is seen in the crossing of four out of nine planetary boundaries (Berg and Cernava, 2022). Increase in the rate of plant extinction, particularly tropical plants, is a reflection of anthropocene. Furthermore, global diet appears homogenous, particularly dependent on certain main crops, thus reducing the intake of certain local crops. This has significantly affected plant-based microbiota (Berg and Cernava, 2022).

Climate change is major planetary boundary capable of adversely affecting global ecosystem. Evidence of climate change include depletion of the ozone layer, global warming, and erratic weather. These invariably alter plant microbiota. In a study on *Prunus padua* aimed at determining the effect of global warming on pathogenic microbes, excessive disease-causing fungi and other infective agents were observed (Liu and He, 2021). Alteration in microbiome in *Sphagnum* was associated with reduced bacterial diversity and a decline in the rate of nitrogen fixation (Carrell et al., 2019). Ozone layer depletion is another factor that results in plant's exposure to UV-B radiation. There are evidence of loss of a variety of native microbes as well as reduction in shoot mass, foliage area, and plant height (Caldwell et al., 2003). The adverse effect of UV-B radiation on microbiomes has also been reported (Berg and Cernava, 2022).

Unwarranted use of nitrogen-based fertilizers could alter microbial-driven biogeochemical cycles, which play a significant role in the maintenance of the ecosystem. Unabsorbed nitrogen by plants is converted to nitrate, which percolates the soil easily (Berg and Cernava, 2022). In a study, certain plant microbiota was altered upon the administration of nitrogen and phosphorus fertilizer. The use of chemical fertilizers has greatly reduced food shortage; it has been shown that their consistent use could eventually increase crop's susceptibility to diseases.

High levels of environmental pollutants could cause dysbiosis in plants. This is of serious concern when these pollutants are of high concentration and linger in the environment. These environmentally persistent pollutants have a tendency of reshaping plant microbiomes. Of note are the toxic metals, antibiotics as well as microplastics that can act together to bring about antibiotic resistance (Feng et al., 2021).

Challenges, future perspectives, and conclusion

Interactions among plant microbiota and the environment vary greatly with a gamut of factors involved. It is recognized that microbiomes play a critical role in plant development. It is hoped that smart microbial consortia, microbiota-favored agricultural practices, and development of novel plans will enhance the application of microbiome in plant health (Berg and Cernava, 2022).

Adequate understanding of plant microbiome will be enhanced by extensive use of multiomics techniques, artificial intelligence, genomic editing, and technologically sophisticated culturing techniques. This will go a long way in boosting plant health and development (Srivastava et al., 2022). It is also obvious that anthropogenic activities, urbanization, increase in global population, and extensive agriculture interfere with plant microbiome. This has to be put in check if possible.

References

Abeles, F.B., Morgan, P.W., Salveit Jr., M.E., 1992. Ethylene in Plant Biology, second ed. Academic Press, New York, NY, USA, p. 41.

Adetunji, C.O., Olaniyan, O.T., Adeyomoye, O., Dare, A., Adeniyi, M.J., Alex, E., Rebezov, M., Garipova, L., Shariati, M.A., 2022a. eHealth, mHealth, and telemedicine for COVID-19 pandemic. In: Pani, S.K., Dash, S., dos Santos, W.P., Chan Bukhari, S.A., Flammini, F. (Eds.), Assessing COVID-19 and Other Pandemics and Epidemics Using Computational Modelling and Data Analysis. Springer, Cham. https://doi.org/10.1007/978-3-030-79753-9_10.

Adetunji, C.O., Olaniyan, O.T., Adeyomoye, O., Dare, A., Adeniyi, M.J., Alex, E., Rebezov, M., Petukhova, E., Ali Shariati, M., 2022b. Machine learning approaches for COVID-19 pandemic. In: Pani, S.K., Dash, S., dos Santos, W.P., Chan Bukhari, S.A., Flammini, F. (Eds.), Assessing COVID-19 and Other Pandemics and Epidemics Using Computational Modelling and Data Analysis. Springer, Cham. https://doi.org/10.1007/978-3-030-79753-9_8.

Adetunji, C.O., Olaniyan, O.T., Adeyomoye, O., Dare, A., Adeniyi, M.J., Enoch, A., Rebezov, M., Isabekova, O., Ali Shariati, M., 2022c. Smart sensing for COVID-19 pandemic. In: Pani, S.K., Dash, S., dos Santos, W.P., Chan Bukhari, S.A., Flammini, F. (Eds.), Assessing COVID-19 and Other Pandemics and Epidemics Using Computational Modelling and Data Analysis. Springer, Cham. https://doi.org/10.1007/978-3-030-79753-9_9.

Adetunji, C.O., Olaniyan, O.T., Adeyomoye, O., Dare, A., Adeniyi, M.J., Enoch, A., Rebezov, M., Petukhova, E., Ali Shariati, M., 2022d. Internet of health things (IoHT) for COVID-19. In: Pani, S.K., Dash, S., dos Santos, W.P., Chan Bukhari, S.A., Flammini, F. (Eds.), Assessing COVID-19 and Other Pandemics and Epidemics Using Computational Modelling and Data Analysis. Springer, Cham. https://doi.org/10.1007/978-3-030-79753-9_5.

Adetunji, C.O., Olaniyan, O.T., Adeyomoye, O., Dare, A., Adeniyi, M.J., Enoch, A., Rebezov, M., Koriagina, N., Ali Shariati, M., 2022e. Diverse techniques applied for effective diagnosis of COVID-19. In: Pani, S.K., Dash, S., dos Santos, W.P., Chan Bukhari, S.A., Flammini, F. (Eds.), Assessing COVID-19 and Other Pandemics and Epidemics Using Computational Modelling and Data Analysis. Springer, Cham. https://doi.org/10.1007/978-3-030-79753-9_3.

Adetunji, C.O., Samuel, M.O., Adetunji, J.B., Oluranti, O.I., 2022f. Corn silk and health benefits. Medical Biotechnology, Biopharmaceutics, Forensic Science and Bioinformatics, first ed.st Edition. Imprint CRC Press, ISBN 9781003178903, p. 12. https://doi.org/10.1201/9781003178903-11. First Published 2022.

Adetunji, C.O., Wilson, N., Olayinka, A.S., Olugbemi, O.T., Akram, M., Laila, U., Samuel, M.O., Oshinjo, A.M., Adetunji, J.B., Okotie, G.E., Esiobu, N.(D.), 2022g. Computational intelligence techniques for combating COVID-19. In: Medical Biotechnology, Biopharmaceutics, Forensic Science and Bioinformatics, first ed.st Edition. Imprint CRC Press, ISBN 9781003178903, p. 12. https://doi.org/10.1201/9781003178903-16. First Published 2022.

Adetunji, C.O., Olugbemi, O.T., Akram, M., Laila, U., Samuel, M.O., Michael Oshinjo, A., Adetunji, J.B., Okotie, G.E., Esiobu, N.(D.), Oyedara, O.O., Adeyemi, F.M., 2022h. Application of computational and bioinformatics techniques in drug repurposing for effective development of potential drug candidate for the management of COVID-19. In: Medical Biotechnology, Biopharmaceutics, Forensic Science and Bioinformatics, first ed.st Edition. Imprint CRC Press, ISBN 9781003178903, p. 14. https://doi.org/10.1201/9781003178903-15. First Published 2022.

Adetunji, C.O., Nwankwo, W., Olayinka, A.S., Olugbemi, O.T., Akram, M., Laila, U., Olugbenga, M.S., Oshinjo, A.M., Adetunji, J.B., Okotie, G.E., Esiobu, N.(D.), 2022i. Machine learning and behaviour modification for COVID-19. In: Medical Biotechnology, Biopharmaceutics, Forensic Science and Bioinformatics, first ed.st Edition. Imprint CRC Press, ISBN 9781003178903, p. 17. https://doi.org/10.1201/9781003178903-17. First Published 2022.

Ansari, M.I., 2018. Plant microbiome and its functional mechanism in response to environmental stress. Int. J. Green Pharm. 12 (01).

Babalola, O.O., Fadiji, A.E., Enagbonma, B.J., Alori, E.T., Ayilara, M.S., Ayangbenro, A.S., 2020. The nexus between plant and plant microbiome: revelation of the networking strategies. Front. Microbiol. 11, 548037. https://doi.org/10.3389/fmicb.2020.548037.

Badri, D.V., Vivanco, J.M., 2009. Regulation and function of root exudates. Plant Cell Environ. 32, 666−681. https://doi.org/10.1111/j.1365-3040.2009.01926.x.

Bais, H.P., Weir, T.L., Perry, L.G., Gilroy, S., Vivanco, J.M., 2006. The role of root exudates in rhizosphere interactions with plants and other organisms. Annu. Rev. Plant Biol. 57, 233−266. https://doi.org/10.1146/annurev.arplant.57.032905.105159.

Berendsen, R.L., Vismans, G., Yu, K., Song, Y., de Jonge, R., Burgman, W.P., et al., 2018. Disease-induced assemblage of a plant-beneficial bacterial consortium. ISME J. 12, 1496−1507. https://doi.org/10.1038/s41396-018-0093-1.

Berg, G., Cernava, T., 2022. The plant microbiota signature of the Anthropocene as a challenge for microbiome research. Microbiome 10, 54. https://doi.org/10.1186/s40168-021-01224-5.

Berg, M., Koskella, B., 2018. Nutrient-and dose-dependent microbiome-mediated protection against a plant pathogen. Curr. Biol. 28 (15), 2487−2492.

Bever, J.D., Platt, T.G., Morton, E.R., 2012. Microbial population and community dynamics on plant roots and their feedbacks on plant communities. Annu. Rev. Microbiol. 66, 265−283. https://doi.org/10.1146/annurev-micro-092611-150107.

Bulgarelli, D., Rott, M., Schlaeppi, K., Ver Loren van Themaat, E., Ahmadinejad, N., Assenza, F., Rauf, P., Huettel, B., Reinhardt, R., Schmelzer, E., Peplies, J., Gloeckner, F.O., Amann, R., Eickhorst, T., Schulze-Lefert, P., 2012. Revealing structure and assembly cues for *Arabidopsis* root-inhabiting bacterial microbiota. Nature 488, 91−95.

Caldwell, M.M., Ballaré, C.L., Bornman, J.F., Flint, S.D., Björn, L.O., Teramura, A.H., et al., 2003. Terrestrial ecosystems, increased solar ultraviolet radiation and interactions with other climatic change factors. Photochem. Photobiol. Sci. 2, 29−38.

Carrell, A.A., Kolton, M., Glass, J.B., Pelletier, D.A., Warren, M.J., et al., 2019. Experimental warming alters the community composition, diversity, and N2 fixation activity of peat moss (Sphagnum fallax) microbiomes. Global Change Biol. 25, 2993−3004.

Chiu, L., Bazin, T., Truchetet, M.E., Schaeverbeke, T., Delhaes, L., Pradeu, T., 2017. Protective microbiota: from localized to long-reaching co-immunity. Front. Immunol. 8, 1678. https://doi.org/10.3389/fimmu.2017.01678.

Durán, P., Jorquera, M., Viscardi, S., Carrion, V.J., de la Luz Mora, M., Pozo, M.J., 2017. Screening and characterization of potentially suppressive soils against gaeumannomyces graminis under extensive wheat cropping by chilean indigenous communities. Front. Microbiol. 8, 1552. https://doi.org/10.3389/fmicb.2017.01552.

Durán, P., Tortella, G., Viscardi, S., Barra, P.J., Carrión, V.J., Mora, M.L., et al., 2018. Microbial community composition in take-all suppressive soils. Front. Microbiol. 9, 2198. https://doi.org/10.3389/fmicb.2018.02198.

Esiobu, N.D., Ogbonna, J.C., Adetunji, C.O., Obembe, O.O., Ezeonu, I.M., Ibrahim, A.B., Ubi, B.E., 2022. Microbiomes and Emerging Applications, first ed.st Edition. Imprint CRC Press, Boca Raton, ISBN 9781003180241, p. 186. https://doi.org/10.1201/9781003180241. First Published 2022. eBook

Published 11 May 2022 eBook ISBN 9781003180241. Subjects Bioscience, Engineering and Technology.

Feng, G., Huang, H., Chen, Y., 2021. Effects of emerging pollutants on the occurrence and transfer of antibiotic resistance genes: a review. J. Hazard Mater. 420, 126602.

Fierer, N., 2017. Embracing the unknown: disentangling the complexities of the soil microbiome. Nat. Rev. Microbiol. 15, 579–590. https://doi.org/10.1038/nrmicro.2017.87.

Finkel, O.M., Castrillo, G., Paredes, S.H., González, I.S., Dangl, J.L., August 2017. Understanding and exploiting plant beneficial microbes. Curr. Opin. Plant Biol. 38, 155–163. https://doi.org/10.1016/j.pbi.2017.04.018.

Glick, B.R., Gamalero, E., 2021. Recent developments in the study of plant microbiomes. Microorganisms 9, 1533. https://doi.org/10.3390/microorganisms9071533.

Hartman, K., Tringe, S.G., 2019. Interactions between plants and soil shaping the root microbiome under abiotic stress. Biochem. J. 476 2705–2724.

Hough, M., McClure, A., Bolduc, B., Dorrepaal, E., Saleska, S., Klepac-Ceraj, V., Rich, V., 2020. Biotic and environmental drivers of plant microbiomes across a permafrost thaw gradient. Front. Microbiol. 11, 796.

Kavamura, V.N., Hayat, R., Clark, I.M., Rossmann, M., Mendes, R., Hirsch, P.R., et al., 2018. Inorganic nitrogen application affects both taxonomical and predicted functional structure of wheat rhizosphere bacterial communities. Front. Microbiol. 9, 1074. https://doi.org/10.3389/fmicb.2018.01074.

Khan, Z., Rho, H., Firrincieli, A., Hung, S.H., Luna, V., Masciarelli, O., et al., 2016. Growth enhancement and drought tolerance of hybrid poplar upon inoculation with endophyte consortia. Curr. Plant Biol. 6, 38–47. https://doi.org/10.1016/j.cpb.2016.08.001.

Kim, B., Song, G.C., Ryu, C.-M., 2016. Root exudation by aphid leaf infestation recruits root-associated Paenibacillus spp. to lead plant insect susceptibility. J. Microbiol. Biotechnol. 26, 549–557.

Knief, C., Delmotte, N., Chaffron, S., Stark, M., Innerebner, G., Wassmann, R., et al., 2012. Metaproteogenomic analysis of microbial communities in the phyllosphere and rhizosphere of rice. ISME J. 6, 1378–1390. https://doi.org/10.1038/ismej.2011.192.

Korenblum, E., Dong, Y., Szymanski, J., Panda, S., Jozwiak, A., Massalha, H., Meir, S., Rogachev, I., Aharoni, A., 2020. Rhizosphere microbiome mediates systemic root metabolite exudation by root-to-root signaling. Proc. Natl. Acad. Sci. USA 117 (7), 3874–3883.

Kozlowski, T.T., 1984. Plant responses to flooding of soil. Bioscience 34, 162–167. https://doi.org/10.2307/1309751.

Kwak, M.-J., Kong, H.G., Choi, K., Kwon, S.-K., Song, J.Y., Lee, J., et al., 2018. Rhizosphere microbiome structure alters to enable wilt resistance in tomato. Nat. Biotechnol. 36, 1100–1109. https://doi.org/10.1038/nbt.4232.

Lazcano, C., Boyd, E., Holmes, G., Hewavitharana, S., Pasulka, A., Ivors, K., 2021. The rhizosphere microbiome plays a role in the resistance to soil-borne pathogens and nutrient uptake of strawberry cultivars under filed conditions. Sci. Rep. 11, 3188.

Lebeis, S.L., Herrera Paredes, S., Lundberg, D.S., Breakfield, N., Gehring, J., McDonald, M., Malfatti, S., Glavina del Rio, T., Jones, C.D., Tringe, S.G., et al., 2015. Salicylic acid modulates colonization of the root microbiome by specific bacterial taxa. Science 349, 860–864.

Liu, Y., He, F., 2021. Warming intensifies soil pathogen negative feedback on a temperate tree. New Phytol. 231, 2297–2307.

Liu, H., Brettell, L.E., Qiu, Z., Singh, B.K., 2020. Microbiome-mediated stress resistance in plants. Trends Plant Sci. 25 (8), 733–743.

Lundberg, D.S., Lebeis, S.L., Paredes, S.H., Yourstone, S., Gehring, J., Malfatti, S., Tremblay, J., Engelbrektson, A., Kunin, V., Rio, T.G.D., Edgar, R.C., 2012. Defining the core *Arabidopsis thaliana* root microbiome. Nature 488 (7409), 86–90.

Lundberg, D.S., Lebeis, S.L., Paredes, S.H., Yourstone, S., Gehring, J., Malfatti, S., Tremblay, J., Engelbrektson, A., Kunin, V., del Rio, T.G., Edgar, R.C., Eickhorst, T., Ley, R.E., Hugenholtz, P., Tringe, S.G., Dangl, J.L., 2012. Defining the core *Arabidopsis thaliana* root microbiome. Nature 488, 86–90.

Malacrinò, A., Mosca, S., Li Destri Nicosia, M.G., Agosteo, G.E., Schena, L., 2022. Plant genotype shapes the bacterial microbiome of fruits, leaves, and soil in olive plants. Plants 11, 613. https://doi.org/10.3390/plants11050613.

Mavrodi, D.V., Mavrodi, O.V., Elbourne, L.D.H., Tetu, S., Bonsall, R.F., Parejko, J., Yang, M., Paulsen, I.T., Weller, D.M., Thomashow, L.S., 2018. Long-term irrigation affects the dynamics and activity of the wheat rhizosphere microbiome. Front. Plant Sci. 9, 345.

Meyer, S.L.F., Halbrendt, J.M., Carta, L.K., Skantar, A.M., Liu, T., Abdelnabby, H.M.E., Vinyard, B.T., 2009. Toxicity of 2,4-diacetylphloroglucinol (DAPG) to plant-parasitic and bacterial-feeding nematodes. J. Nematol. 41, 274–280.

Michael, O.S., Oluranti, O.I., Oshinjo, A.M., Adetunji, C.O., Olaniyi, K.S., Adetunji, J.B., 2022. Microbiome Reshaping and Epigenetic Regulation, first ed.st Edition. Imprint CRC Press, ISBN 9781003180241, p. 22. https://doi.org/10.1201/9781003180241-6. First Published 2022.

Michael, O.S., Oluranti, O.I., Oshinjo, A.M., Adetunji, C.O., Adetunji, J.B., Esiobu, N.(D.), 2022b. Microbiota transplantation, health implications, and the way forward. In: Microbiomes and Emerging Applications, first ed.st Edition. Imprint CRC Press, ISBN 9781003180241, p. 19. https://doi.org/10.1201/9781003180241-5. First Published 2022.

Olaniyan, O.T., Adetunji, C.O., Adeniyi, M.J., Hefft, D.I., 2022a. Machine learning techniques for high-performance computing for IoT applications in healthcare. In: Deep Learning, Machine Learning and IoT in Biomedical and Health Informatics, first ed.st Edition. Imprint CRC Press, ISBN 9780367548445, p. 13. https://doi.org/10.1201/9780367548445-20. First Published 2022.

Olaniyan, O.T., Adetunji, C.O., Adeniyi, M.J., Hefft, D.I., 2022b. Computational intelligence in IoT healthcare, 2022m. In: Deep Learning, Machine Learning and IoT in Biomedical and Health Informatics, first ed.st Edition. Imprint CRC Press, ISBN 9780367548445, p. 13. https://doi.org/10.1201/9780367548445-19. First Published 2022.

Orozco-Mosqueda, M.C., Rocha-Granados, M.C., Glick, B.R., Santoyo, G., 2018. Microbiome engineering to improve biocontrol and plant growth-promoting Mechanisms. Microbiol. Res. 208, 25–31.

Oyedara, O.O., Adeyemi, F.M., Adetunji, C.O., Elufisan, T.O., 2022. Repositioning antiviral drugs as a rapid and cost-effective approach to discover treatment against SARS-CoV-2 infection. In: Medical Biotechnology, Biopharmaceutics, Forensic Science and Bioinformatics, first ed.st Edition. Imprint CRC Press, ISBN 9781003178903, p. 12. https://doi.org/10.1201/9781003178903-10. First Published 2022.

Pang, I.K., Iwasaki, A., 2012. Control of antiviral immunity by pattern recognition and the microbiome. Immunol. Rev. 245, 209–226. https://doi.org/10.1111/j.1600-065X.2011.01073.x.

Rajniak, J., Giehl, R., Chang, E., Murgia, I., von Wirén, N., Sattely, E.S., 2018. Biosynthesis of redox-active metabolites in response to iron deficiency in plants. Nat. Chem. Biol. 14 (5), 442–450. https://doi.org/10.1038/s41589-018-0019-2.

Rezzonico, F., Binder, C., Defago, G., Moenne-Loccoz, Y., 2005. The type III secretion system of biocontrol *Pseudomonas fluorescens* KD targets the phytopathogenic chromista *Pythium ultimum* and promotes cucumber protection. Mol. Plant Microbe Interact. 18, 991–1001.

Round, J.L., Mazmanian, S.K., 2009. The gut microbiota shapes intestinal immune responses during health and disease. Nat. Rev. Immunol. 9, 313–323. https://doi.org/10.1038/nri2515.

Rudrappa, T., Czymmek, K.J., Paré, P.W., Bais, H.P., 2008. Root-secreted malic acid recruits beneficial soil bacteria. Plant Physiol. 148 (3), 1547–1556. https://doi.org/10.1104/pp.108.127613.

Sare, A.R., Stouvenakers, G., Eck, M., Lampens, A., Goormachtig, S., Jijakli, M.H., Massart, S., 2020. Standardization of plant microbiome studies: which proportion of the microbiota is really harvested? Microorganisms 8, 342.

Song, C., Zhu, F., Carrión, V.J., Cordovez, V., 2020. Beyond plant microbiome composition: exploiting microbial functions and plant Traits via integrated Approaches. Front. Bioeng. Biotechnol. 8, 896. https://doi.org/10.3389/fbioe.2020.00896.

Srivastava, A.K., Das, A.K., agannadham, P.T.K., Bora, P., Ansari, F.A., Bhate, R., 2022. Bioprospecting microbiome for soil and plant health management amidst Huanglongbing threat in citrus: a review. Front. Plant Sci. 13, 858842. https://doi.org/10.3389/fpls.2022.858842.

Stringlis, I.A., Yu, K., Feussner, K., de Jonge, R., Van Bentum, S., Van Verk, M.C., Berendsen, R.L., Bakker, P., Feussner, I., Pieterse, C., 2018. MYB72-dependent coumarin exudation shapes root microbiome assembly to promote plant health. Proc. Natl. Acad. Sci. U.S.A. 115 (22), E5213—E5222. https://doi.org/10.1073/pnas.1722335115.

Thoms, D., Liang, Y., Haney, C.H., 2021. Maintaining symbiotic homeostasis: how do plant engage with beneficial microorganisms while at the same time restricting pathogens? Mol. Plant-Microbe Interact. 34 (5), 462—469. https://doi.org/10.1094/MPMI-11-20-0318-FI. Epub Mar 31, 2021. PMID: 33534602.

Trivedi, P., Leach, J.E., Tringe, S.G., Sa, T., Singh, B.K., 2020. Plant—microbiome interactions: from community assembly to plant health. Nat. Rev. Microbiol. 18 (11), 607—621.

Turner, T.R., James, E.K., Poole, P.S., 2013. The plant microbiome. Genome Biol. 14, 209.

Wei, Z., Gu, Y., Friman, V.P., Kowalchuk, G.A., Xu, Y., Shen, Q., Jousset, A., 2019. Initial soil microbiome composition and functioning predetermine future plant health. Sci. Adv. 5 (9), eaaw0759.

Xu, J., Zhang, Y., Zhang, P., Trivedi, P., Riera, N., Wang, Y., Liu, X., Fan, G., Tang, J., Coletta-Filho, H.D., et al., 2018. The structure and function of the global citrus rhizosphere microbiome. Nat. Commun. 9, 4894.

Further reading

Ivanov II, Littman, D.R., 2011. Modulation of immune homeostasis by commensal bacteria. Curr. Opin. Microbiol. 14, 106—114. https://doi.org/10.1016/j.mib.2010.12.003.

CHAPTER 7

Introduction to animal microbiome

Olulope Olufemi Ajayi[1], Charles Oluwaseun Adetunji[2], Olugbenga Samuel Michael[3,4,6] and Juliana Bunmi Adetunji[5]

[1]Department of Biochemistry, Edo State University Uzairue, Edo State, Nigeria; [2]Applied Microbiology, Biotechnology and Nanotechnology Laboratory, Department of Microbiology, and Directorate of Research and Innovation, Edo State University Uzairue, Iyamho, Auchi, Edo State, Nigeria; [3]Cardiometabolic, Microbiome and Applied Physiology Laboratory, Department of Physiology, College of Health Sciences, Bowen University, Iwo, Osun State, Nigeria; [4]Department of Physiology, University of Tennessee Health Science Center, Memphis, TN, United States; [5]Nutritional and Toxicological Research Laboratory, Department of Biochemistry Sciences, Osun State University, Osogbo, Osun State, Nigeria; [6]Department of Medical Pharmacology and Physiology, University of Missouri, Columbia, MO, United States

Introduction

Animals are reservoirs of microorganisms (Song et al., 2019). Certain microbes are beneficial to the host; some are commensals, while some are harmful. The use of microbiome as an index of health and disease as is currently being explored (Song et al., 2019). Ill health is associated with an imbalance in animal microbiota, a situation described as dysbiosis. Dysbiosis could be due to excessive growth of opportunistic microbes, which are usually pathogenic, a significant reduction in or complete absence of the "good" microorganisms, which exert beneficial effects. It may also be attributed to reduction in the variability of microbial population (Song et al., 2019). Use of antibiotics has been associated with dysbiosis. There are instances where antibiotics completely eradicated the good and the pathogenic microbes, concurrently. This occurs not only in humans but also in other lower animals (Suchodolski et al., 2009; Harlow et al., 2013; Costa et al., 2015; Torres-Henderson et al., 2017; Yitbarek et al., 2018; Whittermore et al., 2018).

There are evidence that modulating microbiome could be used in restoring normalcy in dysbiosis and enhancing animal health. Selective targeting, community-level, and culture-level supplementation were identified as possible steps to achieve restoration of microbiome in animals with dysbiosis (Song et al., 2019). These beneficial microorganisms are symbiotic and may be advantageous to animals by supplementing the diet and enhancing immune system (Bahrndorff et al., 2016). This results in general wellness of the animal. Diet and host's genotype are significant factors that affect animal microbiome (Bahrndorff et al., 2016).

A high proportion of microbial population reside in the gastrointestinal tract, particularly the intestine where a cross-link between the host and the microbiota occurs (Esser et al., 2019). Intestinal epithelium is covered by a mucus layer, which primarily contains the mucins that are glycoproteins. The mucins serve as lubricant that protects the

An Introduction to the Microbiome in Health and Diseases
ISBN 978-0-323-91190-0,
https://doi.org/10.1016/B978-0-323-91190-0.00007-2

intestine and also distinguishes the intestinal lumen that contains microbes from the epithelium (Li et al., 2015). Intestinal as well as lung microbiome also promotes mucus production (Esser et al., 2019).

Certain intestinal microbiomes produce indoles that enhance animal health and fertility (Sonowal et al., 17). Microbiomes also play an important role in aging. Report showed that microbiome with less diversity was observed in aged turquoise killifish, which is associated with deficient metabolic processes. An analogous observation was observed in aged mice and human beings (Esser et al., 2019).

Age-associated alterations in the composition of microbial population have been implicated in inflammation and inflammation-dependent diseases (Fransen et al., 2017). *Helicobacter pylori* has been implicated in the etiology of gastric cancer (Bhatt et al., 2017). Dysbiosis is also associated with colorectal cancer. Citrobacter rodentium has also been reported as a risk factor of colorectal cancer (Bhatt et al., 2017). Reports have shown the involvement of dysbiosis in the pathogenesis of intestinal and neurological diseases. There is, however, paucity of information on the role of dysbiosis in metabolic diseases (Dabke et al., 2019).

There are now evidence of a link between dysbiosis and obesity. Reports showed that gut microbiome is extremely susceptible to diet composition (Turnbaugh et al., 2006; David et al., 2014). In a study to compare the effects of plant-based and animal-dependent diets on microbial population in humans, study participants placed on animal-based diet had reduction in both bile-tolerant and fiber-fermenting microbes (David et al., 2014).

It has also been reported that alteration of gut microbiota could disrupt circadian rhythm that can culminate in obesity (Dabke et al., 2019). Alteration in gut microbiome has also been implicated in the etiology of cardiovascular disease. Trimethylamine N-oxide, a proatherosclerotic compound, was produced from choline in animals treated with antibiotics (Dabke et al., 2019).

Microbiome appears to be a panacea to dyslipidemia. Evidence of the association of lipidemia with glucose dysmetabolism have been extensively reported (Matey-Hernandez et a., 2018). Short-chain fatty acids (SCFAs) obtainable from fermentation of dietary fibers are capable of mollifying the effects of chronic dyslipidemia (Canfora et al., 2015). Acetate, propionate, and butyrate are the principal SCFAs produced in the colon, of which acetate accounts for the highest percentage (Dabke et al., 2019).

Rumen and microbiomes

Rumen microbiome has been reported to improve milk production in dairy cows. A number of factors have been implicated in cow milk production, including diet, genetics, and animal management (Xue et al., 2020). Rumen in ruminants is an important organ that allows nutrient derivation from plants. It is possible that microbes in the rumen can enhance milk protein yield. There are also speculations that variability of rumen microbiota may not affect the metabolic function of rumen (Xue et al., 2020).

The anabolism of milk protein involves the breakdown of dietary crude protein, which is used in the biosynthesis of protein of microbes in the rumen. The microbial protein alongside dietary protein not broken down is further broken down into amino acids that are absorbed in the small intestines. Milk protein is therefore synthesized from amino acids transported firstly to the liver and later to the mammary gland (Xue et al., 2020).

The rumen microbes are also important in the biosynthesis of vitamin B, which are important cofactors. For instance, carboxylases utilize biotin as a coenzyme, and reactions involving flavoproteins require riboflavin as coenzyme. Furthermore, the biosynthesis of coenzyme A requires pantothenic acid as coenzyme. It therefore suggests that MPY can be enhanced by the vitamin B produced by rumen microbiota.

Feed effectiveness in ruminant animals depends on the capacity of the rumen microbiome to produce nutrients from digestible feed. Furthermore, the microbiome enhances the conversion of plant fibers to volatile fatty acids, which is the principal source of energy in ruminants (Xue et al., 2022).

Bird gut and microbiomes

Bird gut comprises the gizzard, proventriculus, and the crop, with each section containing unique microbiota this has to do with their respective functions. Food is stored in the crop prior to its digestion, which is initiated in the proventriculus (Deusch et al., 2015). Highly diverse microbial population resides in the intestine of birds, particularly chicken. The bacterial population of intestinal duodenum is the least when compared with the jejunum and ileum.

The microbiota enhances health and development of these animals (Deusch et al., 2015). It is pertinent to say that the gut microbiota helps with synthesis of vitamins, SCFAs, and amino acids. Bird intestine also houses bacteria that produce butyrate that enhances food absorption and nutrient maximization. The effect of diet, gender, and antibiotics on bird GIT microbiota in chicken has been extensively studied. Antibiotics alters bird gut microbiomes (Deusch et al., 2015).

Microbiome in captive and wild animals

Animal captivity is a human interference on the lifestyle of an animal. This could be beneficial and conversely could be detrimental to the well-being of some animals. Animals in captivity receive better diet, protection, and veterinary attention, thus resulting in their extended life span. Quite a number of challenges have, however, been observed in captive animals. This includes dysmetabolism, stress, infertility, and infections. Change of habitat may interfere with microbial population and diversity in captive animals.

Gut microbiome composition is significantly dependent on the environment and diet. This could have systemic affect on the animal. For instance, a similarity in gut microbiome of captive wild horses and domestic horses was reported (Metcalf et al., 2017). Helicobacter gastritis was observed in captive cheetahs, an indication of the link between shift in microbiome and animal health. Pro- and prebiotic effect of diet on microbiome has been reported (Diaz and Reese, 2021).

Regular contact with humans could also interfere with the microbiome of captive animals. The use of chemotherapeutics, particularly antibiotics, rapidly eliminates significant part of the wild microbiota (Ramirez et al., 2020). There is the possibility of direct introduction of human-related microbes due to direct human contact, thereby causing stress known to alter microbial constitution in the captive animal (Willing et al., 2011).

Omics in animal microbiome analysis

A study suggested the consideration of rumen metabolome in the association of feed efficiency with rumen microbiome. In the study, the integration of metatranscriptomics, metabolomics, and metagenomics was employed to determine the role of rumen microbial population in food efficiency in milk-producing cows (Xue et al., 2022). The study suggested that archaea and bacteria are associated with feed efficiency. There was a higher prevalence of *Selenomonas* in the rumen of high efficiency cows relative to low efficiency cows (Xue et al., 2022). A number of *Selenomonas* spp. could produce acetate and lactate from starch. Using metabolomics, about six metabolites of carbohydrate metabolic pathways could predict cow's feed efficiency. Machine learning technique is a vital tool that can be used to enhance the utility of microbiota in predicting animal phenotypes including animal's strength and susceptibility to diseases. This is because this prediction using omics technology is demanding due to the microbial characterization and identification techniques involved. Another pitfall of meta-omics is the high cost implication, which makes it readily unaffordable (Xue et al., 2022).

Meta-omics are used for the study of rumen microbiota. Metagenomics utilizes shortgun sequencing in characterizing possible role of microbiota as a function of their genomes. It enables the analysis of not easily cultivable microbes (Denman et al., 2018). A pitfall of metabolomics is its inability to directly correlate metabolites with microbial spp. This challenge can be overcome by metatranscriptomic, microbial profiling, or metaproteomics (Denman et al., 2018).

Metataxonomics is a more sophisticated technique that gives more detailed information about microbial composition and diversity (Denman et al., 2018). The cost-effectiveness of metataxonomics is one of the reasons why it is commonly used for ruminants (Denman et al., 2018).

Animal microbiome

Certain studies have been carried out on microbiomes in birds. In a study, the micro-biome in the cecum of both wild and captive rock ptarmigans was studied. Rock ptarmigans are birds found in the Arctic and sub-Arctic regions of the world, whose diet changes with the season. Significantly higher bacterial population and methanogenic archaea were found in wild compared with the captive birds (Salgado-Flores et al., 2019). Bacteria that degrade plant secondary metabolites as well as the genes responsible for the conversion of phenolic compounds to pyruvate were found in wild ptarmigans (Salgado-Flores et al., 2019).

Influenza virus A is found in waterducks. In a study, the microbiomes of the cloaca of five species of wild ducks were examined. Absence of consistent flu-like microbiome was observed across the species. Furthermore, two species, northern shovelers and mallards, had a highly diversified microbiome composition (Hird et al., 2018).

In another study carried out on the influence of external environment on gut micro-biome in wild and captive deer mice, it was reported that gut microbiome alpha diversity and composition depend on the immediate environment. Mice in natural habitat had more varied gut microbiome in excess of Helicobacteraceae, Ruminoccocaceae, and Lechnospiraceae spp. The latter two had been associated with maintenance of healthy microbiome and digestion in wild deer mice. There are indications that captive animals may be deficient of these beneficial microbes (Schmidt et al., 2019). Bacteroidetes were the dominant microbiome in captive deer mice.

The effects of season and environment on the gut microbiota were studied in wild red squirrels. Two microbial genera, *Corpococcus* and *Oscillospira*, were observed in the gut of the wild red squirrels. Seasonal variation in gut microbiota in these animals was associated with seasonal dietary pattern. Individuality and maternal effects also play important roles in wild red squirrel gut microbiota (Ren et al., 2017).

Diversity in fecal microbiome and functional genes were reduced in pandas in captivity relative to those in the wild. A reduction in functional ability to metabolize cel-lulose as well as enhanced starch metabolism was observed in captive pandas (Guo et al., 2019). A significant elevation in the activity of amylase and reduced cellulase activity were also observed in the feces of captive pandas in comparison with wild pandas (Guo et al., 2019). It was also reported that microbiomes of captive pandas were enhanced by antibiotic resistance genes, heavy metal tolerance, and virulence factors (Guo et al., 2019).

Seasonal variation in bacterial population was observed in wild fish. Furthermore, the interaction of diet with immune control of bacterial composition in mucus layers indic-ative of host-microbiome association was also observed (Friberg et al., 2019).

The gut microbiota of captive and wild black rhinos were compared to determine taxonomic and functional diversity of microbiome. There was no difference in the alpha diversity; however, significant difference was observed between the wild and captive black rhinos (Gibson et al., 2019).

Furthermore, elevated glycolysis and amino acid synthesis were observed in captive rhino microbiomes indicative of malnutrition in captivity, i.e., not having access to optimal nutrition (Gibson et al., 2019).

Conclusion

It is obvious from this review that microbiome plays a significant role in animal's health and disease. Quite a number of gut microbes are commensals that aid nutrient utilization and absorption, thereby enhancing the growth and development of animals. Conversely, dysbiosis has been implicated in disease conditions in animals. Diet and other environmental factors influence animal microbiome.

References

Bahrndorff, S., Alemu, T., Alemneh, T., Nielsen, J.L., 2016. The microbiome of animals: implications for conservation biology. Int. J. Genomics 2016, 1—7. https://doi.org/10.1155/2016/5304028. Article ID 5304028. PMID: 27195280.

Bhatt, A.P., Redinbo, M.R., Bultman, S.J., July 08, 2017. The role of the microbiome in cancer development and therapy. CA Cancer J. Clin. 67 (4), 326—344. https://doi.org/10.3322/caac.21398.

Canfora, E.E., Jocken, J.W., Blaak, E.E., 2015. Short-chain fatty acids in control of body weight and insulin sensitivity. Nat. Rev. Endocrinol. 11 (10), 577—591. https://doi.org/10.1038/nrendo.2015.128.

Costa, M.C., St€ampfli, H.R., Arroyo, L.G., Allen-Vercoe, E., Gomes, R.G., Weese, J.S., 2015. Changes in the equine fecal microbiota associated with the use of systemic antimicrobial drugs. BMC Vet. Res. 11, 19.

Dabke, K., Hendrick, G., Devkota, S., October 1, 2019. The gut microbiome and metabolic syndrome. J. Clin. Invest. 129 (10), 4050—4057. https://doi.org/10.1172/JCI129194.

David, L.A., et al., 2014. Diet rapidly and reproducibly alters the human gut microbiome. Nature 505 (7484), 559—563. https://doi.org/10.1038/nature12820.

Denman, S.E., Morgavi, D.P., McSweeney, C.S., 2018. Review: the application of omics to rumen microbiota function. Animal 12 (S2), s233—s245.

Deusch, S., Tilocca, B., Camarinha-Silva, A., Seifert, J., 2015. News in livestock research—use of omics-technologies to study the microbiota in the gastrointestinal tract of farm animals. Comput. Struct. Biotechnol. J. 13, 55—63.

Diaz, J., Reese, A.T., 2021. Possibilities and limits for using the gut microbiome to improve captive animal health. Animal Microbiome 3, 89. https://doi.org/10.1186/s42523-021-00155-8.

Esser, D., Lange, J., Marinos, G., Sieber, M., Best, L., Prasse, D., Bathia, J., Rühlemann, M.C., Boersch, K., Jaspers, C., Sommer, F., 2019. Functions of the microbiota for the physiology of animal metaorganisms. J. Innate Immun. 11, 393—404.

Fransen, F., van Beek, A.A., Borghuis, T., Aidy, S.E., Hugenholtz, F., van der Gaast-de Jongh, C., et al., November 2017. Aged gut microbiota contributes to systemical inflammaging after transfer to germ-free mice. Front. Immunol. 8, 1385.

Friberg, I.M., Taylor, J.D., Jackson, J.A., 2019. Diet in the driving seat: natural diet-immunity-microbiome interactions in wild fish. Front. Immunol. 10, 243. https://doi.org/10.3389/fimmu.2019.00243.

Gibson, K.M., Nguyen, B.N., Neumann, L.M., Miller, M., Buss, P., Daniels, S., Ahn, M.J., Crandall, K.A., Pukazhenthi, B., 2019. Gut Microbiome Differences between Wild and Captive Black Rhinoceros—Implications for Rhino health. Sci. Rep. 9, 7570. https://doi.org/10.1038/s41598-019-43875-3.

Guo, W., Mishra, S., Wang, C., Zhang, H., Ning, R., Kong, F., Zeng, B., Zhao, J., Li, Y., 2019. Comparative study of gut microbiota in wild and captive giant pandas (Ailuropoda melanoleuca). Genes 10, 827. https://doi.org/10.3390/genes10100827.

Harlow, B.E., Lawrence, L.M., Flythe, M.D., 2013. Diarrhea-associated pathogens,lactobacilli and cellulolytic bacteria in equine feces: responses to antibiotic challenge. Vet. Microbiol. 166, 225–232.

Hird, S.M., Ganz, H., Eisen, J.A., Boyce, W.M., 2018. The cloacal microbiome of five wild duck species varies by species and influenza A virus infection status. mSphere 3, 1–15. https://doi.org/10.1128/mSphere.00382-18. PMID: 30355662.

Li, H., Limenitakis, J.P., Fuhrer, T., Geuking, M.B., Lawson, M.A., Wyss, M., et al., September 2015. The outer mucus layer hosts a distinct intestinal microbial niche. Nat. Commun. 6 (1), 8292.

Matey-Hernandez, M.L., Williams, F.M.K., Potter, T., Valdes, A.M., Spector, T.D., Menni, C., 2018. Genetic and microbiome influence on lipid metabolism and dyslipidemia. Physiol. Genom. 50 (2), 117–126. https://doi.org/10.1152/physiolgenomics.00053.2017.

Metcalf, J.L., Song, S.J., Morton, J.T., Weiss, S., Seguin-Orlando, A., Joly, F., et al., 2017. Evaluating the impact of domestication and captivity on the horse gut microbiome. Sci. Rep. 7, 15497.

Ramirez, J., Guarner, F., Bustos Fernandez, L., Maruy, A., Sdepanian, V.L., Cohen, H., 2020. Antibiotics as major disruptors of gut microbiota. Front. Cell Infect. Microbiol. 10, 572912. https://doi.org/10.3389/fcimb.2020.572912.

Ren, T., Boutin, S., Humphries, M.M., Dantzer, B., Gorrell, J.C., Coltman, D.W., McAdam, A.G., Wu, M., 2017. Seasonal, spatial, and maternal effects on gut microbiome in wild red squirrels. Microbiome 5, 163.

Salgado-Flores, A., Tveit, A.T., Wright, A.-D., Pope, P.B., Sundset, M.A., 2019. Characterization of the cecum microbiome from wild and captive rock ptarmigans indigenous to Arctic Norway. PLoS One 14 (3), e0213503. https://doi.org/10.1371/journal.pone.0213503.

Schmidt, E., Mykytczuk, N., Schulte-Hostedde, A.I., 2019. Effects of the captive and wild environment on diversity of the gut microbiome of deer mice (Peromyscus maniculatus). ISME J. 13, 1293–1305.

Song, S.J., Woodhams, D.C., Martino, C., Allaband, C., Mu, A., Javorschi-Miller-Montgomery, S., Suchodolski, J.S., Knight, R., 2019. Engineering the microbiome for animal health and conservation. Exp. Biol. Med. 244, 494–504. https://doi.org/10.1177/1535370219830075.

Sonowal, R., Swimm, A., Sahoo, A., Luo, L., Matsunaga, Y., Wu, Z., et al., 2017. Indoles from commensal bacteria extend healthspan. Proc. Natl. Acad. Sci. USA 114 (36), E7506–E7515.

Suchodolski, J.S., Dowd, S.E., Westermarck, E., Steiner, J.M., Wolcott, R.D., Spillmann, T., Harmoinen, J.A., 2009. The effect of the macrolide anti biotictylosin on microbial diversity in the canine small intestine as demonstrated by massive parallel 16S rRNA gene sequencing. BMC Microbiol. 9, 210.

Torres-Henderson, C., Summers, S., Suchodolski, J., Lappin, M.R., 2017. Effect of Enterococcus faecium strain SF68 on gastrointestinal signs and fecal microbiome in cats administered amoxicillin-clavulanate. Top. Companion Anim. Med. 32, 104–108.

Turnbaugh, P.J., Ley, R.E., Mahowald, M.A., Magrini, V., Mardis, E.R., Gordon, J.I., 2006. An obesity-associated gut microbiome with increased capacity for energy harvest. Nature 444 (7122), 1027–1031. https://doi.org/10.1038/nature05414.

Whittemore, J.C., Stokes, J.E., Laia, N.L., Price, J.M., Suchodolski, J.S., 2018. Short and long-term effects of a synbiotic on clinical signs, the fecal microbiome,and metabolomic profiles in healthy research cats receiving clindamycin: a randomized, controlled trial. PeerJ 6, e5130.

Willing, B.P., Russell, S.L., Finlay, B.B., 2011. Shifting the balance: antibiotic effects on host-microbiota mutualism. Nat. Rev. Microbiol. 9, 233–243.

Xue, M.-Y., Sun, H.-Z., Wu, X.-H., Liu, J.-X., Guan, L.L., 2020. Multi-omics reveals that the rumen microbiome and its metabolome together with the host metabolome contribute to individualized dairy cow performance. Microbiome 8, 64. https://doi.org/10.1186/s40168-020-00819-8.

Xue, M.-Y., Xie, Y.-Y., Zhong, Y., Ma, X.-J., Sun, H.-Z., Liu, J.-X., 2022. Integrated meta-omics reveals new ruminal microbial features associated with feed efficiency in dairy cattle. Microbiome 10, 32. https://doi.org/10.1186/s40168-022-01228-9.

Yitbarek, A., Taha-Abdelaziz, K., Hodgins, D.C., Read, L., Nagy, E., Weese, J.S., Caswell, J.L., Parkinson, J., Sharif, S., 2018. Gut microbiota-mediated protection against influenza virus subtype H9N2 in chickens is associated with modulation of the innate responses. Sci. Rep. 8, 13189.

CHAPTER 8

Patents, bioproducts, commercialization, social, ethical, and economic policies on microbiome

Olugbenga Samuel Michael[1,7,8], Juliana Bunmi Adetunji[2],
Ebenezer Olusola Akinwale[3], Olufemi Idowu Oluranti[4], Olulope Olufemi Ajayi[5],
Charles Oluwaseun Adetunji[6], Ayodele Olufemi Soladoye[1] and
Oluwafemi Adebayo Oyewole[9]

[1]Cardiometabolic, Microbiome and Applied Physiology Laboratory, Department of Physiology, College of Health Sciences, Bowen University, Iwo, Osun State, Nigeria; [2]Nutritional and Toxicological Research Laboratory, Department of Biochemistry Sciences, Osun State University, Osogbo, Osun State, Nigeria; [3]Department of Physiology and Biomedical Science, Faculty of Science and Engineering, University of Wolverhampton, Wolverhampton, United Kingdom; [4]Applied and Environmental Research Unit, Department of Physiology, College of Health Sciences, Bowen University, Iwo, Osun State, Nigeria; [5]Department of Biochemistry, Edo University Iyamho, Auchi, Edo State, Nigeria; [6]Applied Microbiology, Biotechnology and Nanotechnology Laboratory, Department of Microbiology, and Directorate of Research and Innovation, Edo State University Uzairue, Iyamho, Auchi, Edo State, Nigeria; [7]Department of Physiology, University of Tennessee Health Science Center, Memphis, TN, United States; [8]Department of Medical Pharmacology and Physiology, University of Missouri, Columbia, MO, United States; [9]Department of Microbiology, Federal University of Technology, Minna, Nigeria

Introduction

Recently, across the globe, there has been explosive information and research on gut microbiota and its interaction with the human health and diseases (Martínez et al., 2021; Afzaal et al., 2022). The microbiome has been broadly studied using interdisciplinary approaches to comprehend the physiological and pathological consequences of the huge microbial populations inside and around us on daily basis (Berg et al., 2020; Gebrayel et al., 2022; Aggarwal et al., 2023). Microbiome regulates physiological activities such as immune system, metabolism, energy production, growth, and endocrine functions among other important regulatory processes influenced by the microbiome (Hou et al., 2022; Colella et al., 2023). Regardless of massive microbiome-related research information overload, there have been serious difficulties and challenges in progressing into products for treatment and prevention of diseases in patients (Schupack et al., 2022; Puschhof and Elinav, 2023). Fecal microbiome transplantation (FMT) has been used crudely for some disease conditions, but major regulated microbiome-based therapeutics are still under clinical trials (Michael et al., 2022; Wu et al., 2022). For the advancement of microbiome from research to clinical applications, there is need for a coalition or partnership among evolving stakeholders in the microbiome field such as inventors, research establishments, and companies to build formidable, robust, consistent, and efficient methods of developing microbiome-based therapeutics and diagnostics that will gain clinical acceptance and utility (Cullen et al., 2020). The hope of

An Introduction to the Microbiome in Health and Diseases
ISBN 978-0-323-91190-0,
https://doi.org/10.1016/B978-0-323-91190-0.00008-4

117

developing microbiome-based therapeutics is improved because of the successful use of FMT in the management of *Clostridium difficile*, inflammatory bowel disease, Chron's disease, and ulcerative colitis (Borody and Campbell, 2011; van Nood et al., 2013). The concept of the FMT proves microbiome-immune cross-talk exists (Duan, 2018; Campbell et al., 2023). Furthermore, the sequencing technological advances made it possible to identify the microbiome compositions, which introduced a hypothetical as well as mechanistic approach to the generation of microbiome therapeutics (Atarashi et al., 2011; Tanoue et al., 2019; Almeida et al., 2019).

These new information and knowledge about the microbiome have transformed the way we think about or perceive microorganisms. The advancement in technology using next-generation sequencing techniques, omics approaches, and bioinformatics has increased our appreciation of microbiome capabilities (Turnbaugh et al., 2007), hence forming part of the stimulant that instigated the microbiome explosion. Microbiome is now a very prevalent area of interest in the skincare, biomedicine, biotechnology, and food and agricultural industries with the global microbiome market worth estimated to be around $1.5 billion by the year 2025.

Diverse microbiome startups began due to tremendous advancements in microbiome-based medical and scientific research over the years. Seres Therapeutics, Rebiotix, Ferring Pharmaceuticals, and Finch Therapeutics are among the many microbiome companies that are working to produce microbiome therapeutics for treatment of diseases. Microbiome therapeutics can be classified as biological concept, bug status, or therapeutic area. When microbiome is classified as bugs, it could be applied as follows: "**bugs as drugs**," "**drugs for bugs**," or "**drugs from bugs**" (FitzGerald and Spek, 2020). Drugs from bugs connote the microbiome therapeutics derived from microbial origins such as bioactive molecules from bacteria from microbiome, metabolites, and compounds. Drugs for bugs imply application of feedstuff for bacteria. This classification involves administration of dietary fibers capable of stimulating the production of metabolites and growth of the microbes. "Bugs as drugs" on the other hand means administration of microbes as therapeutics. Bugs as drugs can also be called live biopharmaceutic products (LBPs) because live microbes are administered as drugs example is seen in FMT and probiotics (FitzGerald and Spek, 2020). This concept is based on the fact that dysbiosis of the microbiome results in disease conditions (Borody and Campbell, 2011; van Nood et al., 2013). Interestingly, some LBPs have scaled far in clinical trials to the third phase in the year 2020, but no LBP has gained approval by the Food and Drug Administration. The main challenges with FMT-based microbiome therapeutics remain getting sufficient clean and healthy donors, sample screening, and storage to ensure that the donor is not transferring another disease to the recipient (DeFilipp et al., 2019). FMT is associated with the administration of frozen fecal microbes as capsule via oral route, nasogastric intubation, and enema. Food and Drug Administration has limited the use of FMT to investigation of new drug application in the United States (FitzGerald and Spek, 2020).

The world is witnessing one of the worst pandemics caused by COVID-19. This is a pointer to the fact that microbes can alter the course of our lives in ways never imagined such that many people now believe that microbes are mainly disease-causing organisms (Feehan and Apostolopoulos, 2021). Also, the idea of infusing, injecting, or taking bugs as medicine does not really appeal to vast majority of human population (Cherniack, 2010; Siddiqui et al., 2023). However, studies and findings as well as some of the many landmark discoveries in the field of microbiome research have started to cause a paradigm shift in the knowledge and perspective of humans on the microbes (Malla et al., 2019; Cullen et al., 2020; Michael et al., 2022; Gebrayel et al., 2022; McGuinness et al., 2024). Biomedical applications of microbiome have offered hope for patients in places where conventional chemical drugs such as antibiotics have failed to manage or treat diseases (Yadav and Chauhan, 2021). The saying that "we are, what we eat" is very important in considering our health in the context of the physiological influences of microbiome (Armet et al., 2022). Diet and nutrition have been shown to affect the microbiome composition (Singh et al., 2017; Makki et al., 2018; Cuevas-Sierra et al., 2019). High-fat diet or Western diet, for instance, has been reported to alter the microbiome homeostasis resulting dysbiosis causing metabolic disorders such as diabetes, obesity, hypertension, liver disease, inflammatory bowel disease, neurodegenerative diseases, and colorectal cancer (Murphy et al., 2015; Martinez et al., 2017; Canale et al., 2021; Kang et al., 2022). Dysbiosis is the alteration of the balance established by the microbial flora or community. This balance is between the beneficial and pathogenic microbes in the gastrointestinal system (Weiss and Hennet, 2017; Shanahan et al., 2021).

There is need for more advocacy and enlightenment on the importance of microbiome so that the general population can understand the promises, risks, hopes, and hazards of using microbiome-based products (Ma et al., 2018; Cordaillat-Simmons et al., 2020). Microbiome-derived products have a lot of unique benefits that conventional drug formulation may not provide (Feng et al., 2020). Transferring the microbial contents from one individual to another for the purpose of treating diseases is a concept that the general population is trying to grapple with because of their perception of microbes as disease-causing organisms (Ser et al., 2021; Michael et al., 2022). Also, the fact that no two individuals have the same microbiome composition is another issue that requires intense discussion (Kurilshikov et al., 2021). There are also issues of sample collection and storage concerns in terms of safety and privacy of microbiome data especially because information on lifestyle, habits, and health of individuals will be available to researchers from their microbiome analysis, and this information could be accessed by other people, thereby violating privacy and safety (Wagner et al., 2016).

The skincare and cosmetic industry are another area that is taking advantage of the microbiome research innovation (Souak et al., 2021). Diverse skincare products are now in the markets that are microbiome-based (Wallen-Russell et al., 2017). The skin acts as the first defensive structure preventing toxic invasion from the environment such as ultraviolet radiation and air pollution (Parrado et al., 2019; Lee and Kim, 2022).

The skin is the largest epithelial surface for interactions with microorganisms (Gallo, 2017). The skin harbors its own microbiome, and it is considered as an essential element of the skin protective structure (Byrd et al., 2018). The microbiome—health connection is well recognized, but the understanding of microbiome inhabiting the skin is just evolving resulting in a revolutionary trend in the cosmetic, beauty, and skincare industry (Mahmud et al., 2022). There is now substantial commercialization of microbiome-based skincare products using the information generated from microbiome research on the skin by the cosmetic industry, and tremendous growth of this novel approach across the globe is speculated. The skin microbiome approach used by the skincare and cosmetics industry includes probiotic, prebiotics, and postbiotics (Puebla-Barragan and Reid, 2021; Al-Smadi et al., 2023). Probiotics are beneficial bacteria used to boost and upsurge the number of good bacteria on the skin while prebiotics are indigestible dietary fibers that aid the growth of probiotics as food (Hardy et al., 2013). Postbiotics are metabolites produced by live bacteria such as polysaccharides, enzymes, peptides, and fatty acids that are beneficial to the health of the skin (Scott et al., 2022; Prajapati et al., 2023).

Microbiome patents

Recently, there is increased desire in understanding how microbiome can be harnessed for human good, either in terms of health, agriculture, cosmetics, etc. (Cullen et al., 2020; Berg et al., 2020). This intensifying interest has led to increased demand to patent discoveries and applications of microbiome. More so, the research process to achieve therapeutic or other beneficial effects is lengthy and expensive. However, obtaining a commercially viable patent comes with the challenge of intense scrutiny by the awarding bodies due to the nature of microbiome-based innovations: There is a potential that they are viewed as natural phenomena. This was particularly recently highlighted by the American Supreme Court decision in the case of *Mayo Collaborative Services versus Prometheus Laboratory and Association for Molecular Pathology versus Myriad Genetics, Inc.* where patent claims were invalidated, overturning long-established United States Patent and Trademark Office ("PTO") policy and disrupting expectations of investors (Association For Molecular Pathology v. Myriad Genetics, Inc. | Supreme Court | US Law | LII/Legal Information Institute, no date; Mayo Collaborative Services v. Prometheus Laboratories, Inc.: 566 U.S. 66 (2012) Justia US Supreme Court Center, no date; Gordon, 2015).

Although, there are claims that this ruling will enable more accessibility to medical information, the potential impact of this in the long run is the discouragement of investment in research, development, and innovation especially in biotechnology and drug development (Statement by NIH Director Francis Collins on U.S. Supreme Court Ruling on Gene Patenting National Institutes of Health (NIH), 2013). Therefore, it stands to reason that workable commercial patents are important to drive innovation and encourage continued research and that the intellectual property of those working in and funding these innovations needs to be adequately protected (Gordon, 2015; Sabatelli et al., 2017; FitzGerald and Spek, 2020).

Numerous patents have been approved on microbiome spanning various areas of its applications such as agriculture, cosmetic, medicine, etc.

Metabolic syndrome

Diabetes and obesity are prevailing debilitating health conditions of global proportions and relevance. They are also risk factors for development of cardiovascular diseases. These conditions have been associated with alteration in microbiome composition resulting in disease pathogenesis. Dysbiosis-associated pathogenesis involved in diabetes and obesity is the basis for microbiome-related therapeutics for the management of metabolic disorders (Parekh et al., 2015). Numerous institutions (e.g., MicroBiome Therapeutics—US9040101, Synlogic Inc—WO2016210384, Nestec SA—US8318150) have been granted patents on microbiome manipulation for the purpose of treating metabolic disorders such as obesity and diabetes (Sabatelli et al., 2017).

Neurological disorders

There is cross-talk between the gut and the brain resulting in bidirectional control of functions and secretions. The microbiome in the gut produces metabolites capable of signal transduction to the brain stimulating production of neurotransmitters and influences some neurological development and functions such as formation of new neurons, myelination, neuroimmune cell maturation, and formation of the blood—brain barrier (Carabotti et al., 2015; Sharon et al., 2016). Therefore, microbiome is a crucial player in the neurological functions and disorders. Targeting the microbiome for management of neurodegenerative diseases, depression, and other psychiatric disorders is indeed very promising prospect, as such patents have been invented by manipulating microbiome composition (Whole Biome—US20150259728, Nubiome—US8927242) (Sabatelli et al., 2017).

Agriculture

The relevance of microbiome in sustainable agricultural practices is an emerging aspect of microbiome application. The fact that plant, soil, and environment have their various microbiomes is interesting. Improvement in crop yield and pest control are ways in which microbiome has been used sustainably as opposed to use of chemical fertilizers and pesticides that have been shown to have detrimental consequences on the environment and human health. Also, the genetic modification to improve crop yield is a method that has its own disadvantages. Therefore, manipulating microbiome to improve growth, yield, and shelf life of farm produce as well as reducing antibiotic resistance in livestock is economically relevant, and some patents have been invented in this area (Biotal—US5718894, Cornell University—WO2016057991) (Shabat et al., 2016; Sabatelli et al., 2017; Orozco-Mosqueda et al., 2018; Ray et al., 2020; Trivedi et al., 2021).

Gastrointestinal and autoimmune disorders

Microbiome manipulation is a very critical factor in the treatment of gastrointestinal diseases and autoimmune disorders because microbiome has the capability to influence innate and adaptive immunity to regulate homeostasis. Microbiome-based products associated with suppression or enhancement of immunological system possess ability to treat inflammatory disorders, and autoimmune diseases have been granted patents (4D Pharma Research Limited—WO2016102950; The University of Tokyo—US9415079) (Mathewson et al., 2016). Inflammatory bowel disease and *Clostridium difficile* have been associated with impaired microbiome homeostasis. Hence, microbiome-based therapies have shown tremendous promise in the management of these gastrointestinal disorders infection (Shreiner et al., 2015). FMT has shown better prospects compared with antibiotics used in the treatment of *Clostridium difficile*, and some patents in this area have been filed and granted (US9511099—Rebiotix; Seres Therapeutics—US9028841; Synthetic Biologics—US9446080).

Cancer immunotherapy

Microbiome influences immunological response to cancer. Microbiome-based cancer immunotherapy is offering hope in the fight against deadly disease that has claimed several lives. The efficacy of cancer immunotherapies has been shown to be enhanced by intestinal microbiome (Gopalakrishnan et al., 2018; Matson et al., 2021). Hence, microbiome modulation to improve cancer immunotherapy is showing great prospects although it is still emerging (Decoy Biosystems—US20160228523; Institut Gustave Roussy—WO2016063263; North Carolina State University—US7981651; Vedanta Biosciences—US20160144014).

Skin care products

The skin being the largest organ in the human body contains its own microbiome, which performs similar function like the gut microbiome mainly preventing invasion of disease-causing microorganism by serving as a physical barrier. However, this skin microbiome barrier can be altered resulting in skin diseases due to imbalance in skin microbiome. The skin microbiome has been shown to have cosmetic and therapeutic potentials (Reisch, 2017). Therefore, cosmetic and pharmaceutical companies are making use of skin microbiome modulation to improve skin care and treat skin diseases (AOBiome—WO2016161285; L'Oreal—US20110064835; Azitra—WO2015184134).

Microbiome biproducts

Microbiota, both human and nonhuman, has been found to biosynthesize a variety of metabolites, which can exert a wide range of effects. The mammalian gut is particularly replete with these microbiotas and, consequently, a host of unexplored biosynthesized bioactive metabolites. With various vital host—microbe and intermicrobe interactions, the bioactivity of these metabolites may range from antibiotic, antiinflammatory, anticancer, and antifungal to immunomodulation and neuromodulation. Hence, they present novel and alternative routes to approaching medical, agricultural, and cosmetic issues (Burja et al., 2001; Lozupone et al., 2012; Jandhyala et al., 2015; Milshteyn et al., 2018; Wang et al., 2019; Ke et al., 2021).

Gut microbiomes produce metabolites either from ingested food (metabolic products such as oligosaccharides, succinate, short-chain fatty acids) or via biosynthetic gene clusters (BGCs) (an organized group of genes responsible for producing various distinct bioactive molecules) (Macfarlane et al., 1992; Lozupone et al., 2012; Lee and Hase, 2014; Rowland et al., 2018; Liu et al., 2020; Wu et al., 2021). Ability to identify microbiome-synthesized genes and infer their chemical structures has led to an expanding pool of identified biomolecules; automation of these processes has further led to the identification of a plethora of biosynthetic gene clusters from the human microbiome. Further to this is the discovery of the so-called cryptic BGCs—BGCs that elude the detection of pharmaceutical testing instruments (Milshteyn et al., 2018; Travin et al., 2020; Rowe and Spring 2021).

Although uncharacterized, BGCs are distinct bioactive molecules that can code for secondary metabolic pathways and produce small molecules; for example, colibactin (genotoxic), microcin (has antibiotic activities), tilivalline (cytotoxic), and the dipeptide aldehydes (protease inhibitors) (Pawłowski, 2008). There are different classifications of BGCs based on the function of the small molecules they code for: terpenoids; saccharides; polyketides (PKs); nonribosomal peptide synthetases (NRPSs); ribosomally synthesized and posttranslationally modified peptides (RiPPs); and hybrid compounds (Dev, 1989; Wagner and Elmadfa, 2003; Pawłowski, 2008; Reen et al., 2015; Hudson and Mitchell, 2018).

Under stress conditions, bacteria have been observed to respond by producing antimicrobial peptides, some of which are later modified posttranslation; hence, the name RiPPs. These have various subgroups with the ones common in humans being lantibiotics, microcins, bacteriocins, thiopeptides, and thiazole/oxazole-modified microcins (TOMMs). RiPPs can hinder the growth of close species of bacteria. Additionally, some RiPPs have structural advantages that lend to usefulness in the synthesis of other small molecules; lasso peptides are an example of peptides posttranslationally modified to have a lasso–shaped structure hence their use as scaffolding buffers to provide strength and stability in the synthesis of other small molecules. With increasing bacterial resistance to established antibiotics, it is imperative to explore and harness the pharmaceutical and therapeutic alternatives that these bioproducts present (Dev, 1989; Wagner and Elmadfa, 2003; Hudson and Mitchell, 2018; Braffman et al., 2019; Rowe and Spring 2021).

Even though the therapeutic potential of microbiome bioproducts is well established, a lot more still needs to be done in further understanding the bioactivity of the molecules and how to harness their benefits. For example, capturing large BGCs on a single clone remains a challenge, and due to low levels of expression in laboratory cultures, many of these metabolites are still yet to be identified (Donia and Fischbach, 2015; Milshteyn et al., 2018). Notwithstanding, with continued research, development, and advances in genetics and synthetic biology, it can be expected that many more vital natural products will be developed.

Social and ethical concerns of microbiome research

The public perception of microorganism is that they cause diseases. Indeed, the recent and ongoing coronavirus-19 (COVID-19) pandemic really showed the world that we share our world with organisms that cannot be seen but powerful to the extent of taking millions of lives across the globe. Education, lifestyle, workplace, societal interactions, and the world as we know it have changed tremendously due to COVID-19. This is the most reawaking call we have had of late that microbes can change our societies and lives even though we live in a microbial world. Microbes are an integral part of our daily living because they are within and outside the human body (Segal et al., 2020; Dhar and Mohanty, 2020; Donati Zeppa et al., 2020).

Microbiome research has shown that the microbes have tremendous benefits that are possible of transforming our world through improvement and innovation in medicine, agriculture, bioengineering, environment, food, and cosmetic industry. The microbiome has a way of interfering with our society and lives because they are an integral part of our environment. In developed world, knowledge of microbial diversity or communities will improve cities' drive to attain a great public health (Shao et al., 2006). Hence, microbiome research will eventually enhance the public perception of the cities' interest in the maintenance and monitoring of the health status of the citizens through improved knowledge of the cities' microbial population (O'Doherty et al., 2016; Hodgetts et al., 2018; Hadrich, 2018).

Microbiome manipulation has been suggested to be the basis of the revolution in the microbiome revolution through the proper understanding of the microbiome dysbiosis, DMT, use of microbiome as biotherapeutics such as probiotics and prebiotics. Most microbiome innovations and therapeutics have not gained approval by the FDA as they are mostly under clinical trial because there are so many reservations about the microbiome therapeutics that needs to be clarified or properly elucidated. Also, societal acceptance of interindividual transference fecal matter for the purpose of treatment of diseases seems vulgar. This is not the only issue with FMT acceptance by the public. There are also safety, privacy, and ethical concerns of using donor microbiome from someone you may not even know. There are also compatibility issues (Park et al., 2017; Cani, 2018; Vargason and Anselmo, 2018).

The microbiome research has a far wide consequence beyond an individual patient or subject. Human microbiome manipulations can have societal and public health relevance because the public can be exposed to risks and harms. It is imperative to be aware that humans are not only afflicted with diseases, but they can also be a host or carrier of pathogens capable of causing diseases (Smith et al., 2004; Francis et al., 2005). In addition, microbiome products or metabolites that have health-promoting consequences in some people can have detrimental effects in others. Furthermore, since microorganisms that offer beneficial effects in some individuals could somehow get transferred to other people and cause them serious damage, it is quite significant to put in consideration the grave societal consequences of microbiome manipulations and therapeutic applications especially as it regards to the potential public relevance of the microbiome research or therapeutics (Nuffield Council on Bioethics, 2007; O'Doherty et al., 2016; Mimee et al., 2016).

There are diverse ongoing changes taking place in the medical space and how medicine is practiced due to the innovation from microbiome research. Numerous social, ethical, and legal implications of microbiome abound that must be addressed to properly guide, engage, and advocate with the society and scientific community to incorporate the consequences and risks associated with microbiome research and innovation. The ethical consideration associated with the conduct of microbiome-based research includes informed consent, participant diversity, privacy, return of research results, and invasiveness of sampling (McGuire et al., 2008). Another aspect of ethical issues to consider is also the impact of the microbiome research on healthcare delivery, perception of the public, commercialization of microbiome-based products, and economic implications (Sharp et al., 2009). There is constant battle to translate the microbiome-based research into commercially relevant products with significant societal relevance where the values of microbiome products are based on the benefits offered by the microbiome innovation (Waldby, 2000).

Westernized societies have been reported to have lower intestinal microbiome diversity in relation to traditional people with local lifestyles. The Amerindians have higher fecal microbial diversity when compared with people living in the United States (Clemente et al., 2015). Hence, people living local habitats unperturbed with industrial pollution and westernized diets, and sedentary lifestyles have rich microbiome diversity suggesting that there are missing microbes in western and industrialized societies (Blaser, 2006). Hence, native people may be a source of microbiome to help restore microbiome that has been lost in western societies. Ethical policies and standards must be the guiding principles when using microbiome from donors for therapy and commercialization. The interest and privacy of the donor must be protected (Ma et al., 2018).

Microbiome commercialization

Microbiome has gained immense relevance due to major scientific discoveries concerning the various applications of microbiome in diverse areas of human life such as agriculture, biomedicine and healthcare, bioengineering, waste management, food industry, cosmetics and skin care, environmental rejuvenation, population growth, and crime investigation. The uniqueness of microbiome and its obvious potentials have necessitated huge investment worth over 3 billion dollars, large-scale production, and manufacturing of microbiome-derived products and applications by major startup biotechnological companies, existing conglomerates, and powerhouse in pharmaceutical, cosmetic and skin care, agriculture, and food industries. Interestingly, funding for microbiome research has gone up in the past decade with the United States spearheading the microbiome research revolution with over two billion dollars. Commercialization of microbiome includes microbiome-related technological and instrumentation advances, microbiome-derived therapeutics, as well as retail or end user products such as probiotics. The recent tremendous investments in basic microbiome research have provided results that enlightens major policy players and the world at large about the crucial roles played by microbiome in the human body as well as her environment. Modern microbiome characterization and sequencing technologies have made it possible to access the composition and function of the human microbiome using next-generation sequencing techniques, metagenomics, and metabolomics techniques resulting in deep understanding of microbiome—host relationship especially in terms of disease pathogenesis and health maintenance (Gilbert et al.,2016, 2018; Shendure et al., 2017; Quince et al., 2017; Lagier et al., 2018; Uzbay, 2019; Zhang et al., 2019).

FMT is one of the breakthrough applications of microbiome research that is likely to be the first to gain FDA approval for clinical application in the management of *Clostridium difficile*, ulcerative colitis, and inflammatory bowel syndrome. Translational application of microbiome research findings from the bench to bed side has been very challenging because many microbiome-therapeutics are in clinical trials with some in phase 3, phase 2, and others in phase 1 (Fernández and Rodrguez, 2019; Bamforth, 2019). The approval of safe clinical application by FDA is just a first step in the process of developing a full commercially viable product. There are other major logistic and product development challenges that have to do with large-scale industrial manufacturing processes and quality assurance. For instance, growing microbes is very sensitive and requires technological soundness and skills to control the needed medium for the microbes to thrive especially when large commercial quantities of microbes with therapeutic potentials are needed. The processes involved in the production or growth of commercial quantities of microbes must be economical and extremely consistent. Technological expertise is favored above ecological principles in selection of most strains of probiotics to be commercially produced possibly due to difficulties in growing large quantities of the microbes that have ecological suitability. Essentially, therapeutically relevant microbes might be the most difficult to grow requiring further ground-breaking efforts to achieve the commercialization of the microbes of relevance (McBurney et al., 2019).

Economic impact on microbiome

Microbiome is economically relevant for its role in disease management, improved crop production, environmental cleansing, waste recycling, improved soil quality, sustainable agriculture, and metabolic engineering. Since microbiome can be reshaped by many factors such as diet, drugs, disease, and exposure to genetic or environmental stressors, it becomes evident that microbiome varies among individuals, community ethnic groups, and countries. Interestingly, microbiome of developed countries was reported to show capacity to induce obesity compared with microbiome community of developing or low-income countries (De Filippo et al., 2010).

The socioeconomic condition of an individual has far-reaching implications on the microbiome diversity and integrity. Socioeconomic status influences the type or quality of diets an individual consumes, mode of delivery, antibiotics usage, sedentary lifestyle, and length of breastfeeding, and this generally has implication of the development of the intestinal microbiome. This is a factor that is not usually put into consideration when discussing microbiome research and application. Urbanization and economic growth resulting in sedentary lifestyle have high tendency in reshaping the microbiome, which has associated with increased incidence of metabolic disorders in Western countries, suggesting a connection between the microbiome, metabolic disorder, and economic status (He et al., 2018).

Amaral and colleagues demonstrated that social relationships alter microbiome constitution via microbial sharing (Amaral et al., 2017). Physicosocial environmental interactions may influence the composition of intestinal microbiome as depicted in a study that similar microbiome composition exists between cohabiting individuals, whereas individuals living separately have distinctly varied microbial composition. In addition, there are microbiome modifications observed in elderly when they move to nursing homes from their community residence (Claesson et al., 2012; Lax et al., 2014).

Societal status and income level influence vulnerability to diverse kinds of diseases. Poverty or low socioeconomic standing is a predisposing factor for many infections or diseases. Low income is strongly associated with poor health outcomes; this is corroborated by a study that demonstrated that more hospitalization for infection in children living in poor overcrowded neighborhoods compared with children living in wealthy and less crowded neighborhood (Braveman et al., 2010; McDonough et al., 2011; Yousey-Hindes and Hadler, 2012). The microbiome constitution is subject to environmental modifications, which affect immunological, metabolic, and neurological regulations in the host. Hence, political, socioeconomic status, or income level, which determines environment and standard of living of individuals, are factors responsible for altered microbiome composition resulting in detrimental health outcomes. Environmental and economic influences on gut microflora play significant involvement in the determination of the state of health of individuals. Healthcare policies must put microbiome into consideration in the determination of public health because of the modulatory role of socioeconomic status on intestinal microbiome (Amato et al., 2021).

Microbiome research has given us tremendous insights into the scope and potentials of microbiome if fully utilized in every aspect of human life. However, translation of research findings into commercially viable products and applications is the next challenge to over in the field of microbiome. Microbiome-derived products, applications, and technology are projected to be major contributors to economic growth globally. To effectively transition from the bench to the bedside or to the market, microbiome research needs to move from largely descriptive to one that is mechanistically driven to explain cause and effects and vice versa, which will increase the availability of the necessary knowledge and data for microbiome-derived applications and technologies (Meisner et al., 2021).

Economic potentials of microbiome are immense when fully utilized from diverse areas of its relevance from agriculture, climate, waste management, beauty and cosmetics, engineering to healthcare industry. Microbiome can be employed to achieve desired functions as a source for sustainable, circular bioeconomy. Microbiome is capable of food and agricultural transformation needed to tackle the problem of hunger globally. The applications of microbiome to improve soil quality as biofertilizer, pest control, crop protection, and reversal of loss biodiversity are major areas of economic advantage for the world to explore strategically to solve hunger problem in a sustainable manner (D'Hondt et al., 2021).

Community education and enlightenment about microbiome and its applications is very crucial to the acceptance of microbiome products and applications. Massive community engagement, international advocacy, and strategic frameworks that will enhance the education of individuals from primary schools to tertiary institutions, religious institutions, traditional rulers, policymakers, and political office holders must be put in place to ensure that there is proper understanding of the microbiome and its applications for economic empowerment and sustainable development (Meisner et al., 2021).

Conclusion

Microbiome has been shown to have tremendous potential to benefit humans in diverse ways ranging from its application in agriculture, cosmetics and skincare, food industry, biomedicine, bioengineering, climate and environmental sector, etc. However, extreme care is necessary, and regulatory policies are very pertinent when interpreting or applying microbiome research findings as either therapeutic or for commercial purposes. This is because microbiome can be a double-edged sword with both beneficial and harmful effects. The ongoing COVID-19 demonstrates importance and dangerous effects of microorganisms. Hence, applying microbiome products or their commercialization must be done within proper regulatory framework that will ascertain or validate safety. Also, there is need to scale up advocacy policies that will educate communities on the importance of microbiome as well as the dangers to create adequate awareness of the gains and risks of using microbiome products.

References

Afzaal, M., Saeed, F., Shah, Y.A., Hussain, M., Rabail, R., Socol, C.T., Hassoun, A., Pateiro, M., Lorenzo, J.M., Rusu, A.V., Aadil, R.M., September 26, 2022. Human gut microbiota in health and disease: unveiling the relationship. Front. Microbiol. 13, 999001. https://doi.org/10.3389/fmicb.2022.999001.

Aggarwal, N., Kitano, S., Puah, G.R.Y., Kittelmann, S., Hwang, I.Y., Chang, M.W., January 11, 2023. Microbiome and human health: current understanding, engineering and enabling technologies. Chem. Rev. 123 (1), 31–72. https://doi.org/10.1021/acs.chemrev.2c00431.

Al-Smadi, K., Leite-Silva, V.R., Filho, N.A., Lopes, P.S., Mohammed, Y., December 4, 2023. Innovative approaches for maintaining and enhancing skin health and managing skin diseases through microbiome-targeted strategies. Antibiotics (Basel) 12 (12), 1698. https://doi.org/10.3390/antibiotics12121698.

Almeida, A., Mitchell, A.L., Boland, M., Forster, S.C., Gloor, G.B., Tarkowska, A., Lawley, T.D., Finn, R.D., 2019. A new genomic blueprint of the human gut microbiota. Nature 568 (7753), 499–504.

Amaral, W.Z., Lubach, G.R., Proctor, A., Lyte, M., Phillips, G.J., Coe, C.L., 2017. Social influences on prevotella and the gut microbiome of young monkeys. Psychosom. Med. 79, 888–897.

Amato, K.R., Arrieta, M.C., Azad, M.B., Bailey, M.T., Broussard, J.L., Bruggeling, C.E., Claud, E.C., Costello, E.K., Davenport, E.R., Dutilh, B.E., Swain Ewald, H.A., Ewald, P., Hanlon, E.C., Julion, W., Keshavarzian, A., Maurice, C.F., Miller, G.E., Preidis, G.A., Segurel, L., Singer, B., Subramanian, S., Zhao, L., Kuzawa, C.W., 2021. The human gut microbiome and health inequities. Proc. Natl. Acad. Sci. U. S. A. 118 (25) e2017947118.

Armet, A.M., Deehan, E.C., O'Sullivan, A.F., Mota, J.F., Field, C.J., Prado, C.M., Lucey, A.J., Walter, J., June 8, 2022. Rethinking healthy eating in light of the gut microbiome. Cell Host Microbe 30 (6), 764–785. https://doi.org/10.1016/j.chom.2022.04.016.

Association for Molecular Pathology v. Myriad Genetics, Inc. | Supreme Court | US Law | LII/Legal Information Institute [no date]. Available from: https://www.law.cornell.edu/supremecourt/text/12-398 [Accessed 7 July 2021].

Atarashi, K., Tanoue, T., Shima, T., Imaoka, A., Kuwahara, T., Momose, Y., Cheng, G., Yamasaki, S., Saito, T., Ohba, Y., Taniguchi, T., Takeda, K., Hori, S., Ivanov, I.I., Umesaki, Y., Itoh, K., Honda, K., 2011. Induction of colonic regulatory T cells by indigenous Clostridium species. Science 331 (6015), 337–341.

Bamforth, M., 2019. Supporting Commercialization of Live Biotherapeutic Product for Microbiomebased Therapies. Pharma's Almanac. https://www.pharmasalmanac.com/articles/supporting-commercialization-of-live-biotherapeutic-products-for-microbiome-based-therapies.

Berg, G., Rybakova, D., Fischer, D., Cernava, T., Vergès, M.C., Charles, T., Chen, X., Cocolin, L., Eversole, K., Corral, G.H., Kazou, M., Kinkel, L., Lange, L., Lima, N., Loy, A., Macklin, J.A., Maguin, E., Mauchline, T., McClure, R., Mitter, B., Ryan, M., Sarand, I., Smidt, H., Schelkle, B., Roume, H., Kiran, G.S., Selvin, J., Souza, R.S.C., van Overbeek, L., Singh, B.K., Wagner, M., Walsh, A., Sessitsch, A., Schloter, M., June 30, 2020. Microbiome definition revisited: old concepts and new challenges. Microbiome 8 (1), 103. https://doi.org/10.1186/s40168-020-00875-0.

Blaser, M.J., October 2006. Who are we? Indigenous microbes and the ecology of human diseases. EMBO Rep. 7 (10), 956—960.

Borody, T.J., Campbell, J., December 2011. Fecal microbiota transplantation: current status and future directions. Expet Rev. Gastroenterol. Hepatol. 5 (6), 653—655. https://doi.org/10.1586/egh.11.71.

Braffman, N.R., Piscotta, F.J., Hauver, J., Campbell, E.A., Link, A.J., Darst, S.A., 2019. Structural mechanism of transcription inhibition by lasso peptides microcin J25 and capistruin. Proc. Natl. Acad. Sci. USA 116, 1273—1278. https://doi.org/10.1073/pnas.1817352116.

Braveman, P.A., Cubbin, C., Egerter, S., Williams, D.R., Pamuk, E., 2010. Socioeconomic disparities in health in the United States: what the patterns tell us. Am. J. Publ. Health 100, S186—S196.

Burja, A.M., Banaigs, B., Abou-Mansour, E., Grant Burgess, J., Wright, P.C., 2001. Marine cyanobacteria—a prolific source of natural products. Tetrahedron 57, 9347—9377. https://doi.org/10.1016/S0040-4020(01)00931-0.

Byrd, A.L., Belkaid, Y., Segre, J.A., March 2018. The human skin microbiome. Nat. Rev. Microbiol. 16 (3), 143—155. https://doi.org/10.1038/nrmicro.2017.157.

Campbell, C., Kandalgaonkar, M.R., Golonka, R.M., Yeoh, B.S., Vijay-Kumar, M., Saha, P., January 20, 2023. Crosstalk between gut microbiota and host immunity: impact on inflammation and immunotherapy. Biomedicines 11 (2), 294. https://doi.org/10.3390/biomedicines11020294.

Canale, M.P., Noce, A., Di Lauro, M., Marrone, G., Cantelmo, M., Cardillo, C., Federici, M., Di Daniele, N., Tesauro, M., April 1, 2021. Gut dysbiosis and western diet in the pathogenesis of essential arterial hypertension: a narrative review. Nutrients 13 (4), 1162. https://doi.org/10.3390/nu13041162.

Cani, P.D., 2018. Human gut microbiome: hopes, threats and promises. Gut 67, 1716—1725.

Carabotti, M., Scirocco, A., Maselli, M.A., Severi, C., 2015. The gut-brain axis: interactions between enteric microbiota, central and enteric nervous systems. Ann. Gastroenterol. 28 (2), 203—209.

Cherniack, E.P., July 2010. Bugs as drugs, Part 1: insects: the "new" alternative medicine for the 21st century? Alternative Med. Rev. 15 (2), 124—135.

Claesson, M.J., Jeffery, I.B., Conde, S., Power, S.E., O'Connor, E.M., Cusack, S., Harris, H.M.B., Coakley, M., Lakshminarayanan, B., O'Sullivan, O., et al., 2012. Gut microbiota composition correlates with diet and health in the elderly. Nature 488, 178.

Clemente, J.C., Pehrsson, E.C., Blaser, M.J., Sandhu, K., Gao, Z., Wang, B., Magris, M., Hidalgo, G., Contreras, M., Noya-Alarcón, Ó., Lander, O., McDonald, J., Cox, M., Walter, J., Oh, P.L., Ruiz, J.F., Rodriguez, S., Shen, N., Song, S.J., Metcalf, J., Knight, R., Dantas, G., Dominguez-Bello, M.G., April 3, 2015. The microbiome of uncontacted Amerindians. Sci. Adv. 1 (3), e1500183.

Colella, M., Charitos, I.A., Ballini, A., Cafiero, C., Topi, S., Palmirotta, R., Santacroce, L., July 28, 2023. Microbiota revolution: how gut microbes regulate our lives. World J. Gastroenterol. 29 (28), 4368—4383. https://doi.org/10.3748/wjg.v29.i28.4368.

Cordaillat-Simmons, M., Rouanet, A., Pot, B., September 2020. Live biotherapeutic products: the importance of a defined regulatory framework. Exp. Mol. Med. 52 (9), 1397—1406. https://doi.org/10.1038/s12276-020-0437-6.

Cuevas-Sierra, A., Ramos-Lopez, O., Riezu-Boj, J.I., Milagro, F.I., Martinez, J.A., January 1, 2019. Diet, gut microbiota, and obesity: links with host genetics and epigenetics and potential applications. Adv. Nutr. 10 (Suppl_1), S17—S30. https://doi.org/10.1093/advances/nmy078.

Cullen, C.M., Aneja, K.K., Beyhan, S., Cho, C.E., Woloszynek, S., Convertino, M., McCoy, S.J., Zhang, Y., Anderson, M.Z., Alvarez-Ponce, D., Smirnova, E., Karstens, L., Dorrestein, P.C., Li, H., Sen Gupta, A., Cheung, K., Powers, J.G., Zhao, Z., Rosen, G.L., February 19, 2020. Emerging priorities for microbiome research. Front. Microbiol. 11, 136. https://doi.org/10.3389/fmicb.2020.00136.

D'Hondt, K., Kostic, T., McDowell, R., Eudes, F., Singh, B.K., Sarkar, S., Markakis, M., Schelkle, B., Maguin, E., Sessitsch, A., 2021. Microbiome innovations for a sustainable future. Nat Microbiol 6 (2), 138—142.

De Filippo, C., Cavalieri, D., Di Paola, M., Ramazzotti, M., Poullet, J.B., Massart, S., et al., 2010. Impact of diet in shaping gut microbiota revealed by a comparative study in children from Europe and rural Africa. Proc. Natl. Acad. Sci. U. S. A. 107 (33), 14691—14696.

DeFilipp, Z., Bloom, P.P., Torres Soto, M., Mansour, M.K., Sater, M.R.A., Huntley, M.H., Turbett, S., Chung, R.T., Chen, Y.B., Hohmann, E.L., 2019. Drug-resistant E. coli bacteremia transmitted by fecal microbiota transplant. N. Engl. J. Med. 381 (21), 2043—2050.

Dev, S., 1989. Terpenoids. In: Rowe, J.W. (Ed.), Natural Products of Woody Plants: Chemicals Extraneous to the Lignocellulosic Cell Wall, Springer Series in Wood Science. Springer, Berlin, Heidelberg, pp. 691–807. https://doi.org/10.1007/978-3-642-74075-6_19.

Dhar, D., Mohanty, A., 2020. Gut microbiota and Covid-19- possible link and implications. Virus Res. 285, 198018.

Donati Zeppa, S., Agostini, D., Piccoli, G., Stocchi, V., Sestili, P., 2020. Gut microbiota status in COVID-19: an unrecognized player? Front. Cell. Infect. Microbiol. 10, 576551.

Donia, M.S., Fischbach, M.A., 2015. Small molecules from the human microbiota. Science 349, 1254766. https://doi.org/10.1126/science.1254766.

Duan, M., May 28, 2018. Microbiota and immune cell crosstalk: dialogues driving health and disease. Clin. Transl. Immunol. 7 (5), e1020. https://doi.org/10.1002/cti2.1020.

Feehan, J., Apostolopoulos, V., July 2021. Is COVID-19 the worst pandemic? Maturitas 149, 56–58. https://doi.org/10.1016/j.maturitas.2021.02.001.

Feng, W., Liu, J., Ao, H., Yue, S., Peng, C., September 14, 2020. Targeting gut microbiota for precision medicine: focusing on the efficacy and toxicity of drugs. Theranostics 10 (24), 11278–11301. https://doi.org/10.7150/thno.47289.

Fernández, C.R., 22 Jan. 2019. No Guts, No Glory: How Microbiome Research Is Changing Medicine. Labbiotech (Web).

FitzGerald, M.J., Spek, E.J., 2020. Microbiome therapeutics and patent protection. Nat. Biotechnol. 38 (7), 806–810. https://doi.org/10.1038/s41587-020-0579-z.

Francis, L.P., Battin, M.P., Jacobson, J.A., Smith, C.B., Botkin, J., 2005. How infectious diseases got left out—and what this omission might have meant for bioethics. Bioethics 19 (4), 307–322.

Gallo, R.L., June 2017. Human skin is the largest epithelial surface for interaction with microbes. J. Invest. Dermatol. 137 (6), 1213–1214. https://doi.org/10.1016/j.jid.2016.11.045.

Gebrayel, P., Nicco, C., Al Khodor, S., Bilinski, J., Caselli, E., Comelli, E.M., Egert, M., Giaroni, C., Karpinski, T.M., Loniewski, I., Mulak, A., Reygner, J., Samczuk, P., Serino, M., Sikora, M., Terranegra, A., Ufnal, M., Villeger, R., Pichon, C., Konturek, P., Edeas, M., March 7, 2022. Microbiota medicine: towards clinical revolution. J. Transl. Med. 20 (1), 111. https://doi.org/10.1186/s12967-022-03296-9.

Gilbert, J.A., Quinn, R.A., Debelius, J., Xu, Z.Z., Morton, J., Garg, N., Jansson, J.K., Dorrestein, P.C., Knight, R., 2016. Microbiome-wide association studies link dynamic microbial consortia to disease. Nature 535 (7610), 94–103.

Gilbert, J.A., Blaser, M.J., Caporaso, J.G., Jansson, J.K., Lynch, S.V., Knight, R., 2018. Current understanding of the human microbiome. Nat. Med. 24 (4), 392–400.

Gopalakrishnan, V., Helmink, B.A., Spencer, C.N., Reuben, A., Wargo, J.A., 2018. The influence of the gut microbiome on cancer, immunity, and cancer immunotherapy. Cancer Cell 33 (4), 570–580.

Gordon, J., 2015. The impact of Myriad and Mayo: will advancements in the biological sciences be spurred or disincentivized? (or was biotech patenting not complicated enough?). Cold Spring Harbor Perspect. Med. 5 (5). https://doi.org/10.1101/cshperspect.a020917.

Hadrich, D., 2018. Microbiome research is becoming the key to better understanding health and nutrition. Front. Genet. 9, 212.

Hardy, H., Harris, J., Lyon, E., Beal, J., Foey, A.D., May 29, 2013. Probiotics, prebiotics and immunomodulation of gut mucosal defences: homeostasis and immunopathology. Nutrients 5 (6), 1869–1912. https://doi.org/10.3390/nu5061869.

He, Y., Wu, W., Wu, S., Zheng, H.M., Li, P., Sheng, H.F., Chen, M.X., Chen, X.H., Ji, G.Y., Zheng, Z.D.X., Mujagond, P., Chen, X.J., Rong, Z.H., Chen, P., Lyu, L.Y., Wang, X., Xu, J.B., Wu, C.B., Yu, N., Xu, Y.J., Yin, J., Raes, J., Ma, W.J., Zhou, H.W., 2018. Linking gut microbiota, metabolic syndrome and economic status based on a population-level analysis. Microbiome 6, 172.

Hodgetts, T., Grenyer, R., Greenhough, B., McLeod, C., Dwyer, A., Lorimer, J., 2018. The microbiome and its publics: a participatory approach for engaging publics with the microbiome and its implications for health and hygiene. EMBO Rep. 19 (6), e45786.

Hou, K., Wu, Z.X., Chen, X.Y., Wang, J.Q., Zhang, D., Xiao, C., Zhu, D., Koya, J.B., Wei, L., Li, J., Chen, Z.S., April 23, 2022. Microbiota in health and diseases. Signal Transduct. Targeted Ther. 7 (1), 135. https://doi.org/10.1038/s41392-022-00974-4.

Hudson, G.A., Mitchell, D.A., 2018. RiPP antibiotics: biosynthesis and engineering potential. Curr. Opin. Microbiol. 45, 61−69. https://doi.org/10.1016/j.mib.2018.02.010.

Jandhyala, S.M., Talukdar, R., Subramanyam, C., Vuyyuru, H., Sasikala, M., Reddy, D.N., 2015. Role of the normal gut microbiota. World J. Gastroenterol. 21, 8787−8803. https://doi.org/10.3748/wjg.v21.i29.8787.

Kang, G.G., Trevaskis, N.L., Murphy, A.J., Febbraio, M.A., December 30, 2022. Diet-induced gut dysbiosis and inflammation: key drivers of obesity-driven NASH. iScience 26 (1), 105905. https://doi.org/10.1016/j.isci.2022.105905.

Ke, J., Wang, B., Yoshikuni, Y., 2021. Microbiome engineering: synthetic biology of plant-associated microbiomes in sustainable agriculture. Trends Biotechnol. 39, 244−261. https://doi.org/10.1016/j.tibtech.2020.07.008.

Kurilshikov, A., Medina-Gomez, C., Bacigalupe, R., Radjabzadeh, D., Wang, J., Demirkan, A., Le Roy, C.I., Raygoza Garay, J.A., Finnicum, C.T., Liu, X., Zhernakova, D.V., Bonder, M.J., Hansen, T.H., Frost, F., Rühlemann, M.C., Turpin, W., Moon, J.Y., Kim, H.N., Lüll, K., Barkan, E., Shah, S.A., Fornage, M., Szopinska-Tokov, J., Wallen, Z.D., Borisevich, D., Agreus, L., Andreasson, A., Bang, C., Bedrani, L., Bell, J.T., Bisgaard, H., Boehnke, M., Boomsma, D.I., Burk, R.D., Claringbould, A., Croitoru, K., Davies, G.E., van Duijn, C.M., Duijts, L., Falony, G., Fu, J., van der Graaf, A., Hansen, T., Homuth, G., Hughes, D.A., Ijzerman, R.G., Jackson, M.A., Jaddoe, V.W.V., Joossens, M., Jørgensen, T., Keszthelyi, D., Knight, R., Laakso, M., Laudes, M., Launer, L.J., Lieb, W., Lusis, A.J., Masclee, A.A.M., Moll, H.A., Mujagic, Z., Qibin, Q., Rothschild, D., Shin, H., Sørensen, S.J., Steves, C.J., Thorsen, J., Timpson, N.J., Tito, R.Y., Vieira-Silva, S., Völker, U., Völzke, H., Võsa, U., Wade, K.H., Walter, S., Watanabe, K., Weiss, S., Weiss, F.U., Weissbrod, O., Westra, H.J., Willemsen, G., Payami, H., Jonkers, D.M.A.E., Arias Vasquez, A., de Geus, E.J.C., Meyer, K.A., Stokholm, J., Segal, E., Org, E., Wijmenga, C., Kim, H.L., Kaplan, R.C., Spector, T.D., Uitterlinden, A.G., Rivadeneira, F., Franke, A., Lerch, M.M., Franke, L., Sanna, S., D'Amato, M., Pedersen, O., Paterson, A.D., Kraaij, R., Raes, J., Zhernakova, A., February 2021. Large-scale association analyses identify host factors influencing human gut microbiome composition. Nat. Genet. 53 (2), 156−165. https://doi.org/10.1038/s41588-020-00763-1.

Lagier, J.C., Dubourg, G., Million, M., Cadoret, F., Bilen, M., Fenollar, F., Levasseur, A., Rolain, J.M., Fournier, P.E., Raoult, D., 2018. Culturing the human microbiota and culturomics. Nat. Rev. Microbiol. 16 (9), 540−550.

Lax, S., Smith, D.P., Hampton-Marcell, J., Owens, S.M., Handley, K.M., Scott, N.M., Gibbons, S.M., Larsen, P., Shogan, B.D., Weiss, S., et al., 2014. Longitudinal analysis of microbial interaction between humans and the indoor environment. Science 345, 1048−1052.

Lee, W.-J., Hase, K., 2014. Gut microbiota−generated metabolites in animal health and disease. Nat. Chem. Biol. 10, 416−424. https://doi.org/10.1038/nchembio.1535.

Lee, H.J., Kim, M., October 28, 2022. Skin barrier function and the microbiome. Int. J. Mol. Sci. 23 (21), 13071. https://doi.org/10.3390/ijms232113071.

Liu, Z., Dai, X., Zhang, H., Shi, R., Hui, Y., Jin, X., Zhang, W., Wang, L., Wang, Q., Wang, D., Wang, J., Tan, X., Ren, B., Liu, X., Zhao, T., Wang, J., Pan, J., Yuan, T., Chu, C., Lan, L., Yin, F., Cadenas, E., Shi, L., Zhao, S., Liu, X., 2020. Gut microbiota mediates intermittent-fasting alleviation of diabetes-induced cognitive impairment. Nat. Commun. 11, 855. https://doi.org/10.1038/s41467-020-14676-4.

Lozupone, C.A., Stombaugh, J.I., Gordon, J.I., Jansson, J.K., Knight, R., 2012. Diversity, stability and resilience of the human gut microbiota. Nature 489, 220−230. https://doi.org/10.1038/nature11550.

Ma, Y., Chen, H., Lan, C., Ren, J., May 2018. Help, hope and hype: ethical considerations of human microbiome research and applications. Protein Cell 9 (5), 404−415. https://doi.org/10.1007/s13238-018-0537-4.

Macfarlane, G.T., Gibson, G.R., Cummings, J.H., 1992. Comparison of fermentation reactions in different regions of the human colon. J. Appl. Bacteriol. 72, 57−64. https://doi.org/10.1111/j.1365-2672.1992.tb04882.x.

Mahmud, M.R., Akter, S., Tamanna, S.K., Mazumder, L., Esti, I.Z., Banerjee, S., Akter, S., Hasan, M.R., Acharjee, M., Hossain, M.S., Pirttilä, A.M., 2022 . Impact of gut microbiome on skin health: gut-skin axis observed through the lenses of therapeutics and skin diseases. Gut Microb. 14 (1), 2096995. https://doi.org/10.1080/19490976.2022.2096995.

Makki, K., Deehan, E.C., Walter, J., Bäckhed, F., June 13, 2018. The impact of dietary fiber on gut microbiota in host health and disease. Cell Host Microbe 23 (6), 705–715. https://doi.org/10.1016/j.chom.2018.05.012.

Malla, M.A., Dubey, A., Kumar, A., Yadav, S., Hashem, A., Abd Allah, E.F., January 7, 2019. Exploring the human microbiome: the potential future role of next-generation sequencing in disease diagnosis and treatment. Front. Immunol. 9, 2868. https://doi.org/10.3389/fimmu.2018.02868.

Martinez, K.B., Leone, V., Chang, E.B., March 4, 2017. Western diets, gut dysbiosis, and metabolic diseases: are they linked? Gut Microb. 8 (2), 130–142. https://doi.org/10.1080/19490976.2016.1270811.

Martínez, J.E., Vargas, A., Pérez-Sánchez, T., Encío, I.J., Cabello-Olmo, M., Barajas, M., August 24, 2021. Human microbiota network: unveiling potential crosstalk between the different microbiota ecosystems and their role in health and disease. Nutrients 13 (9), 2905. https://doi.org/10.3390/nu13092905.

Mathewson, N.D., Jenq, R., Mathew, A.V., et al., 2016. Corrigendum: gut microbiome-derived metabolites modulate intestinal epithelial cell damage and mitigate graft-versus-host disease. Nat. Immunol. 17 (10), 1235.

Matson, V., Chervin, C.S., Gajewski, T.F., 2021. Cancer and the microbiome-influence of the commensal microbiota on cancer, immune responses, and immunotherapy. Gastroenterology 160 (2), 600–613.

McBurney, M.I., Davis, C., Fraser, C.M., et al., 2019. Establishing what constitutes a healthy human gut microbiome: state of the science, regulatory considerations, and future directions. J. Nutr. 149 (11), 1882–1895.

McDonough, P., Duncan, G.J., Williams, D., House, J., 2011. Income dynamics and adult mortality in the United States, 1972 through 1989. Am. J. Publ. Health 87, 1476–1483.

McGuinness, A.J., Stinson, L.F., Snelson, M., Loughman, A., Stringer, A., Hannan, A.J., Cowan, C.S.M., Jama, H.A., Caparros-Martin, J.A., West, M.L., Wardill, H.R., Australasian Human Microbiome Research Network (AHMRN) Committee, January 2024. From hype to hope: considerations in conducting robust microbiome science. Brain Behav. Immun. 115, 120–130. https://doi.org/10.1016/j.bbi.2023.09.022.

McGuire, A.L., Colgrove, J., Whitney, S.N., Diaz, C.M., Bustillos, D., Versalovic, J., 2008. Ethical, legal, and social considerations in conducting the human microbiome project. Genome Res. 18, 1861–1864.

Meisner, A., Wepner, B., Kostic, T., van Overbeek, L.S., Bunthof, C.J., de Souza, R.S.C., Olivares, M., Sanz, Y., Lange, L., Fischer, D., Sessitsch, A., Smidt, H., MicrobiomeSupport Consortium, 2021. Calling for a systems approach in microbiome research and innovation. Curr. Opin. Biotechnol. 73, 171–178.

Michael, O.S., Oluranti, O.I., Oshinjo, A.M., Adetunji, C.O., Adetunji, J.B., Esiobu, N.D., 2022. Microbiota transplantation, health implications, and the way forward. Microbiomes and Emerging Applications 79–97.

Milshteyn, A., Colosimo, D.A., Brady, S.F., 2018. Accessing bioactive natural products from the human microbiome. Cell Host Microbe 23, 725–736. https://doi.org/10.1016/j.chom.2018.05.013.

Mimee, M., Citorik, R.J., Lu, T.K., 2016. Microbiome therapeutics - advances and challenges. Adv. Drug Deliv. Rev. 105 (Pt A), 44–54.

Murphy, E.A., Velazquez, K.T., Herbert, K.M., September 2015. Influence of high-fat diet on gut microbiota: a driving force for chronic disease risk. Curr. Opin. Clin. Nutr. Metab. Care 18 (5), 515–520. https://doi.org/10.1097/MCO.0000000000000209.

Nuffield Council on Bioethics, 2007. Public Health: Ethical Issues. London. http://www.nuffieldbioethics.org/wp-content/uploads/2014/07/Publichealth-ethical-issues.pdf.

Orozco-Mosqueda, M.D.C., Rocha-Granados, M.D.C., Glick, B.R., Santoyo, G., 2018. Microbiome engineering to improve biocontrol and plant growth-promoting mechanisms. Microbiol. Res. 208, 25–31.

O'Doherty, K.C., Virani, A., Wilcox, E.S., 2016. The human microbiome and public health: social and ethical considerations. Am. J. Publ. Health 106 (3), 414–420.

Parekh, P.J., Balart, L.A., Johnson, D.A., 2015. The influence of the gut microbiome on obesity, metabolic syndrome and gastrointestinal disease. Clin. Transl. Gastroenterol. 6 (6), e91. https://doi.org/10.1038/ctg.2015.16.

Park, L., Mone, A., Price, J.C., Tzimas, D., Hirsh, J., Poles, M.A., Malter, L., Chen, L.A., 2017. Perceptions of fecal microbiota transplantation for *Clostridium difficile* infection: factors that predict acceptance. Ann. Gastroenterol. 30 (1), 83—88.

Parrado, C., Mercado-Saenz, S., Perez-Davo, A., Gilaberte, Y., Gonzalez, S., Juarranz, A., July 9, 2019. Environmental stressors on skin aging. Mechanistic insights. Front. Pharmacol. 10, 759. https://doi.org/10.3389/fphar.2019.00759.

Pawłowski, K., 2008. Uncharacterized/hypothetical proteins in biomedical 'omics' experiments: is novelty being swept under the carpet? Brief. Funct. Genomics 7, 283—290. https://doi.org/10.1093/bfgp/eln033.

Prajapati, N., Patel, J., Singh, S., Yadav, V.K., Joshi, C., Patani, A., Prajapati, D., Sahoo, D.K., Patel, A., December 19, 2023. Postbiotic production: harnessing the power of microbial metabolites for health applications. Front. Microbiol. 14, 1306192. https://doi.org/10.3389/fmicb.2023.1306192 eCollection 2023.

Puebla-Barragan, S., Reid, G., February 26, 2021. Probiotics in cosmetic and personal care products: trends and challenges. Molecules 26 (5), 1249. https://doi.org/10.3390/molecules26051249.

Puschhof, J., Elinav, E., March 17, 2023. Human microbiome research: growing pains and future promises. PLoS Biol. 21 (3), e3002053. https://doi.org/10.1371/journal.pbio.3002053.

Quince, C., Walker, A.W., Simpson, J.T., Loman, N.J., Segata, N., 2017. Shotgun metagenomics, from sampling to analysis. Nat. Biotechnol. 35 (9), 833—844.

Ray, P., Lakshmanan, V., Labbé, J.L., Craven, K.D., 2020. Microbe to microbiome: a paradigm shift in the application of microorganisms for sustainable agriculture. Front. Microbiol. 11, 622926.

Reen, F.J., Romano, S., Dobson, A.D.W., O'Gara, F., 2015. The sound of silence: activating silent biosynthetic gene clusters in marine microorganisms. Mar. Drugs 13, 4754—4783. https://doi.org/10.3390/md13084754.

Reisch, M.S., 2017. The microbiome comes to cosmetics. C&EN 95 (19), 30—34.

Rowe, S.M., Spring, D.R., 2021. The role of chemical synthesis in developing RiPP antibiotics. Chem. Soc. Rev. 50, 4245—4258. https://doi.org/10.1039/D0CS01386B.

Rowland, I., Gibson, G., Heinken, A., Scott, K., Swann, J., Thiele, I., Tuohy, K., 2018. Gut microbiota functions: metabolism of nutrients and other food components. Eur. J. Nutr. 57, 1—24. https://doi.org/10.1007/s00394-017-1445-8.

Sabatelli, A.D., Vincent, N.G., Puleo, D.E., 2017. Patenting the microbiome: trends, challenges and insights. Pharm. Pat. Anal. 6 (6), 273—282. https://doi.org/10.4155/ppa-2017-0028.

Schupack, D.A., Mars, R.A.T., Voelker, D.H., Abeykoon, J.P., Kashyap, P.C., January 2022. The promise of the gut microbiome as part of individualized treatment strategies. Nat. Rev. Gastroenterol. Hepatol. 19 (1), 7—25. https://doi.org/10.1038/s41575-021-00499-1.

Scott, E., De Paepe, K., Van de Wiele, T., November 4, 2022. Postbiotics and their health modulatory biomolecules. Biomolecules 12 (11), 1640. https://doi.org/10.3390/biom12111640.

Segal, J.P., Mak, J.W.Y., Mullish, B.H., Alexander, J.L., Ng, S.C., Marchesi, J.R., 2020. The gut microbiome: an under-recognised contributor to the COVID-19 pandemic? Ther. Adv. Gastroenterol. 13, 1756284820974914.

Ser, H.L., Letchumanan, V., Goh, B.H., Wong, S.H., Lee, L.H., May 13, 2021. The use of fecal microbiome transplant in treating human diseases: too early for poop? Front. Microbiol. 12, 519836. https://doi.org/10.3389/fmicb.2021.519836.

Shabat, S.K., Sasson, G., Doron-Faigenboim, A., et al., 2016. Specific microbiome-dependent mechanisms underlie the energy harvest efficiency of ruminants. ISME J. 10 (12), 2958—2972.

Shanahan, F., Ghosh, T.S., O'Toole, P.W., January 2021. The healthy microbiome-what is the definition of a healthy gut microbiome? Gastroenterology 160 (2), 483—494. https://doi.org/10.1053/j.gastro.2020.09.057.

Shao, M., Tang, X., Zhang, Y., Li, W., 2006. City clusters in China: air and surface water pollution. Front. Ecol. Environ. 4, 353—361.

Sharon, G., Sampson, T.R., Geschwind, D.H., Mazmanian, S.K., 2016. The central nervous system and the gut microbiome. Cell 167 (4), 915–932. https://doi.org/10.1016/j.cell.2016.10.027.

Sharp, R.R., Achkar, J.P., Brinich, M.A., Farrell, R.M., 2009. Helping patients make informed choices about probiotics: a need for research. Am. J. Gastroenterol. 104, 809–813.

Shendure, J., Balasubramanian, S., Church, G.M., Gilbert, W., Rogers, J., Schloss, J.A., Waterston, R.H., 2017. DNA sequencing at 40: past, present and future. Nature 550 (7676), 345–353.

Shreiner, A.B., Kao, J.Y., Young, V.B., 2015. The gut microbiome in health and in disease. Curr. Opin. Gastroenterol. 31 (1), 69–75.

Siddiqui, S.A., Li, C., Aidoo, O.F., Fernando, I., Haddad, M.A., Pereira, J.A.M., Blinov, A., Golik, A., Câmara, J.S., April 29, 2023. Unravelling the potential of insects for medicinal purposes - a comprehensive review. Heliyon 9 (5), e15938. https://doi.org/10.1016/j.heliyon.2023.e15938.

Singh, R.K., Chang, H.W., Yan, D., Lee, K.M., Ucmak, D., Wong, K., Abrouk, M., Farahnik, B., Nakamura, M., Zhu, T.H., Bhutani, T., Liao, W., April 8, 2017. Influence of diet on the gut microbiome and implications for human health. J. Transl. Med. 15 (1), 73. https://doi.org/10.1186/s12967-017-1175-y.

Smith, C.B., Battin, M.P., Jacobson, J.A., Francis, L.P., Botkin, J.R., Asplund, E.P., et al., 2004. Are there characteristics of infectious diseases that raise special ethical issues? Develop. World Bioeth. 4 (1), 1–16.

Souak, D., Barreau, M., Courtois, A., André, V., Duclairoir Poc, C., Feuilloley, M.G.J., Gault, M., April 27, 2021. Challenging cosmetic innovation: the skin microbiota and probiotics protect the skin from UV-induced damage. Microorganisms 9 (5), 936. https://doi.org/10.3390/microorganisms9050936.

Statement by NIH Director Francis Collins on U.S, 2013. Supreme Court Ruling on Gene Patenting | National Institutes of Health (NIH). Available from: https://www.nih.gov/about-nih/who-we-are/nih-director/statements/statement-nih-director-francis-collins-us-supreme-court-ruling-gene-patenting. (Accessed 7 July 2021).

Tanoue, T., Morita, S., Plichta, D.R., Skelly, A.N., Suda, W., Sugiura, Y., Narushima, S., Vlamakis, H., Motoo, I., Sugita, K., Shiota, A., Takeshita, K., Yasuma-Mitobe, K., Riethmacher, D., Kaisho, T., Norman, J.M., Mucida, D., Suematsu, M., Yaguchi, T., Bucci, V., Inoue, T., Kawakami, Y., Olle, B., Roberts, B., Hattori, M., Xavier, R.J., Atarashi, K., Honda, K., 2019. A defined commensal consortium elicits CD8 T cells and anti-cancer immunity. Nature 565 (7741), 600–605.

Travin, D.Y., Bikmetov, D., Severinov, K., 2020. Translation-targeting RiPPs and where to find them. Front. Genet. 0. https://doi.org/10.3389/fgene.2020.00226.

Trivedi, P., Mattupalli, C., Eversole, K., Leach, J.E., 2021. Enabling sustainable agriculture through understanding and enhancement of microbiomes. New Phytol. 230 (6), 2129–2147.

Turnbaugh, P.J., Ley, R.E., Hamady, M., Fraser-Liggett, C.M., Knight, R., Gordon, J.I., 2007. The human microbiome project. Nature 449, 804–810.

Uzbay, T., 2019. Germ-free animal experiments in the gut microbiota studies. Curr. Opin. Pharmacol. 49, 6–10.

van Nood, E., Vrieze, A., Nieuwdorp, M., Fuentes, S., Zoetendal, E.G., de Vos, W.M., Visser, C.E., Kuijper, E.J., Bartelsman, J.F., Tijssen, J.G., Speelman, P., Dijkgraaf, M.G., Keller, J.J., 2013. Duodenal infusion of donor feces for recurrent *Clostridium difficile*. N. Engl. J. Med. 368 (5), 407–415.

Vargason, A.M., Anselmo, A.C., 2018. Clinical translation of microbe-based therapies: current clinical landscape and preclinical outlook. Bioeng. Transl. Med. 3 (2), 124–137.

Wagner, K.-H., Elmadfa, I., 2003. Biological relevance of terpenoids. Ann. Nutr. Metab. 47, 95–106. https://doi.org/10.1159/000070030.

Wagner, J., Paulson, J.N., Wang, X., Bhattacharjee, B., Corrada, B.H., 2016. Privacy-preserving microbiome analysis using secure computation. Bioinformatics 32, 1873–1879.

Waldby, C., 2000. The Visible Human Project: Informatic Bodies and Posthuman Medicine. Routledge, New York.

Wallen-Russell, C., Wallen-Russell, S., 2017. Meta analysis of skin microbiome: new link between skin microbiota diversity and skin health with proposal to use this as a future mechanism to determine whether cosmetic products damage the skin. Cosmetics 4, 14. https://doi.org/10.3390/cosmetics4020014.

Wang, L., Ravichandran, V., Yin, Y., Yin, J., Zhang, Y., 2019. Natural products from mammalian gut microbiota. Trends Biotechnol. 37, 492–504. https://doi.org/10.1016/j.tibtech.2018.10.003.

Weiss, G.A., Hennet, T., August 2017. Mechanisms and consequences of intestinal dysbiosis. Cell. Mol. Life Sci. 74 (16), 2959–2977. https://doi.org/10.1007/s00018-017-2509-x.

Wu, J., Wang, K., Wang, X., Pang, Y., Jiang, C., 2021. The role of the gut microbiome and its metabolites in metabolic diseases. Protein Cell 12, 360–373. https://doi.org/10.1007/s13238-020-00814-7.

Wu, J., Lv, L., Wang, C., February 28, 2022. Efficacy of fecal microbiota transplantation in irritable bowel syndrome: a meta-analysis of randomized controlled trials. Front. Cell. Infect. Microbiol. 12, 827395. https://doi.org/10.3389/fcimb.2022.827395.

Yadav, M., Chauhan, N.S., November 15, 2021. Microbiome therapeutics: exploring the present scenario and challenges. Gastroenterol Rep (Oxf). 10. https://doi.org/10.1093/gastro/goab046 goab046.

Yousey-Hindes, K.M., Hadler, J.L., 2012. Neighborhood socioeconomic status and influenza hospitalizations among children: new Haven County, Connecticut, 2003–2010. Am. J. Publ. Health 101, 1785–1789.

Zhang, X., Li, L., Butcher, J., Stintzi, A., Figeys, D., 2019. Advancing functional and translational microbiome research using metaomics approaches. Microbiome 7 (1), 154.

CHAPTER 9

Introduction to gut microbiome and epigenetics: Their role in polycystic ovary syndrome pathogenesis

Ayomide Michael Oshinjo[1,9], Olugbenga Samuel Michael[1,10,11], Lawrence Dayo Adedayo[2], Charles Oluwaseun Adetunji[3], Bamidele Olubayode[1], Juliana Bunmi Adetunji[4], Olaniyan Amos Morakinyo[5], Ebenezer Olusola Akinwale[6], Olulope Olufemi Ajayi[7], Funmileyi Olubajo Awobajo[8], Ayodele Olufemi Soladoye[1] and Oluwafemi Adebayo Oyewole[12]

[1]Cardiometabolic, Microbiome and Applied Physiology Laboratory, Department of Physiology, College of Health Sciences, Bowen University, Iwo, Osun State, Nigeria; [2]Neurophysiology Research Unit, Department of Physiology, College of Health Sciences, Bowen University, Iwo, Osun State, Nigeria; [3]Applied Microbiology, Biotechnology and Nanotechnology Laboratory, Department of Microbiology, and Directorate of Research and Innovation, Edo State University Uzairue, Iyamho, Auchi, Edo State, Nigeria; [4]Nutritional and Toxicological Research Laboratory, Department of Biochemistry Sciences, Osun State University, Osogbo, Osun State, Nigeria; [5]Department of Anatomy, College of Health Sciences, Bowen University, Iwo, Nigeria; [6]Department of Physiology and Biomedical Science, Faculty of Science and Engineering, University of Wolverhampton, Wolverhampton, United Kingdom; [7]Department of Biochemistry, Edo University Iyamho, Auchi, Nigeria; [8]Department of Physiology, College of Medicine, University of Lagos, Idiaraba, Nigeria; [9]Faculty of Biochemistry and Molecular Medicine, University of Oulu, Oulu, Finland; [10]Department of Physiology, University of Tennessee Health Science Center, Memphis, TN, United States; [11]Department of Medical Pharmacology and Physiology, University of Missouri, Columbia, MO, United States; [12]Department of Microbiology, Federal University of Technology, Minna, Nigeria

Introduction

Polycystic ovary syndrome (PCOS) is a complex endocrine disease that occurs in about 8%—13% of premenopausal reproductive-aged, and this prevalence may exceed 20% depending on demography and underlying comorbidities (Bozdag et al., 2016; Neven et al., 2018). According to a recent PCOS guide, there is an average gap of approximately 2 years and three physicians visit between the initial consultation and the final diagnosis (Teede et al., 2018). The disorder includes a constellation of features including but not limited to ano-/oligovulation, amenorrhea, hyperandrogenism, metabolic dysfunction, chronic inflammation, and polycystic ovary morphology (Patel et al., 2018; Neven et al., 2018). Although the etiology of this disorder is unclear, research has shown that it may be multifactorial arising from genetic, epigenetic, environmental, and lifestyle factors (Day et al., 2018; Dapas et al., 2019; Kulkarni et al., 2019; Jacobsen et al., 2019). Reproductive and metabolic abnormalities observed in PCOS are also common in offspring of PCOS patients regardless of sex (Crisosto et al., 2017; Torchen et al., 2019). About 60%—70% of the daughters of PCOS patients exhibit PCOS-like traits at the onset of adolescence, with the phenotype exaggerated/dampened by lifestyle and

An Introduction to the Microbiome in Health and Diseases
ISBN 978-0-323-91190-0,
https://doi.org/10.1016/B978-0-323-91190-0.00009-6

dietary conditions (Crisosto et al., 2019; Risal et al., 2019). Identified genetic alterations are responsible for less than 10% of PCOS prevalence and heritability, and epigenetics may explain the intergenerational and transgenerational transmission of PCOS (Azziz, 2016; Dunaif, 2016; Escobar-Morreale, 2018). Epigenetics describes the study of inheritable alterations in gene expression arising with no modification of the genetic code. The mechanisms that mediate epigenetic changes include methylation of DNA molecules, modification of histone proteins, and noncoding RNA interactions (Noble, 2015). These mechanisms are involved in cellular adaptation to internal and external influences, and they allow for controlled expression of specific proteins necessary for survival. Epigenetic modifications are often maintained throughout the cell cycle and division and thus, transferred to offspring (Perez and Lehner, 2019). This explains why epigenetics is thought to underlie the pathophysiology and inheritance of numerous abnormal human conditions including PCOS, the focus of this review (Cheng et al., 2018; Stoccoro et al., 2018; Zarzour et al., 2019; Vázquez-Martínez et al., 2019; Jung et al., 2020). The human gut serves as home to a diverse and abundant population of microbial organisms. Its microbiota is a community of prokaryotes, eukaryotes, and viruses, which colonize the lumen and mucosa of the human gastrointestinal tract from the mouth to the anus, with varying abundance and diversity in each segment of the tract. It functions as a supplementary organ playing active roles in digestion, absorption, immune function, neural activity, and general metabolism (De vadder et al., 2018; Wang and Zhao, 2018). Most microbiota activities are mediated via its metabolites, e.g., vitamins (Roager and Dragsted, 2019). This explains the link between altered gut communities, i.e., dysbiosis and the pathophysiology of various diseases (Perez et al., 2020). This relationship has been extensively observed in both PCOS disease models and patients (Liu et al., 2017; Thackray, 2019). The gut dysbiotic state results in abnormal levels of gut metabolites, which are considered as critical modulators of epigenetic mechanisms (Gerhauser, 2018). This review unravels the mechanisms by which gut dysbiosis and its accompanying irregularity in gut-derived metabolites underlie the epigenetic alterations observed in PCOS patients.

Overview of PCOS

PCOS has a widespread effect on various systems in the human body. Women diagnosed with PCOS tend to have abnormal features such as excess androgens, hyperinsulinemia, dyslipidemia, and obesity. It is expected that these abnormalities are a result of alterations in the functions of various physiological systems.

Insulin resistance

Glucose is an essential substrate for metabolism in the human body. Maintenance of normal glucose levels depends on a careful scale between insulin secretion and insulin sensitivity. Impaired cellular response to insulin results in compensatory hyperinsulinemia, a condition in which there is an unusual increase in the insulin concentration necessary for the production of its metabolic effects. Hyperinsulinemia is closely linked to insulin resistance, which describes the decreased responsiveness to target tissues to normal insulin concentrations. Insulin resistance (IR) appears to be central to the metabolic dysfunction found in most women with PCOS across all phenotypes, with the estimated prevalence of IR in obese patients being around 70% and 30% in lean patients (Randeva et al., 2012). Even with normal age and weight controls, PCOS patients have heightened likelihood of having IR and glucose intolerance (Cassar et al., 2016). Ethnicity also contributes to the prevalence of IR. Hispanic patients have elevated prevalence of hyperglycaemia and IR in comparison with non-Hispanic white women (Cassar et al., 2016). There are several theories around the origin of IR and compensatory hyperinsulinemia in PCOS. The mainstream proposed mechanism is a defect in insulin signaling after binding with its receptors. This is caused by hyperserine phosphorylation of insulin receptor and insulin receptor substrate-1. These changes are responsible for metabolic disruption in ovarian cells and insulin binding cells (Anagnostis et al., 2018). Hyperinsulinemia often results due to increased basal secretion of insulin and decreased clearance by the liver, which in turn alters androgen production as insulin stimulates ovarian steroidogenesis.

Sex hormones

Two out of the three diagnostic criteria (NIH and Androgen Excess Society) explicitly require the presence of hyperandrogenemia for PCOS to be clinically diagnosed (Dunaif and Zawadski, 1996; Azziz et al., 2006). Epidemiological studies have estimated hyperandrogenemia to be present in about 90% of patients diagnosed with PCOS (Lizneva et al., 2016). Interestingly, the severity of cardiometabolic symptoms and elevated androgen levels in PCOS patients are two positively correlated factors even when matched with PCOS patients with normal androgen levels (Jones et al., 2012; Kim et al., 2014). However, the pathway by which androgens alter metabolic functions is relatively unknown. In most women with PCOS, androgen excess arises from ovarian steroidogenesis, but in about 20%—30%, excess androgen is secreted from the adrenal gland. Thus, plasma concentrations of androgens are significantly elevated in women with PCOS (Huang et al., 2010). A study has also demonstrated increased steroidogenesis in subcutaneous adipocytes of PCOS patients (O'Reilly et al., 2015). There is also a positive correlation between excess androgens and hyperinsulinemia. As stated earlier, insulin directly activates ovarian androgen secretion or indirectly through the release of luteinizing hormone (LH) (Moret et al., 2009; Lim et al., 2017). Hyperandrogenemia can also arise from hyperinsulinemia by reducing the liver synthesis of sex hormone—binding globulin (SHBG), thus elevating levels of unbound circulating androgens. Other possible causes of hyperandrogenemia include intrauterine growth retardation, sedentary lifestyle, diet, and in utero exposure to excess androgens (Abott et al., 2019).

Lipid metabolism

Dyslipidemia is a common metabolic disruption in PCOS patients (Macut et al., 2013). These patients often present with unusual lipid profiles such as elevated levels of triglycerides (Ghaffazard et al., 2016; Pergialiotis et al., 2018). Recently, an experiment showed that high-fat diet—fed mice had altered ovarian morphology and produced metabolic patterns commonly found in PCOS (Patel and Shah, 2018). Obesity can produce such an impact on the ovaries via organelle level (endoplasmic reticulum and mitochondria) disruptions and cellular apoptosis (Broughton and Moley, 2017). PCOS phenotypes appear to have varying lipid patterns due to different androgen levels, suggesting that hyperandrogenemia may be implicated in dyslipidemia (Spalkowska et al., 2018). However, a study has shown that genes responsible for the production of steroids and lipids may also facilitate the synthesis of androgens, which are steroid hormones (Pan et al., 2018). Oxidative stress is also a common feature in PCOS and has been closely linked with dyslipidemia (Imran et al., 2018). It was demonstrated that hypercholesterolemia led to elevated serum levels of triglycerides, decreasing sirtuin 1 (S1RT1) and ultimately triggering oxidative stress in the liver (Bonomini et al., 2018). PCOS patients who show a degree of dyslipidemia, specifically hypercholesterolemia, tend to have a higher BMI, fasting insulin levels, significant insulin resistance than other PCOS patients with normal lipid levels (Pergialiotis et al., 2018). Dyslipidemia in PCOS also exacerbates the tendency of developing cardiovascular disease (Kumar et al., 2017). Therefore, dyslipidemia in PCOS is interwoven with other metabolic alterations and worsens the overall pathology of PCOS.

Ovulation and fertility

Anovulatory women with PCOS are just as many or even more than their counterparts (March et al., 2010; Yildiz et al., 2012). Although there is a lack of clarity regarding the reason for this, available data indicate that anovulation is not a product of ovarian androgen excess. This is because hyperandrogenism is sometimes found in women with neither PCOS nor anovulation (Guastella et al., 2010; Carmina, 2015). It has been suggested that anovulation in PCOS may occur due to disruption in early follicle development. Mason et al. (1994) showed that follicular granulosa cells extracted from anovulatory PCOS patients elicited excessive estradiol response to follicle-stimulating hormone (FSH) compared with extracts from ovulatory PCOS patients. A later experiment by Willis et al. (1998) also points that only small follicular granulosa cells from anovulatory PCOS patients responded to the effect of luteinizing hormone (LH) in certain instances while only granulosa cells from large dominant follicles responded to LH in ovulatory women with PCOS. Thus, the abnormal response of smaller follicles to LH in anovulatory PCOS women could lead to an arrest of follicular development and lack of ovulation. There is no genetic basis for the ovulatory differences in PCOS patients, and thus, anovulation may be linked with multiple factors (Cui et al., 2013). A significant reduction in BMI is often enough to convert anovulatory PCOS women into ovulatory PCOS women and vice versa (Carmina, 2015). With anovulation being at the top, followed closely by decreased oocyte quality and endometrial receptivity, it is no surprise that PCOS is often accompanied by subfertility. Close associations have been

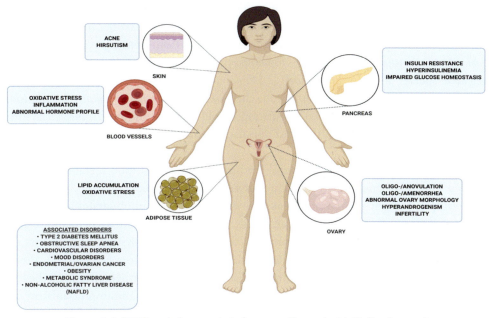

Figure 9.1 PCOS and characteristic features. *(Created with BioRender.com.)*

made between oligomenorrhea, obesity, and reproductive ability. Obesity coupled with insulin resistance also causes reproductive dysfunction. Lifestyle modification, particularly weight loss, remains the most reliable option to ameliorate subfertility in women with PCOS (Ecklund and Usadi, 2015) (Fig. 9.1).

Gut microbiota dysbiosis in polycystic ovary syndrome

Normal gut microbiota

The human skin, gastrointestinal tract, respiratory system, oral cavity, and vagina serve as home to various microorganisms such as protozoa, arche, fungi, and bacteria, and these organisms are regarded as "microbiota" (Quigley, 2017). The gut serves as home to $10^{13}-10^{14}$ microbial organisms, roughly 10 times the human cells population and the genes are about 150 times the number of genes in human genetic makeup (Perez et al., 2020). Although the human gastrointestinal tract was originally considered to be sterile at birth, new evidence has shown that colonization may have been initiated during the intrauterine life (Collado et al., 2016). However, a bulk of studies still support the original thought that colonization begins at the birth via the route of delivery and is influenced by diet, location, antibiotics, etc. (Dogra et al., 2015; Cong et al., 2016). The early gastrointestinal microbial composition begins to mirror the normal adult composition by age 2—3. However, a study has reported gut microbiota changes even in adolescence

(Hollister et al., 2015). The gut microbiota then remains relatively stable till old age when it becomes altered due to diet and other factors (Claesson et al., 2012). Gut microbiota in humans is made up of 5 major bacterial phyla, namely, *Proteobacteria* phylum, *Verrucomicrobia* phylum, *Actinobacteria* phylum, *Firmicutes* phylum, and *Bacteroidetes* phylum. Gram-negative species include *Proteobacteria*, *Verrucomicrobia*, *Bacteroidetes*, and *Actinobacteria* phyla, while gram-positive species include the *Firmicutes* phylum. *Firmicutes* and *Bacteroidetes* form almost 90% of the total population, while *Proteobacteria* and *Actinobacteria* form about 10%. *Verrucomicrobia* has the least population among the five phyla (Ottman et al., 2012; Pascale et al., 2018).

This microbial community is critical for human health and food metabolism. Thus, this is a mutually beneficial relationship between the microbiota and the human host. For example, the gut microbiota breaks down indigestible carbohydrates and peptides into metabolites such as short-chain fatty acids (SCFAs) and branched-chain amino acids (BCAAs), respectively, which are ultimately absorbed into the systemic circulation and used as a source of cellular energy (Wang and Zhao, 2018; Canfora et al., 2019). Propionate, butyrate, and acetate are the primary SCFAs formed via fermentation, and they are important due to their anticarcinogenic, antiinflammatory, and immunomodulatory actions. They facilitate the synthesis and release of protective cytokines and chemokines. Moreover, acetate plays a role in lipid metabolic pathways and propionate contributes to glucose synthesis. Butyrate is also known for causing increased apoptosis of colon cancer cells. The gut microbiota even exhibits a paradoxical action by having an antimicrobial effect of inhibiting the production of harmful lipopolysaccharides (LPS) and peptidoglycans. The microbiota also interferes with the bodily defense by giving signals for the growth and maturation of T-helper (type 1, 2, and 17) cells and T cells (Surana and Kasper, 2014; Carmody and Turnbaugh, 2014; Falluca et al., 2014; Mayer et al., 2015). In addition, it is important for the gut intestinal wall through the secretion of mucins and regulation of tight junctions (Jia et al., 2018; De vadder et al., 2018). This cumulative impact of the normal gut microbiota for host health is seen in cases of microbial imbalance termed "dysbiosis," a critical factor underlying the pathophysiology cardiometabolic diseases and degenerative conditions.

Altered gut microbial communities in PCOS

Gut dysbiosis can be defined as "a disruption in the balance, diversity, and function of symbiotic intestinal microbial communities precipitated by changes in the biological milieu and manifested as biological dysregulation resulting in further biological milieu disruption, disease development, and/or presentation of a wide range of symptoms" (Perez et al., 2020). A few years ago, the discussion about a possible relationship between PCOS and gut dysbiosis was initiated by two Australian researchers (Tremellen and Pearce, 2012). The correlation between ecological changes in the gut microbial community and the metabolic disruptions as well as clinical presentations in PCOS patients has been recorded (Liu et al., 2017). Although the interaction between PCOS and gut dysbiosis can be regarded as noncausative, there appears to be a strong link between the duo. PCOS patients have reduced alpha (α) diversity and modified beta (β) diversity, which provide an explanation for the occurrence of inflammation and likelihood of obesity in the condition (Walters et al., 2014; Liu et al., 2017). α-diversity describes the amount of various species present in the microbial community or individual sample, while β-diversity describes the similarity between the microbial community or individual samples (Fig. 9.2).

In 2016, a study demonstrated that letrozole administration in peripubertal mice decreased α-diversity of the gut microbiota and relatively changed bacteria abundance, most of which belonged to the Firmicutes and Bacteroidetes phyla. These results were reported in comparison with a control placebo group and independent of significant dietary change. Some of the bacteria, which showed a change in relative abundance, have been recorded in other mouse metabolic disease models (Kelley et al., 2016). In a later but similar study using rats, it was shown that letrozole-induced PCOS rats possessed a relatively elevated abundance of *Prevotella* and reduced abundance of *Ruminococcus*, *Clostridium*, and *Lactobacillus* when compared with controls. Interestingly, *Prevotella* has been associated with inflammatory conditions such as periodontitis and chronic sinusitis, and chronic inflammation is a key feature of PCOS. *Lactobacillus* also helps immunity by producing SCFAs and contributing to a healthy balance of microbiota (Guo et al., 2016).

In a study of using stool microbiome, 24 women with PCOS were found to relatively reduce bacterial abundance in three taxa in comparison with the controls (Lindheim et al., 2017). These findings provide an insight into the metabolic disruptions present in PCOS as bacteria from phylum *Tenericutes* have been indicated to have a higher abundance and diversity in healthy individuals in comparison with patients with metabolic syndrome. Also, the relative abundance of bacteria from the family S24-7 was found to decrease in high-fat diet—fed mice (Evans et al., 2014; Rizk and Thackray, 2020). Similar altered gut microbial composition was reported between obese PCOS patients, and the obese controls compared with the nonobese controls, suggesting a microbial relationship between PCOS and obesity. They also reported a higher degree of dysbiosis in obese PCOS patients among the participants. Increased abundance of gram-negative bacteria arising from the *Escherichia/Shigella* and *Bacteroides* genera and a decrease in protective *Akkermansia* were also reported in PCOS patients (Liu et al., 2017). These bacteria are linked to increased production of LPS, which cause metabolic disruptions such as obesity. *Akkermansia* also plays a protective role by repairing the intestinal mucosal layer and reducing metabolic endotoxemia (Everard et al., 2013). Similarly, Torres et al. demonstrated decreased α-diversity and modified β-diversity in PCOS patients in comparison with normal controls. In addition, there is a recorded correlation between hyperandrogenism and the gut biodiversity. Increased colonic abundance of *Faecalibacterium prausnitzii*, *Bacteroides coprophilus*, *Porphyromonas* spp., *Blautia* spp., and a decreased

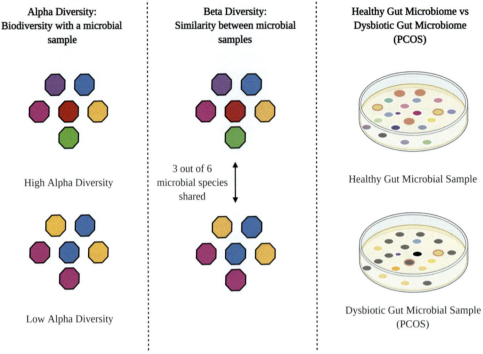

Note: Each different colour represents a different microbial species

Figure 9.2 Dysbiotic gut microbiome composition in PCOS. *(Created with BioRender.com.)*

abundance of *Ruminococcus bromii*, *Odoribacter* spp., *Anaerococcus* spp., and *Roseburia* spp., were also observed. *Porphyromonas* is associated with decreased intestinal mucosal permeability and gut dysbiosis, while an increased abundance of *Blautia* has been noted in conditions such as glucose intolerance and diabetes (Egshatyan et al., 2016; Torres et al., 2018). An animal study using the prenatal androgen exposure (PNA) model observed increased abundance of Nocardiaceae and Clostridiaceae bacterial family and decreased abundance of *Clostridium*, *Akkermansia*, *Bacteroides*, and *Lactobacillus* was reported in fecal microbial profile. Nocardiaceae and Clostridiaceae have been previously associated with steroid hormone synthesis (Sherman et al., 2018). Conversely, a positive association was detected between α-diversity and testosterone concentrations as well as the ratio of circulating testosterone to circulating estradiol. In addition, estradiol levels were inversely associated with α-diversity. Explaining these results, the researchers indicated the possibility of a mutual causality between microbial alteration and sex hormones regulation (Insenser et al., 2018). It also reported decreased β-diversity in both obese PCOS patients and controls in addition to a relatively increased abundance of *Catenibacterium* and *Kandleria* genera in PCOS patients compared with both nonhyperandrogenic male and female controls. The abundance of *Kandleria* was positively associated with levels of free androstenedione in all participants. *Catenibacterium* has also been positively associated with autoimmune conditions and infections, and a study in which mice were fed with prebiotics that resulted in decreased abundance of this genus, amelioration of mild inflammation and fat accumulation was reported (Everard et al., 2011; Insenser et al., 2018). Furthermore, in a research comparing non–insulin-resistant PCOS (NIR-PCOS) patients and insulin-resistant PCOS (IR-PCOS) patients to each other and against healthy controls, Prevotellaceae was reported to have decreased abundance having in IR-PCOS women accompanied with increased abundance of *Bacteroides*. *Bacteroides* was noted to have a positive correlation with increased sex hormones levels, inflammation, and degree of insulin resistance, while *Prevotellaceae* showed a negative correlation (Zeng et al., 2019). Recently, "beneficial" bacteria such as the *Faecalibacterium prausnitzii* and *Bifidobacterium* sp. were significantly decreased in PCOS. *Faecalibacterium prausnitzii* and *Bifidobacterium* sp. are regarded as beneficial bacteria for their activities around microbial balance, promotion of innate immunity, and production of SCFAs (Table 9.1).

Table 9.1 Recent studies linking epigenetic mechanisms and PCOS.

Study design	Key findings when PCOS group is compared with healthy controls	References
Histone modification		
United States case–control prenatal androgen-treated sheep model/control (n = 5/n = 4–5) intramuscular administration of 100 mg testosterone propionate twice weekly from day 30 to day 90 of gestation	↑ hepatic HDAC2 and HDAC3 expression, ↓ hepatic HDAC1 expression, ↑ HDAC1 expression in epicardiac adipocytes, ↓ HDAC3 expression in visceral adipocytes, ↑ sirtuin1 expression in subcutaneous adipocytes, ↑ acetylation at H3K9ac and H3K27ac (hepatic)	Guo et al. (2020)
United States case–control prenatal androgen-treated sheep model/control (n = 3/n = 3–4) intramuscular administration of 100 mg of 1.2 mg/kg testosterone propionate twice weekly from day 30 to day 90 of gestation	No difference in H3K27ac, H3K4me3 and H3K9ac (gene inducing histone marks) as well as H3K9me3 and H3K27me3 (gene repressing marks) in ovaries of day 90 sheep, upregulation of H3K27ac, H3K4me3, and H3K9ac marks in ovaries of 2-year-old adult sheep, downregulation of diacylglycerol kinase (DGKG) and adrenergic receptor genes, ADRA1A and ADRA2A in ovaries of 2-year-old adult sheep	Sinha et al. (2020)

Continued

Table 9.1 Recent studies linking epigenetic mechanisms and PCOS.—cont'd

Study design	Key findings when PCOS group is compared with healthy controls	References
Iran case—control human participants women with PCOS/healthy controls (n = 12/n = 12)	Hypomethylation of H3K9 of promoter PII in cumulus cells, no significant difference in H3K9 methylation at PI.3 and PI.4 in cumulus cells	Hosseini et al. (2019)
miRNA expression		
India case—control human participants women with PCOS/healthy controls (n = 20/n = 20)	↓ miR-24 expression in serum, no significant difference in miR-29a and miR-502-3p expression in serum	Nanda et al. (2020)
China case—control human participants insulin-resistant PCOS (IR-PCOS) women/control PCOS (PCOS Con) women (n = 26/n = 24)	59 differentially expressed miRNAs (↓ 37 miRNAs and ↑ 22 miRNAs) in the cumulus cells of PCOS-IR women	Hu et al. (2020)
China case-control human participants insulin-resistant PCOS (IR-PCOS) women/Control PCOS (PCOS Con) women (n = 68/n = 44)	↓ miR-204 expression in granulosa cells of PCOS-IR women	Jiang et al. (2020a,b)
China case—control hCG-induced rat granulosa cells/control (n = 15/n = 15)	Upregulation of miR-139, miR-21b, miR-9119, miR-126, and miR-556 in ovarian cortexes	Ding et al. (2020)
United States case—control human participants women with PCOS/healthy controls (n = 7/n = 7)	Eighteen differentially expressed miRNA (13 upregulated and 5 downregulated) in theca interna tissue	McAllister et al. (2019)
China case—control human participants ovarian cortical tissues from PCOS patients/healthy controls (n = 20/n = 20)	Overexpression of miR-155 in ovarian cortical tissue	Xia and Zhao (2020)
China case—control human participants ovarian cortical tissues from PCOS patients/healthy controls (n = 5/n = 5)	Upregulation of 7 miRNAs and downregulation of 5 miRNAs Increased miR-130b expression	Jiang et al. (2020a,b)

Table 9.1 Recent studies linking epigenetic mechanisms and PCOS.—cont'd

Study design	Key findings when PCOS group is compared with healthy controls	References
DNA methylation		
The United Kingdom case—control human participants granulosa lutein cells from PCOS patients/healthy controls (n = 16/n = 16)	106 differentially methylated CpG sites	Makrinou et al. (2020)
China case—control human participants subcutaneous adipose tissue samples from PCOS patients/healthy controls (n = 10/n = 10)	1275 differentially expressed genes and 556 differentially methylated genes	Liu et al. (2020)
India case—control human participants cumulus granulosa cells from PCOS patients/healthy controls (n = 35/n = 38)	Differential methylation of 6486 CpG sites (2977 hypomethylated and 2509 hypermethylated)	Sagvekar et al. (2019)
Australia cross-sectional study human participants immune cells from PCOS patients/healthy controls (n = 17/n = 17)	DNA hypomethylation in monocytes and T helpers	Hiam et al. (2019)
United States case—control prenatal androgen-treated sheep model/control(n = 5/n = 4—5) intramuscular administration of 100 mg Testosterone Propionate twice weekly from day 30 to day 90 of gestation	↑ DNMT1, DNMT3A, and DNMT3B in visceral adipose tissue; ↑ DNMT1 and DNMT3A in subcutaneous adipose tissue; ↑ DNMT3A in perirenal, subcutaneous, and epicardial adipose tissue; ↑ DNMT1 and DNMT3A in liver ↑ DNMT3A in muscle	Guo et al. (2020)

Epigenetics mechanisms and gut microbiota

Epigenetics is the inheritable alteration in genetic expression without any change to the DNA sequence. It allows cells to successfully adapt to their internal or external environment without DNA sequence restructuring. This is important in numerous cellular activities such as aging, chromosomal silencing, and cell differentiation. Epigenetic changes occur via various mechanisms, which can be synergic in driving changes (Moutinho and Esteller, 2017).

DNA methylation

Pathological conditions ranging from cancer to polycystic ovary syndrome have been linked to aberrant methylation (Muse et al., 2020).

Interaction with gut microbiota

The gut microbiota can be considered as a key epigenetic controller. The link between DNA methylation patterns and host microbiota dominance is highlighted by Kumar and colleagues. The genome-wide methylation study using hematological samples from eight pregnant women revealed that a *Firmicutes* dominant gut microbiota resulted in the hypermethylation of 568 genes and hypomethylation of 245 genes when compared with *Bacteroidetes* and *Proteobacteria* dominant gut microbiota. A portion of the affected genes has been correlated with the incidence of cardiovascular and metabolic dysfunction, inflammatory mediation, and obesity (Kumar et al., 2014). In a study using *Bacteroidetes* and *Firmicutes* ratio (BFR) in obese individuals, a strong variation was reported between the adipose tissue and blood of high BFR and low BFR groups. The genome-wide methylation analysis indicated the differential methylation of 258 genes in blood and adipose samples. There was also hypomethylation and overexpression of HDAC7 and IGF2BP2 genes, which are relevant in glucose homeostasis and energy metabolism, respectively, in the individuals with low BFR when compared with those with high BFR (Ramos-Molina et al., 2019). Overexpression of HDAC7 in pancreatic islet cells of human and rats has been correlated with the β-cell disruption and impairment of insulin secretion (Daneshpajooh et al., 2017). In addition, there is a vast amount of gut microbial metabolites, which can alter methylation patterns present in the host epigenome. Several metabolites such as folate, vitamin B12, SCFAs, and EGCGs are capable of influencing DNA methylation via interactions with the one-carbon metabolic pathway or key enzymes that catalyze methylation reactions (Bhat and Kapila, 2017). A change in diet leading to gut dysbiosis alters the availability of microbial metabolites and thus results in epigenetic modifications. For metabolites involved in one-carbon metabolism, they contribute to the formation of S-adenosylmethionine, the primary methyl donor utilized by DNMTs and HMTs. B vitamins have also been reported to serve as cofactors to key regulatory enzymes in DNA methylation (Yamada et al., 2015; Bhat and Kapila, 2017). Gurwara and colleagues showed that the reduced consumption of one-carbon metabolism-related B vitamins led to increased pathogenic bacteria and reduced butyrate-producing bacteria (Gurwara et al., 2019). This study further highlights the relationship between healthy microbiota, its metabolites, and one-carbon metabolism, which is necessary for DNA methylation. Furthermore, SCFAs and their contribution to the Kreb's cycle via acetyl-CoA may produce intermediates such as NAD+, fumarate, and succinate, which influence DNA methylation. α-Ketoglutarate is also utilized as a cosubstrate by ten–eleven translocation (TET) enzymes for DNA demethylation. These enzymes may be inhibited by succinate and fumarate due to structural similarity with substrate (An et al., 2017; Miro-Blanch and Yanes, 2019). A whole-genome bisulfite sequencing has demonstrated the dependence o + f microbiota-related DNA methylation reactions on TET2 and TET3 enzymes, particularly concerning acute inflammation in the intestinal epithelium (Ansari et al., 2020). Apart from their HDAC inhibitory activity, SCFAs may influence epigenetic modifications via DNA methylation. SCFA supplementation in obese mice ameliorates the adiponectin and resistin expression via DNA methylation alterations in their promoter regions. This effect was facilitated by decreased expression of DMNT1, DMNT3a, and DMNT3b and prevention of the interaction between these proteins and the promoter regions (Lu et al., 2018).

Histone modification

Histone proteins form the backbone of nuclear chromatin and regulate the packing of DNA. There are five families of histone proteins that play a role in epigenetic regulation. All except H1 and H5 form the octamer of the nucleosome. H1 closes the entry and exit points of DNA around the nucleosome, while H5 is an isoform of H1. Histones undergo multiple posttranslational modifications (Ye et al., 2017; Garcia-Gimenez et al., 2019). Methylation and acetylation are the most studied modifications, and they are capable of altering the physical features of chromatin. Chromatin can be converted into euchromatin (loosely packed) and heterochromatin (tightly packed), increasing or decreasing gene transcription, respectively (Moutinho and Esteller, 2017). Acetylation and deacetylation reactions are catalyzed by histone acetyltransferases (HATs) and histone deacetylases (HDACs). There are 18 HDACs enzymes divided into four major classes based on their cellular location and mechanism of action. On the other hand, HATs are classified into three broad families—MYST (MORF, MOF, Tip60, MOZ, HBO1), GNAT (HAT1, GCN5, PCAF), and p300/CBP. Both HATs and HDACs can act on other protein substrates such as p53, β-catenin, Hsp 90, and Rb, which are relevant in cancer therapy (Narita et al., 2019). Histone acetylation activates gene transcription by promoting loose histone—DNA interactions (euchromatin), while deacetylation represses gene transcription by tightening chromatin structure (heterochromatin). Histone methylation occurs on nitrogen atoms present in lysine and arginine residues on primarily histones H3 and H4. At histone H3, lysines 4 (H3K4), 9 (H3K9), 20(H3K20), 27(H3K27), 36 (H3K36) and 79(H3K79) as well as arginines 2 (H3R2), 8(H3R8), 17 (H3R17), 26 (H3R26), 128 (H3R128), 129 (H3R129), 131 (H3R131), and 134 (H3R134) can undergo mono/di/trimethylation. The same applies to methylated arginine 3 at histone H4 (H4R3) and H2A(H2AR3) (Gavin et al., 2016; Poulin et al., 2016; Zhang et al., 2016). Lysine methylation and demethylation are catalyzed by six main families of histone lysine methyltransferase complex (KMT 1−6) and histone lysine demethylases (KDM1−6), respectively. The enzymatic methylation of arginine occurs via a subclass of protein arginine methyltransferases (PRMTs). The histone methylation effect on gene expression may vary according to the location and current methylation as it influences chromatin architecture and interacts with RNAs and transcription factors (Jambhekar et al., 2019; Zhao and Shilatifard et al., 2019).

Interaction with gut microbiota

Gut metabolites also influence epigenetic regulation via modification of histone proteins around which DNA wraps (Gerhauser, 2018; Qin and Wade, 2018). Among gut metabolites, SCFAs are the most studied concerning the modification of histone proteins. SCFAs, mainly butyrate, have been reported as HDAC inhibitors, explaining its relevance in cancer and immunological therapy. HDAC activity is associated with conditions such as schizophrenia, Alzheimer's, schizophrenia, drug addiction, etc. (Volmar and Wahlestedt, 2015; Liu et al., 2018). Butyrate with an efficiency of 80% is a strong HDAC inhibitor. This is closely followed by propionate with about 60% inhibitory efficiency. Acetate has been reported to show little to no HDAC inhibitory ability (Soliman and Rosenberger, 2011; Kasubuchi et al., 2015; Liu et al., 2018). Single-dose systemic administration of butyrate was shown to induce histone hyperacetylation (arising from HDAC inhibition) in the frontal cortex and hippocampus. Similarly, chronic intraperitoneal administration of butyrate facilitated increased memory and learning ability via chromatin remodeling in mice with brain atrophy and wild-phenotype mice. Intrahippocampal and systemic butyrate also enhances extinction of fear through histone acetylation (Fischer et al., 2007; Stafford et al., 2012). Likewise, dietary fibers that promote SCFA production may stimulate the acetylation of histone proteins (Carrer et al., 2017; Krautkramer et al., 2016, 2017).

Noncoding RNA regulation

Noncoding RNAs (ncRNAs) are capable of mediating processes relevant in gene expression and epigenetic modifications. These RNAs that do not undergo translation may play either housekeeping or regulatory roles. Regulatory RNAs are grouped into two based on size—short ncRNAs and long ncRNAs (min. length of 200 nucleotides) (Wei et al., 2017). This review emphasizes miRNAs, a short ncRNA, as its aberrance is closely associated with PCOS (Chen et al., 2019a,b). miRNAs are endogenic partial hairpin-like double-stranded molecules, which comprise of 18—25 nucleotides and have body-wide distribution in tissues and fluids (Yao et al., 2019). The biogenesis of miRNAs involves their gene transcription by RNA polymerase II and posttranscriptional activation of pre-miRNAs by Drosha and Dicer ribonucleases. miRNAs have been reported to target about 30% of the entire human genetic composition (Lewis et al., 2005; Valinezhad et al., 2014). They alter genetic expression by binding to 3′ untranslated region (UTR) of a specific gene, leading to repression or activation of its activity. miRNAs contribute to a wide range of cellular activities ranging from differentiation to apoptosis.

A reciprocal relationship exists between miRNAs and epigenetic regulation as miRNAs target enzymes, which play a role in epigenetic modification (e.g., DMNTs and HDACs). Likewise, other epigenetic modifications also control miRNA activity (Yao et al., 2019; Chen et al., 2019a,b). This underlies the miRNA-epigenetic cycle, which is currently being explored to develop novel therapies for different diseases (Chen et al., 2019a,b; Tirronen et al., 2019; Handa et al., 2019; Churov et al., 2019).

Role of diet in microbiota composition and epigenetics

The gut microbiome is affected by numerous factors, including diet (Bassis et al., 2019). Diet is central to modulating the abundance of different microbial species in the gut and regulating their functions (Leeming et al., 2019). High-fat—low-fiber diet consumption has been implicated in the pathogenesis of many conditions ranging from cardiovascular diseases to inflammatory bowel disease (Gentile and Weir, 2018). Interestingly, diet is capable of inducing both short-term (24—48 hours) and long-term effects on gut microbial ecology (David et al., 2014; Rodriguez et al., 2019). This is a result of microbes metabolizing ingested food constituents as fuel for their cellular activities. Thus, a change in the diet influences the abundance of certain species by altering available substrates for microbial metabolism. Dietary components such as fats, proteins, macronutrients, and micronutrients have individual effects on the microbial ecology in the gut. Sonnenburg et al. described the group of carbohydrates readily metabolized by gut microbiota as "microbiota-accessible carbohydrates (MACs)." Changes in dietary MACs over multiple generations were reported to cause a transgenerational gut microbial shift in terms of composition and activity (Sonnenburg et al., 2016). Two population-based research studies indicate the divergence of gut microbiome between individuals based in agrocentric communities and others in industrial settings. In the Hadza-Tanzania population, a community of hunters and gatherers increased consumption diet, including berries and honey, was positively associated with better microbial abundance and diversity when compared with urban Italian counterparts. The Hadza community possessed a microbial community, which aided in the digestion and extraction of nutrients from ingested plant fibers, consistent with their foraging habits (Schnorr et al., 2014). Across Malawi, Venezuela, and the United States, a similar variation is highlighted in the fecal microbial community of 531 individuals from 150 families (Yatsunenko et al., 2012).

Consumption of MACs also showcases the symbiotic link of the gut microbiome and its human host. The bacterial fermentation of MACs (e.g., starch) produces SCFAs, which contribute to processes such as inflammation, apoptosis, cellular proliferation, immunomodulation, etc. SCFAs, particularly acetate, butyrate, and propionate, also inhibit HDACs, a key enzyme in the epigenetic modification of histones (Dalile et al., 2019). This is discussed in a subsequent section of this chapter. In the same vein, some MACs (e.g., fructooligosaccharides and inulins) can enhance the population of beneficial bacterial species (Ho et al., 2019). The increase of these microbial species has been closely associated with better cardiometabolic health indices in diabetes and PCOS (Shamasbi et al., 2020; Ho et al., 2019). Also, dietary fats influence gut microbial diversity. The protection of germ-free mice from the metabolic dysfunction induced by increased high-fat diet consumption indicates gut microbial involvement in lipid metabolic pathways (Rabot et al., 2010). However, the cholesterol content of the diet is a noteworthy factor as it mediates this relationship (Kübeck et al., 2016). Ramos-Romero et al. observed that male rats on a high-fat diet demonstrated an increased abundance of *Escherichia coli* and Enterobacterales communities. This was accompanied by increased inflammatory mediators (IL-6 and PGE2) and lipopolysaccharides (LPS), resulting in insulin resistance (Ramos-Romero et al., 2017). The interaction of LPS with Toll-like receptor 4 may highlight the mediative activity of the microbiota for the metabolic outcomes of a high-fat diet. This interaction is closely linked with obesity, sepsis, and metabolic syndrome (Munford, 2016).

Influence of gut metabolites on epigenetics

Numerous gut metabolites have been recorded to influence the epigenetic modification of genes. They achieve this feat by interacting with key enzymes, which serve as epigenetic regulators or altering the pool of substrates required for epigenetic reactions (Bhat and Kapila, 2017). These metabolites include SCFAs, phenol compounds, fats, polyamines, and vitamins.

Short-chain fatty acids

SCFAs are saturated carboxylic acids with a chain length between one and six carbon atoms. The abundant SCFAs are acetate (C2), butyrate (C3), and propionate (C4), while valerate (C5), formate (C1), and caproate (C6) are formed in lesser quantities (Cummings et al., 1987). SCFAs are derived mainly from gut anaerobic microbial breakdown of dietary fibers (Macfarlane and Macfarlane, 2003; Dalile et al., 2019). These dietary fibers include inulin, oats bran, cellulose, pectin, resistant starch, and others (Basson et al., 2016). In the colon, acetate, butyrate, and propionate are distributed, respectively, in the approximate molar ratio 60:20:20; howbeit, this proportion is dependent on host, diet, and microbiota diversity (Cummings et al., 1987). Pyruvate, hydrogen and carbon dioxide, and methanoate can serve as source molecules for the formation of acetate via different pathways (Ragsdale and Pierce, 2008; Koh et al., 2016). Butyrate is synthesized from butyryl-CoA via two metabolic pathways (Louis et al., 2004, 2010). Also, certain microbes (e.g., *Anaerostipes* spp.) can produce butyrate from acetate and lactate (Louis et al., 2014; Koh et al., 2016). Interestingly, propionate can be generated from substrates such as methylmalonyl-CoA, lactate, or deoxyhexose sugars (e.g., rhamnose) via three pathways—succinate, acrylate, and propanediol, respectively (Reichardt et al., 2014). SCFAs can act as modulators of epigenetic activity via HDAC inhibition. Butyrate is the most studied SCFA due to its highly potent inhibition of HDAC. It limits the expression of proinflammatory genes (CCL2, IL-6, and IL-8) in vascular endothelial tissues even when stimulated by LPS (Chriett et al., 2019). Butyrate-mediated HDAC inhibitory activity interacts with some oncogenic signaling pathways (STAT3 and VEGF). In cancer cells, butyrate is not used as an energy substrate; rather, it accumulates in the nucleus and inhibits HDAC. This results in heightened histone acetylation and thus causes a reduction in cancer cell proliferation (Eckschlager et al., 2017; Chen et al., 2019a,b). Recently, it was demonstrated that butyrate inhibits cellular division and induces apoptosis in colorectal cancerous cells (Cao et al., 2019). Also, histone modifications that arise from HDAC inhibition have been implicated in metabolic disorders. Olaniyi and colleagues reported that acetate supplementation ameliorates cardiometabolic features in diabetic Wistar rats. The inhibition of HDAC by acetate prevented cardiac tissue damage and suppressed parameters, which indicated oxidative stress and cardiometabolic dysfunction in the animal model (Olaniyi et al., 2020). This study hints that the interplay between acetate and epigenetic alterations may contribute to the pathogenesis of metabolic diseases.

Polyphenol compounds

Polyphenols are phytochemicals widely present in fruits, vegetables, and herbs. These compounds can be grouped into three main classes—phenolic compounds, flavonoids, and stilbenoids (Brglez Mojzer et al., 2016). Flavonoids form the majority of polyphenol compounds, and they are classified based on chemical structure. They account for about 60% of polyphenols with two or more aromatic rings. Stilbenoids form a smaller group of polyphenols that contain polyhydroxystilbenes, while phenolic compounds that constitute about 30% of polyphenols can be subdivided into hydroxybenzoic and hydroxycinnamic acids (Brglez Mojzer et al., 2016; Arora et al., 2019). Gut microbiota, e.g., Bacteroides species metabolize polyphenols to produce some metabolites associated with epigenetic modifications found in cancer, diabetes, and cardiovascular diseases (Rasines-Perea and Teissedre et al., 2017; Arora et al., 2019). Although a large number of polyphenols and their metabolites limit this research area, the interaction between certain polyphenols and epigenetic regulators has been documented. For instance, epigallocatechin-3-gallate (EGCG), a polyphenol, was observed to reactivate TIMP-3. TIMP-3 codes for metalloproteinase inhibitor 3, a protein that limits prostate cancer progression. EGCG administration led to the trimethylation of H3K27, increased acetylation of H3K9/18, and decreased HDAC Class I activity (Deb et al., 2019). Similarly, EGCG inhibits cellular proliferation and mediate apoptosis in acute promyelocytic leukemia (APL). In this study, EGCG was reported to mediate hyperacetylation of histone H3 and acetylation of H4 at lysine 14 (H3K14 and H4K14). In addition, EGCG was shown to downregulate key epigenetic regulatory enzymes such as HDAC1, HDAC2, and DMNT1 (Borutinskaitė et al., 2018). EGCG mediates its biological outcomes through its metabolites rather than the parent compound as studies have shown that these metabolites have longer half-lives and higher concentrations in vitro when compared with the parent compound (Pervin et al., 2019). Ellagitannins are also polyphenols that double as epigenetic modifiers. Ellagitannins are found in fruits such as strawberries, pomegranates, and blackberries as well as nuts such as almonds. They are hydrolyzed and metabolized by the gut microbiota into ellagic acid and urolithins, respectively (Huller and Fu, 2014; DaSilva et al., 2019). The treatment of colon cancer cells with ellagic acid led to the differential expression of 857 genes (363 upregulated and 494 downregulated), most of which underlie cellular activities including proliferation and apoptosis (Zhao et al., 2017). Ellagic acid has also been reported to target HDAC6-related pathways in cancer cells (Lernoux et al., 2018).

Dietary fats

The absorption of dietary fats such as triacylglycerol, sterols, and fatty acids usually takes place in the small intestine. However, about 7% of dietary fat passes through the colon for excretion. This portion of dietary fats is metabolized by the gut microbiota (Gabert et al., 2011; Hullar and Fu, 2014). In a 6-month-long study, supplementation of omega-3 polyunsaturated fatty acids (n-3 PUFAs) in obese and overweight participants interacted with pathways related to lipid metabolism, cardiovascular health, development of type 2 diabetes, and inflammatory responses via epigenetic alterations. In blood leukocytes, about 308 CpG sites (231 genes) presented with different methylation patterns with hypomethylation and hypermethylation in 22 and 286 sites, respectively (Tremblay et al., 2017). Likewise, in a study using animal colorectal cancer model, n-3 PUFA was reported to produce its anticarcinogenic effects via DNA hydroxymethylation (Huang et al., 2019). Another study suggests that n-3 PUFA may protect cardiomyocytes from the effects of myocardial infarction via interaction with apoptosis miRNA and related genes (Kura et al., 2019). Flores-sierra and colleagues documented the DNA hypermethylation effect of elaidate, a transfatty acid in human monocytes and its detectable epigenetic modification of offspring genome (Flores-sierra et al., 2016).

Polyamines

Polyamines are low-molecular-weight and positively charged aliphatic amines present in mammals. The gut microbiota serves as the major source of polyamines in the colon (Tofalo et al., 2019). They include spermine, spermidine, cadaverine, and putrescine. These amines are closely related to cellular functions such as DNA, RNA and protein synthesis, chromatin remodeling, and interaction with receptor and ion channels. Outside the cell, polyamines are antiinflammatory, and they maintain the intestinal barrier via promotion of mucus secretion and upregulation of tight junction proteins (Igarashi and Kashiwagi, 2019). Due to their cellular importance, polyamine analogs have been developed to target cell growth and suppress tumor activity via epigenetic mechanisms. Polyamines have been reported to interact with the activity of HATs and HDACs as well as bind to miRNAs (Damiani and Wallace, 2018; Sakamoto et al., 2020). Polyamine analog treatment of human myeloid leukemia cells led to the inhibition of LSD-1. This resulted in increased methylation (mono and di) of H3K4 and reexpression of e-cadherin gene, which is commonly repressed in acute myeloid leukemia (Murray-Stewart et al., 2014). Also, a recent review has highlighted the role of spermine and spermidine metabolism in the modulation of DMNT activity and regulation of DNA methylation aberrance, further associating it with prolongation of life (Soda, 2020).

Vitamins

In addition to vitamins derived from exogenous sources, gut microbes play a role in the availability of vitamins such as vitamin K and most water-soluble vitamins (riboflavin, folate, pyroxidine, pantothenic acid, biotin, thiamine, and cobalamin). These vitamins, particularly folate, have been reported to mediate epigenetic changes. The role of folate in one-carbon metabolism allows it to contribute to DNA and histone methylation reactions (Said and Nexo, 2018). Recently, a large epigenome-wide study involving 5841 volunteers reported significant DNA methylation changes associated with folate intake in blood leukocytes. The metaanalysis revealed 74 folate-associated differentially methylated regions and 6 differentially methylated positions, which are related to genes concerned with cancer development, cell proliferation, and centrosome formation (Mandaviya et al., 2019).

Similarly, aberrant miRNA expression in cancer, metabolic, and neurodegenerative conditions has been linked to folate deficiency (Beckett et al., 2017). Concerning the B vitamins, they contribute to both one-carbon and folate metabolism. In the aforementioned epigenome-wide study, vitamin B12 intake was correlated with 29 differentially methylated regions linked to 48 genes (Mandaviya et al., 2019). In another study, vitamin B12 deficiency in adipocytes derived from pregnant women led to changes in circulating miRNAs. In addition to increased expression of adipogenesis-related genes, adipocytes also showed increased lipogenesis and triglyceride accumulation. This study suggests that vitamin B12 deficiency may underlie maternal obesity and metabolic dysfunction (Adaikalakoteswari et al., 2017).

Furthermore, Arreguín and colleagues evaluated the impact of the supplementation of two dietary forms of vitamin A, β-carotene, and retinyl ester on gene translation and transcription in white adipose tissue of Wistar rats. The study reported that retinyl ester treatment hypermethylated RBP4, PPARG2, and hypomethylated PCNA promoters. Meanwhile, hypermethylation of PCNA and hypomethylation of RBP4 promoters were observed in the β-carotene group. This study highlights the impact of vitamin A supplementation on white adipose tissue in early postnatal life (Arreguín et al., 2018).

Transgenerational inheritance of PCOS: role of epigenetics

Although there is no widely accepted consensus on the specific mechanisms involved of PCOS inheritance, there is a general agreement about the PCOS traits can be inherited by the offspring of women with PCOS. Genetic, epigenetic, and environmental (external and in utero) conditions are thought to contribute to the transmission of PCOS phenotype to offspring (Gorsic et al., 2019; Dapas et al., 2019). Heightened likelihood of metabolic (insulin resistance) and reproductive parameters present in PCOS is expressed in both male and female offspring (Crisosto et al., 2017; Torchen et al., 2019). Also, the risk of psychiatric conditions has also been indicated in such offspring (Cesta et al., 2016). A genome-wide association study (GWAS) has highlighted susceptibility loci, which contribute to PCOS phenotypes (irrespective of diagnostic criteria). These loci were closely linked to excessive androgen secretion and altered gonadotropin levels in PCOS patients (Day et al., 2018). Overexpression of an identified susceptibility locus, DENND1A (DENN/MADD domain containing 1A) in ovarian theca cells of PCOS patients, was central to the development of hyperandrogenism (McAllister et al., 2014). However, these loci cannot account for more than 10% of PCOS inheritance (Aziz et al., 2016). This indicates that other factors, including epigenetic modifications, may underlie the heritability of PCOS. The interaction between epigenetic modifications and the identified loci may result in gene activation and/or gene silencing, which contributes to the pathophysiology of PCOS. Hypomethylation of genes related to steroid biosynthesis altered their expression in PCOS patients and promoted increased synthesis of sex hormones. This was suggested to contribute to the hyperandrogenic features seen in PCOS. Similar investigations have described the aberrance of methylation patterns and resulting abnormal genetic expression in fat tissue and skeletal muscle of PCOS patients. These transcriptional and epigenetic modifications are indicated to underlie the PCOS-associated metabolic irregularities (Kokosar et al., 2016; Nilsson et al., 2018). This begs the question of whether aberrant DNA methylation and other epigenetic marks can be transmitted across multiple generations. For intergenerational transmission, the parent phenotype (F0) must be seen in the F1 and F2 generations, while transgenerational inheritance involves the display of this phenotype in the F2 and F3 generation in males and female offspring, respectively (Perez and Lehner, 2019). This brings us to the developmental origins of health and disease (DOHaD) hypothesis, which proposes that in utero and early infant conditions of an individual might lead to (re)programming of fetal physiology and predisposition to metabolic disorders (Lacagnina, 2020). This suggests that in utero environment might influence epigenetic modification of fetal DNA, perhaps in anticipating of the external environment. This might contribute to the inheritance of PCOS phenotype as gestational conditions in PCOS differ from the norm. PCOS patients have excess androgen concentrations compared with healthy participants, and this increase extends into pregnancy. This, in addition to anti-Mullerian hormone concentration, alters placenta signaling and nutrient delivery (Maliquo et al., 2015; Tata et al., 2018). Increased anogenital distance, a marker of prenatal androgen (PNA) levels, has been observed in offspring (regardless of sex) of PCOS women (Barrett et al., 2018; Glintborg et al., 2019). Combined with epigenetic changes in the adipose tissue of PNA-exposed rhesus monkey and evidence from the tissue-specific DNA methylation patterns in PCOS patients, there is a solid background for the DOHaD hypothesis in the origin of PCOS (Xu et al., 2011; Vázquez-Martínez et al., 2019). While there is research literature supporting the intergenerational transmission of PCOS traits, there are limited studies on the transgenerational transmission of these traits. This may be due to the difficulty of maintaining such a longitudinal study across successive generations (F3 for females, F2 for males) in women with PCOS. For transgenerational epigenetic alterations to be recorded, epigenetic modifications must appear in the germ cells of three generations of offspring (irrespective of somatic alterations). In a study by Saben et al., a high-fat—high-sugar diet (before conception till weaning)-induced maternal obesity resulted in transgenerational oocyte mitochondrial dysfunction in female offspring. This increased the likelihood of insulin resistance and predisposition to metabolic syndrome in F1, F2, and F3 generations placed on the control diet (Saben et al., 2016). Concerning PCOS, Risal et al. demonstrated the impact of epigenetics in the transgenerational inheritance of PCOS phenotypes using a fat diet and PNA exposure mice model (mimicking lean and obese PCOS phenotypes). Except for appropriate controls, obesity and glucose dysregulation was induced in the late gestation stage of F0 generation dams using a high-fat diet, in addition to daily subcutaneous injection of dihydrotestosterone (DHT). The anogenital distance was increased in female offspring across all three generations (F1 to F3) despite the normalcy of androgen concentrations in their adulthood. Alterations in adipocyte morphology, energy metabolism, estrous cycles, and hepatic triglyceride levels were maintained in the preandrogenized lineage till the F3 generation (Risal et al., 2019). Interestingly, the molecular and structural analysis of mitochondria present in MII oocytes of the preandrogenized lineage (F1 to F3) reveals additional alterations in mitochondrial DNA, number, and morphology. Finally, the identification of differential expression genes, which play central roles in steroid synthesis, DNA repair, and glucose homeostasis in the MII oocytes of the three generations (F1—F3), suggests the underlying role of epigenetic modification of germ cells in this transgenerational inheritance of PCOS-like characteristics (Risal et al., 2019).

Interplay between epigenetic mechanisms and PCOS

Histone modification and polycystic ovary syndrome

There is a dearth of research on histone modifications in polycystic ovary syndrome. However, certain papers have elucidated specific epigenetic changes in the histone proteins present in both humans and animal models of PCOS. In a recent study using prenatally androgenized sheep which mimic the metabolic features seen in PCOS, hepatic HDAC1 expression was largely decreased, while HDAC2 (significant) and HDAC3 expression were increased (Guo et al., 2020). In adipose samples (subcutaneous, visceral, perirenal, and epicardiac), there was increased HDAC1 expression in epicardiac adipocytes and decreased HDAC3 expression in visceral adipocytes with no change in that of HDAC2 in all samples. Subcutaneous adipocytes also showed increased expression of Sirtuin1, while significant increase in demethylase enzyme KDM1A, and a nonsignificant increase in SMYD3 and EZH2 expression was observed in visceral adipocytes. Obtained liver samples showed significantly increased acetylation at H3K9ac and H3K27ac (Guo et al., 2020). These observed changes were due to the prenatal androgen treatment. The increased hepatic histone acetylation indicates a decrease in the activity of HDACs, a feature of hepatic steatosis seen in individuals with excessive alcohol consumption or obesity. This change may explain the occurrence of hepatic steatosis in prenatal androgen sheep PCOS model. Also, increased acetylation at H3K9ac and H3K27ac may mediate an increase in expression of PPARA, an important lipid-associated transcription factor in the muscle and liver. Aberrant PPARA expression is linked to NAFLD and lipid metabolic dysfunction (Kersten and Stienstra, 2017; Guo et al., 2020). Thus, histone-related epigenetic changes may produce tissue-specific metabolic disorders seen in the pathophysiology of PCOS.

Another study using a similar model established the relationship between histone modification and differential gene expression in sheep ovary. In an examination of gene inducing and repressing histone marks across experimental groups, no difference was reported in the ovaries of day 90 fetal groups sheep (control and testosterone-treated). However, in the 2-year-old adult ovaries, except for the gene repressing marks, all marks were reported to be upregulated (Sinha et al., 2020). These altered gene marks were shown to interact with genes concerned with inflammation, cellular signaling, and steroid synthesis with the latter being the most significantly affected. Likewise, about 22 inflammation-associated genes were upregulated in the ovaries of 2-year-old testosterone-treated adult sheep compared with appropriate controls. This finding further emphasizes inflammation as a cardinal factor in the pathogenesis of PCOS. Diacylglycerol kinase (DGKG) was also reported to be downregulated in the tissue mentioned before (Sinha et al., 2020), and its downregulation may precede insulin insensitivity in PCOS. Also, adrenergic receptor genes, ADRA1A and ADRA2A, which contribute to the normal growth of follicles and function of ovaries, were shown to be downregulated in prenatal androgen sheep PCOS model (Sinha et al., 2020). Thus, prenatal androgen exposure produces epigenetic changes via histone modifications that result in changes in expression of normal ovarian genes and may lead to abnormal ovarian morphology and ovarian dysfunction typically seen in PCOS. In the same vein, histone modification activity has been observed in the aberrant expression of the aromatase gene, CYP19A1 in PCOS. The cumulus cells showed significant hyperacetylation of promoter regions of the aromatase gene (Hosseini et al., 2019). In summary, these studies indicate a need for further investigation into the contribution of histone modification in PCOS.

MiRNA expression in polycystic ovary syndrome

Recent research has highlighted the aberrant expression of miRNA in fluids, cells, and tissues in PCOS. A recent study attempts to correlate miRNA expression with parameters related to metabolism and the endocrine system. MiR-24 expression was significantly reduced in PCOS patients versus controls, while miR-29a and miR-502-23p expression were comparable across both groups (Nanda et al., 2020). MiR-24 expression had an inverse relationship with insulin, TG, HOMA-IR, testosterone, LH-FSH ratio, FIRI, glucose and BMI, and positively associated with HDL-C levels. Interestingly, miR-24 underlies androgen biosynthesis and estradiol regulation. Its biosynthetic role arises due to its interaction with CYP11A1, a key enzyme in sex steroid synthesis, and its regulatory role is a result of its interaction with TGF-β signaling. Thus, decreased miR-24 levels could result in increased concentrations of CYP11A1and abnormal TGF-β signaling, leading to hyperandrogenemia and dysfunctional follicular growth, respectively (Nanda et al., 2020). The findings also suggest the contribution of miR-24 to hyperglycemia and insulin resistance in PCOS as its expression was inversely associated with insulin and glucose concentrations as well as FIRI and HOMA-IR. This miRNA might be a key player in normal glucose homeostasis. With a focus on insulin resistance, Hu and colleagues highlighted the miRNAs that control the MAPK pathway in insulin-resistant PCOS patients using obtained cumulus cells (Hu et al., 2020). There was increased expression of 22 miRNAs and decreased expression of 37 miRNAs, a total of 59 differentially expressed miRNAs in the PCOS-IR group compared with the PCOS control group. Notably, miR-612 is thought to target Rap1b for MAPK regulation. This is supported by a significant rise in miR-612, accompanied by a severe fall in the expression of Rap1b mRNA and protein (Hu et al., 2020). This study highlights the plausible contribution of miRNAs to the development of insulin-resistant PCOS (Hu et al., 2020). This is further reinforced by a study which reported decreased expression of miR-204 and increased expression of HMGB1, TLR4, and NF-κB p65 in the granulosa cells of PCOS-IR women relative to both healthy and PCOS-NIR controls (Jiang et al., 2020a,b). Moreover, cellular models, which allowed for increased expression of miR-204 and downregulation of HMGB1, demonstrated suppression of the NF-κB p65 pathway mediating decreased insulin resistance and testosterone levels. Thus, upregulation of miR-204 may be considered as a therapeutic tool in ameliorating insulin resistance in PCOS (Jiang et al., 2020a,b). This requires further research to conclusively establish the relationship between the miR-204/HMGB1 axis and insulin resistance in PCOS.

Another study has investigated the miRNA profile in insulin-treated granulosa cells from PCOS patients. MiR-139, miR-21b, miR-9119, miR-126, and miR-556 were observed to be upregulated in PCOS (Ding et al., 2020). MiR-9119 expression was noted to be significantly elevated in ovarian granulosa cells even when treated with insulin. Further investigation showed that miR-9119 prevents cellular proliferation and mediates apoptosis via negative regulation of DICER expression. This, in turn, positively regulates NF-κB/p65 (Ding et al., 2020). The role of miR-9119 in this cross-talk hints at the possibility of using this miRNA as an early biomarker in PCOS. Moreover, the hyperandrogenic feature in PCOS has been correlated with miR-130b-3p expression in PCOS theca cells. McAllister et al. identified 18 differentially expressed miRNA, 13 upregulated, and 5 downregulated in the condition (McAllister et al., 2019). There was an inverse relationship between the decreased expression of miR-130b-3p and the expression of CYP17A1, which mediate thecal androgen production (McAllister et al., 2019). The pathway targeting and regulatory action of miR-130b-3p provides valuable understanding into the pathophysiology of hyperandrogenemia in PCOS.

Aberrant miRNA expression in PCOS has also been linked with the likelihood of carcinogenesis and tumorigenesis. Xia and Zhao observed the overexpression of miR-155 in ovarian cortical cells obtained from PCOS patients compared with controls (Xia and Zhao, 2020). The expression of this miRNA promoted cellular proliferation, invasion, and migration in KGN (ovarian granulosa-like) cell lines. This is thought to arise due to its interaction with a tumor suppressor and apoptosis regulator, PDCD4. The inverse association between miR-155 and PDCD4 mediates the stimulation of the PI3K/AKT and JNK signaling pathway, which is closely associated with cancer development and metastasis (Xia and Zhao, 2020). In addition, differential miRNA expression has been linked with abnormal oocyte growth and development. Microarray analysis of ovarian cortical tissue showed upregulation of seven miRNAs and downregulation of five miRNAs in PCOS patients (Jiang et al., 2020a,b). The study then focused on the relationship between an upregulated miRNA, miR-130b and underexpressed gap junction protein, connexin 43 (cx43) in ovarian cortical and luteinized granulosa cells in women with PCOS. MiR-130b is thought to bind to the 3′UTR region of cx43 and facilitate its decreased levels. Decreased cx43 expression has been linked to atresia and stunted growth of ovarian follicles. This suggests that cx43 plays an important role in intercellular communication between cells in the ovary (Jiang et al., 2020a,b). Thus, PCOS patients with reduced cx43 levels have lower gap junctional intercellular communication in the ovaries, leading to abnormal ovarian morphology and dysfunction.

DNA methylation and polycystic ovary syndrome

DNA methylation is the most widely studied epigenetic modification concerning the pathogenesis of PCOS. Abnormal methylation patterns have been recorded in the PCOS patients, their offspring as well as experimental models of PCOS. DNA methylation studies have provided an insight into the heterogeneity of PCOS. Jacobsen et al. reported an association between clinical subphenotypes of PCOS and the differential methylation of genomic regions (Jacobsen et al., 2019). A recent epigenome-wide study in granulosa lutein cells of PCOS patients highlighted 106 differentially methylated CpG sites compared with normal controls. 88 of these sites were associated with genes concerned with metabolism, reproduction, inflammation, immune response, and endocrine regulation (Makrinou et al., 2020). The role of these epigenetic alterations is further reinforced by a study by Liu et al., which correlates methylation profiles with adipocyte genetic expression in PCOS. Initial analysis revealed 1275 differentially expressed genes and 556 differentially methylated genes in PCOS patients (Liu et al., 2020). Both differential methylation and expression of 499 genes were reported, in addition to a negative relationship between the expression and methylation levels of 237 of these genes. One of such genes, LEP, codes for the polypeptide hormone, leptin. Leptin is released mainly by adipose tissue with increased levels in the serum of PCOS patients (Liu et al., 2020). Leptin concentration has been associated with BMI, metabolic dysfunction and abnormal glucose homeostasis. Hypomethylation of LEP recorded in this study led to increased gene expression and possibly elevated leptin concentrations via interaction with Jak-STaT pathway in women with PCOS. Also, this study highlights the differential methylation and expression of F2R, SPP1, IL12B, and RBP4, which may contribute to the abnormal apoptosis and cellular growth arrest of follicles in PCOS (Liu et al., 2020). A closer examination of the PCOS ovarian DNA methylome is seen in a recent study using cumulus cells from women with PCOS. Results showed differential methylation of 6486 CpG sites (2977 hypomethylated and 2509 hypermethylated) associated with 3403 genes (Sagvekar et al., 2019). Some of the differentially methylated miRNAs have been linked to obesity, abnormal steroid biosynthesis, impaired glucose homeostasis, and folliculogenesis. Similarly, pathway analysis implicated seven differentially expressed genes in the development of excessive androgen secretion, premature apoptosis, and abnormal oocyte growth, resulting in irregularities in the typical PCOS ovary (Sagvekar et al., 2019). This study suggests that epigenetics may underlie defects in follicular development preceding ovarian dysfunction in women with PCOS. This is further validated by a cross-sectional study, which showed that DNA hypomethylation in cumulus cells obtained from PCOS patients triggered a series of reaction, leading to the pathogenic upregulation of CYP19A1, the aromatase gene (Hosseini et al., 2019). In a new direction, another cross-sectional study highlights widespread and specific methylation patterns in monocytes and T helper cells from women with PCOS. DNA hypomethylation was recorded in both monocytes and helper cells from PCOS patients (Hiam et al., 2019). In T helper cells, the differentially methylated genes were related to its immune function and interestingly, reproductive function. Furthermore, global methylation pattern in helper cells was inversely associated with anti-Mullerian hormone (AMH) levels (Hiam et al., 2019). AMH levels have been reported to exhibit a direct relationship with sex steroids and LH, and an inverse relationship with glucose concentration and FSH. The interaction between AMH and FSH inhibits follicular growth and selection and predisposes the individual to the polycystic ovary morphology. Similarly, global methylation in helper cells was negatively correlated with free testosterone levels, a marker of hyperandrogenism (Hiam et al., 2019). These findings imply that epigenetic modifications in immune cells may underlie reproductive characteristics of PCOS. Concerning metabolism, a case−control methylation analysis study using pluripotent stem cells demonstrated the overactivation of the CREB signaling pathway in PCOS (Huang et al., 2019). The CREB pathway has been reported to contribute to lipid metabolism and glucose homeostasis, and its altered regulation may explain the metabolic disruption seen in PCOS. Also, this pathway has been shown to interact with aromatase activity, and thus, its hyperactivation may mediate endocrine aberrations in PCOS (Huang et al., 2019). The prenatal androgen sheep model also showed increased key methylation enzymes such as DNMT1, DNMT3A and DNMT3B in visceral adipose tissue, DNMT1 and DNMT3A in subcutaneous adipose tissue, DNMT3A in perirenal, subcutaneous, and epicardial adipose tissue, DNMT1 and DNMT3A in the liver, and DNMT3A in the muscle (Guo et al., 2020). These hepatic and muscular changes are thought to lead to lipid deposition and insulin resistance in these tissues. Thus, impaired glucose metabolism and lipid accumulation may arise due to DNA methylation−induced changes in PCOS (Guo et al., 2020). In addition, epigenetic links between increased cancer incidence and PCOS have been studied. Jiao et al. compared methylation patterns in PCOS patients with or without menstrual irregularities. The results revealed lower DNA methylation in PCOS patients with amenorrhea compared with PCOS controls without amenorrhea. Aberrant expression of oncogenic genes associated with apoptosis, cellular proliferation, and stress was also noted in PCOS patients with irregular menstrual cycles, giving them a predisposition to ovarian cancer (Jiao et al., 2019).

Summary and conclusion

Gut microbiome−mediated epigenetic alterations are crucial contributors to PCOS pathogenesis. Epigenetic modifications are instrumental to PCOS progression from intrauterine life to adulthood (Fig. 9.3). Coupled with the global obesity pandemic, the constellation of metabolic, endocrine, and reproductive dysfunction in PCOS is alarming and increases the risk of cardiovascular disease. PCOS patients with dysbiotic gut communities might transfer an epigenetic predisposition to this disorder to their offspring. Moreover, evidence suggests that the epigenetic transmission of PCOS may be intergenerational and transgenerational. There is need for more research elucidating the interplay between gut metabolites such as SCFAs, folate and choline, and epigenetic changes observed in PCOS. Identification of miRNAs, methylation patterns, and histone modifications involved in this interplay will highlight landmarks in the diagnosis and prognosis of PCOS. The differential expression of these epigenetic regulators might also serve as an important risk indicator for the metabolic, reproductive, and endocrine sequelae accompanying PCOS.

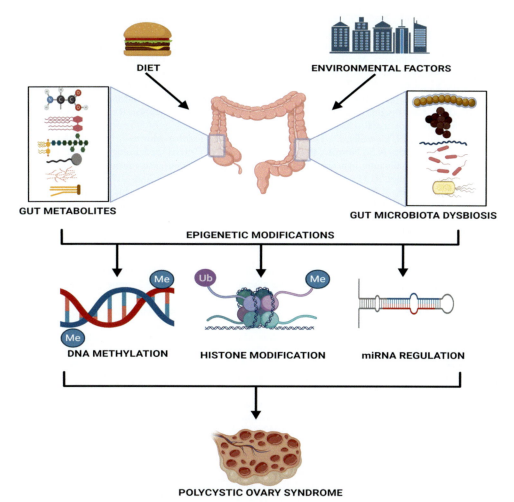

Figure 9.3 Schematic diagram showing gut microbiome modulatory role in PCOS through epigenetical modification. *(Created with BioRender.com.)*

References

Abbott, D.H., Dumesic, D.A., Levine, J.E., 2019. Hyperandrogenic origins of polycystic ovary syndrome—implications for pathophysiology and therapy. Expet Rev. Endocrinol. Metabol. 14 (2), 131—143. https://doi.org/10.1080/17446651.2019.1576522.

Adaikalakoteswari, A., Vatish, M., Alam, M.T., Ott, S., Kumar, S., Saravanan, P., 2017. Low vitamin B12 in pregnancy is associated with adipose-derived circulating miRs targeting PPARγ and insulin resistance. J. Clin. Endocrinol. Metabol. 102 (11), 4200—4209. https://doi.org/10.1210/jc.2017-01155.

Arora, I., Sharma, M., Tollefsbol, T.O., 2019. Combinatorial epigenetics impact of polyphenols and phytochemicals in cancer prevention and therapy. Int. J. Mol. Sci. 20 (18), 4567. https://doi.org/10.3390/ijms20184567.

Arreguín, A., Ribot, J., Mušinović, H., von Lintig, J., Palou, A., Bonet, M.L., 2018. Dietary vitamin A impacts DNA methylation patterns of adipogenesis-related genes in suckling rats. Arch. Biochem. Biophys. 650, 75—84. https://doi.org/10.1016/j.abb.2018.05.009.

An, J., Rao, A., Ko, M., 2017. TET family dioxygenases and DNA demethylation in stem cells and cancers. Exp. Mol. Med. 49 (4), e323. https://doi.org/10.1038/emm.2017.5.

Anagnostis, P., Tarlatzis, B.C., Kauffman, R.P., 2018. Polycystic ovarian syndrome (PCOS): long-term metabolic consequences. Metabolism 86, 33–43. https://doi.org/10.1016/j.metabol.2017.09.016.

Ansari, I., Raddatz, G., Gutekunst, J., Ridnik, M., Cohen, D., Abu-Remaileh, M., Tuganbaev, T., Shapiro, H., Pikarsky, E., Elinav, E., 2020. The microbiota programs DNA methylation to control intestinal homeostasis and inflammation. Nat. Microbiol. 5 (4), 610–619. https://doi.org/10.1038/s41564-019-0659-3.

Azziz, R., 2016. New insights into the genetics of polycystic ovary syndrome. Nat. Rev. Endocrinol. 12 (2), 74–75. https://doi.org/10.1038/nrendo.2015.230.

Azziz, R., Carmina, E., Dewailly, D., Diamanti-Kandarakis, E., Escobar-Morreale, H.F., Futterweit, W., Janssen, O.E., Legro, R.S., Norman, R.J., Taylor, A.E., 2006. Criteria for defining polycystic ovary syndrome as a predominantly hyperandrogenic syndrome: an androgen excess society guideline. J. Clin. Endocrinol. Metabol. 91 (11), 4237–4245. https://doi.org/10.1210/jc.2006-0178.

Barrett, E.S., Hoeger, K.M., Sathyanarayana, S., Abbott, D.H., Redmon, J.B., Nguyen, R.H.N., Swan, S.H., 2018. Anogenital distance in newborn daughters of women with polycystic ovary syndrome indicates fetal testosterone exposure. J. Dev. Orig. Health Dis. 9 (3), 307–314. https://doi.org/10.1017/s2040174417001118.

Bassis, C.M., 2019. Live and diet by your gut microbiota. mBio 10 (5). https://doi.org/10.1128/mBio.02335-19.

Basson, A., Trotter, A., Rodriguez-Palacios, A., Cominelli, F., 2016. Mucosal interactions between genetics, diet, and microbiome in inflammatory bowel disease. Front. Immunol. 7. https://doi.org/10.3389/fimmu.2016.00290.

Beckett, E.L., Veysey, M., Lucock, M., 2017. Folate and microRNA: bidirectional interactions. Clin. Chim. Acta 474, 60–66. https://doi.org/10.1016/j.cca.2017.09.001.

Bhat, M.I., Kapila, R., 2017. Dietary metabolites derived from gut microbiota: critical modulators of epigenetic changes in mammals. Nutr. Rev. 75 (5), 374–389. https://doi.org/10.1093/nutrit/nux001.

Bonomini, F., Favero, G., Rodella, L.F., Moghadasian, M.H., Rezzani, R., 2018. Melatonin modulation of sirtuin-1 attenuates liver injury in a hypercholesterolemic mouse model. BioMed Res. Int. 2018, 1–9. https://doi.org/10.1155/2018/7968452.

Borutinskaitė, V., Virkšaitė, A., Gudelytė, G., Navakauskienė, R., 2018. Green tea polyphenol EGCG causes anti-cancerous epigenetic modulations in acute promyelocytic leukemia cells. Leuk. Lymphoma 59 (2), 469–478. https://doi.org/10.1080/10428194.2017.1339881.

Bozdag, G., Mumusoglu, S., Zengin, D., Karabulut, E., Yildiz, B.O., 2016. The prevalence and phenotypic features of polycystic ovary syndrome: a systematic review and meta-analysis. Hum. Reprod. 31 (12), 2841–2855. https://doi.org/10.1093/humrep/dew218.

Brglez Mojzer, E., Knez Hrnčič, M., Škerget, M., Knez, Ž., Bren, U., 2016. Polyphenols: extraction methods, antioxidative action, bioavailability and anticarcinogenic effects. Molecules 21 (7), 901. https://doi.org/10.3390/molecules21070901.

Broughton, D.E., Moley, K.H., 2017. Obesity and female infertility: potential mediators of obesity's impact. Fertil. Steril. 107 (4), 840–847. https://doi.org/10.1016/j.fertnstert.2017.01.017.

Canfora, E.E., Meex, R.C.R., Venema, K., Blaak, E.E., 2019. Gut microbial metabolites in obesity, NAFLD and T2DM. Nat. Rev. Endocrinol. 15 (5), 261–273. https://doi.org/10.1038/s41574-019-0156-z.

Cao, M., Zhang, Z., Han, S., Lu, X., 2019. Butyrate inhibits the proliferation and induces the apoptosis of colorectal cancer HCT116 cells via the deactivation of mTOR/S6K1 signaling mediated partly by SIRT1 downregulation. Mol. Med. Rep. 19 (5), 3941–3947. https://doi.org/10.3892/mmr.2019.10002.

Carmina, E., 2015. Reproductive system outcome among patients with polycystic ovarian syndrome. Endocrinol Metab. Clin. N. Am. 44 (4), 787–797. https://doi.org/10.1016/j.ecl.2015.07.006.

Carmody, R.N., Turnbaugh, P.J., 2014. Host-microbial interactions in the metabolism of therapeutic and diet-derived xenobiotics. J. Clin. Invest. 124 (10), 4173–4181. https://doi.org/10.1172/jci72335.

Carrer, A., Parris, J.L.D., Trefely, S., Henry, R.A., Montgomery, D.C., Torres, A., Viola, J.M., Kuo, Y.-M., Blair, I.A., Meier, J.L., 2017. Impact of a high-fat diet on tissue acyl-CoA and histone acetylation levels. J. Biol. Chem. 292 (8), 3312–3322. https://doi.org/10.1074/jbc.m116.750620.

Cassar, S., Misso, M.L., Hopkins, W.G., Shaw, C.S., Teede, H.J., Stepto, N.K., 2016. Insulin resistance in polycystic ovary syndrome: a systematic review and meta-analysis of euglycaemic—hyperinsulinaemic clamp studies. Hum. Reprod. 31 (11), 2619—2631. https://doi.org/10.1093/humrep/dew243.

Cesta, C.E., Månsson, M., Palm, C., Lichtenstein, P., Iliadou, A.N., Landén, M., 2016. Polycystic ovary syndrome and psychiatric disorders: co-morbidity and heritability in a nationwide Swedish cohort. Psychoneuroendocrinology 73, 196—203. https://doi.org/10.1016/j.psyneuen.2016.08.005.

Chen, B., Xu, P., Wang, J., Zhang, C., 2019a. The role of MiRNA in polycystic ovary syndrome (PCOS). Gene 706, 91—96. https://doi.org/10.1016/j.gene.2019.04.082.

Chen, J., Zhao, K.-N., Vitetta, L., 2019b. Effects of intestinal microbial—elaborated butyrate on oncogenic signaling pathways. Nutrients 11 (5), 1026. https://doi.org/10.3390/nu11051026.

Cheng, Z., Zheng, L., Almeida, F.A., 2018. Epigenetic reprogramming in metabolic disorders: nutritional factors and beyond. J. Nutr. Biochem. 54, 1—10. https://doi.org/10.1016/j.jnutbio.2017.10.004.

Chriett, S., Dąbek, A., Wojtala, M., Vidal, H., Balcerczyk, A., Pirola, L., 2019. Prominent action of butyrate over β-hydroxybutyrate as histone deacetylase inhibitor, transcriptional modulator and anti-inflammatory molecule. Sci. Rep. 9 (1). https://doi.org/10.1038/s41598-018-36941-9.

Churov, A., Summerhill, V., Grechko, A., Orekhova, V., Orekhov, A., 2019. MicroRNAs as potential biomarkers in atherosclerosis. Int. J. Mol. Sci. 20 (22), 5547. https://doi.org/10.3390/ijms20225547.

Claesson, M.J., Jeffery, I.B., Conde, S., Power, S.E., O'Connor, E.M., Cusack, S., Harris, H.M., Coakley, M., Lakshminarayanan, B., O'Sullivan, O., Fitzgerald, G.F., Deane, J., O'Connor, M., Harnedy, N., O'Connor, K., O'Mahony, D., van Sinderen, D., Wallace, M., Brennan, L., Stanton, C., Marchesi, J.R., Fitzgerald, A.P., Shanahan, F., Hill, C., Ross, R.P., O'Toole, P.W., 2012. Gut microbiota composition correlates with diet and health in the elderly. Nature 488 (7410), 178—184. https://doi.org/10.1038/nature11319.

Collado, M.C., Rautava, S., Aakko, J., Isolauri, E., Salminen, S., 2016. Human gut colonisation may be initiated in utero by distinct microbial communities in the placenta and amniotic fluid. Sci. Rep. 6 (1). https://doi.org/10.1038/srep23129.

Cong, X., Xu, W., Janton, S., Henderson, W.A., Matson, A., McGrath, J.M., Maas, K., Graf, J., 2016. Gut microbiome developmental patterns in early life of preterm infants: impacts of feeding and gender. PLoS One 11 (4), e0152751. https://doi.org/10.1371/journal.pone.0152751.

Crisosto, N., Echiburú, B., Maliqueo, M., Luchsinger, M., Rojas, P., Recabarren, S., Sir-Petermann, T., 2017. Reproductive and metabolic features during puberty in sons of women with polycystic ovary syndrome. Endocr. Connect. 6 (8), 607—613. https://doi.org/10.1530/ec-17-0218.

Crisosto, N., Ladrón de Guevara, A., Echiburú, B., Maliqueo, M., Cavada, G., Codner, E., Paez, F., Sir-Petermann, T., 2019. Higher luteinizing hormone levels associated with antimüllerian hormone in postmenarchal daughters of women with polycystic ovary syndrome. Fertil. Steril. 111 (2), 381—388. https://doi.org/10.1016/j.fertnstert.2018.10.011.

Cui, L., Zhao, H., Zhang, B., Qu, Z., Liu, J., Liang, X., Zhao, X., Zhao, J., Sun, Y., Wang, P., 2013. Genotype—phenotype correlations of PCOS susceptibility SNPs identified by GWAS in a large cohort of Han Chinese women. Hum. Reprod. 28 (2), 538—544. https://doi.org/10.1093/humrep/des424.

Cummings, J.H., Pomare, E.W., Branch, W.J., Naylor, C.P., Macfarlane, G.T., 1987. Short chain fatty acids in human large intestine, portal, hepatic and venous blood. Gut 28 (10), 1221—1227. https://doi.org/10.1136/gut.28.10.1221.

Dalile, B., Van Oudenhove, L., Vervliet, B., Verbeke, K., 2019. The role of short-chain fatty acids in microbiota—gut—brain communication. Nat. Rev. Gastroenterol. Hepatol. 16 (8), 461—478. https://doi.org/10.1038/s41575-019-0157-3.

Damiani, E., Wallace, H.M., 2018. Polyamines and Cancer, pp. 469—488. https://doi.org/10.1007/978-1-4939-7398-9_39.

Daneshpajooh, M., Bacos, K., Bysani, M., Bagge, A., Ottosson Laakso, E., Vikman, P., Eliasson, L., Mulder, H., Ling, C., 2017. HDAC7 is overexpressed in human diabetic islets and impairs insulin secretion in rat islets and clonal beta cells. Diabetologia 60 (1), 116—125. https://doi.org/10.1007/s00125-016-4113-2.

Dapas, M., Sisk, R., Legro, R.S., Urbanek, M., Dunaif, A., Hayes, M.G., 2019. Family-based quantitative trait meta-analysis implicates rare noncoding variants in DENND1A in polycystic ovary syndrome. J. Clin. Endocrinol. Metabol. 104 (9), 3835—3850. https://doi.org/10.1210/jc.2018-02496.

DaSilva, N.A., Nahar, P.P., Ma, H., Eid, A., Wei, Z., Meschwitz, S., Zawia, N.H., Slitt, A.L., Seeram, N.P., 2019. Pomegranate ellagitannin-gut microbial-derived metabolites, urolithins, inhibit neuroinflammation in vitro. Nutr. Neurosci. 22 (3), 185—195. https://doi.org/10.1080/1028415X.2017.1360558.

David, L.A., Maurice, C.F., Carmody, R.N., Gootenberg, D.B., Button, J.E., Wolfe, B.E., Ling, A.V., Devlin, A.S., Varma, Y., Fischbach, M.A., 2014. Diet rapidly and reproducibly alters the human gut microbiome. Nature 505 (7484), 559—563. https://doi.org/10.1038/nature12820.

Day, F., Karaderi, T., Jones, M.R., Meun, C., He, C., Drong, A., Kraft, P., Lin, N., Huang, H., Broer, L., 2018. Large-scale genome-wide meta-analysis of polycystic ovary syndrome suggests shared genetic architecture for different diagnosis criteria. PLoS Genet. 14 (12), e1007813. https://doi.org/10.1371/journal.pgen.1007813.

De Vadder, F., Grasset, E., Mannerås Holm, L., Karsenty, G., Macpherson, A.J., Olofsson, L.E., Bäckhed, F., 2018. Gut microbiota regulates maturation of the adult enteric nervous system via enteric serotonin networks. Proc. Natl. Acad. Sci. U.S.A. 115 (25), 6458—6463. https://doi.org/10.1073/pnas.1720017115.

Deb, G., Shankar, E., Thakur, V.S., Ponsky, L.E., Bodner, D.R., Fu, P., Gupta, S., 2019. Green tea-induced epigenetic reactivation of tissue inhibitor of matrix metalloproteinase-3 suppresses prostate cancer progression through histone-modifying enzymes. Mol. Carcinog. 58 (7), 1194—1207. https://doi.org/10.1002/mc.23003.

Ding, Y., He, P., Li, Z., 2020. MicroRNA-9119 regulates cell viability of granulosa cells in polycystic ovarian syndrome via mediating Dicer expression. Mol. Cell. Biochem. 465 (1—2), 187—197. https://doi.org/10.1007/s11010-019-03678-6.

Dogra, S., Sakwinska, O., Soh, S.-E., Ngom-Bru, C., Brück, W.M., Berger, B., Brüssow, H., Lee, Y.S., Yap, F., Chong, Y.-S., 2015. Dynamics of infant gut microbiota are influenced by delivery mode and gestational duration and are associated with subsequent adiposity. mBio 6 (1). https://doi.org/10.1128/mbio.02419-14.

Dunaif, A., Finegood, D.T., 1996. Beta-cell dysfunction independent of obesity and glucose intolerance in the polycystic ovary syndrome. J. Clin. Endocrinol. Metabol. 81 (3), 942—947. https://doi.org/10.1210/jcem.81.3.8772555.

Dunaif, A., 2016. Perspectives in polycystic ovary syndrome: from hair to eternity. J. Clin. Endocrinol. Metabol. 101 (3), 759—768. https://doi.org/10.1210/jc.2015-3780.

Ecklund, L.C., Usadi, R.S., 2015. Endocrine and reproductive effects of polycystic ovarian syndrome. Obstet. Gynecol. Clin. North Am. 42 (1), 55—65.

Eckschlager, T., Plch, J., Stiborova, M., Hrabeta, J., 2017. Histone deacetylase inhibitors as anticancer drugs. Int. J. Mol. Sci. 18 (7), 1414. https://doi.org/10.3390/ijms18071414.

Egshatyan, L., Kashtanova, D., Popenko, A., Tkacheva, O., Tyakht, A., Alexeev, D., Karamnova, N., Kostryukova, E., Babenko, V., Vakhitova, M., 2016. Gut microbiota and diet in patients with different glucose tolerance. Endocr. Connect. 5 (1), 1—9. https://doi.org/10.1530/EC-15-0094.

Escobar-Morreale, H.F., 2018. Polycystic ovary syndrome: definition, aetiology, diagnosis and treatment. Nat. Rev. Endocrinol. 14 (5), 270—284. https://doi.org/10.1038/nrendo.2018.24.

Evans, C.C., LePard, K.J., Kwak, J.W., Stancukas, M.C., Laskowski, S., Dougherty, J., Moulton, L., Glawe, A., Wang, Y., Leone, V., 2014. Exercise prevents weight gain and alters the gut microbiota in a mouse model of high fat diet-induced obesity. PLoS One 9 (3), e92193. https://doi.org/10.1371/journal.pone.0092193.

Everard, A., Belzer, C., Geurts, L., Ouwerkerk, J.P., Druart, C., Bindels, L.B., Guiot, Y., Derrien, M., Muccioli, G.G., Delzenne, N.M., 2013. Cross-talk between Akkermansia muciniphila and intestinal epithelium controls diet-induced obesity. Proc. Natl. Acad. Sci. USA 110 (22), 9066—9071. https://doi.org/10.1073/pnas.1219451110.

Everard, A., Lazarevic, V., Derrien, M., Girard, M., Muccioli, G.G., Neyrinck, A.M., Possemiers, S., Van Holle, A., François, P., de Vos, W.M., 2011. Responses of gut microbiota and glucose and lipid metabolism to prebiotics in genetic obese and diet-induced leptin-resistant mice. Diabetes 60 (11), 2775—2786. https://doi.org/10.2337/db11-0227.

Fallucca, F., Porrata, C., Fallucca, S., Pianesi, M., 2014. Influence of diet on gut microbiota, inflammation and type 2 diabetes mellitus. First experience with macrobiotic Ma-Pi 2 diet. Diabetes Metab. Res. Rev. 30 (S1), 48–54. https://doi.org/10.1002/dmrr.2518.

Fischer, A., Sananbenesi, F., Wang, X., Dobbin, M., Tsai, L.-H., 2007. Recovery of learning and memory is associated with chromatin remodelling. Nature 447 (7141), 178–182. https://doi.org/10.1038/nature05772.

Flores-Sierra, J., Arredondo-Guerrero, M., Cervantes-Paz, B., Rodríguez-Ríos, D., Alvarado-Caudillo, Y., Nielsen, F.C., Wrobel, K., Wrobel, K., Zaina, S., Lund, G., 2016. The trans fatty acid elaidate affects the global DNA methylation profile of cultured cells and in vivo. Lipids Health Dis. 15 (1). https://doi.org/10.1186/s12944-016-0243-2.

Gabert, L., Vors, C., Louche-Pélissier, C., Sauvinet, V., Lambert-Porcheron, S., Drai, J., Laville, M., Désage, M., Michalski, M.-C., 2011. 13C tracer recovery in human stools after digestion of a fat-rich meal labelled with [1,1,1-13C3]tripalmitin and [1,1,1-13C3]triolein. Rapid Commun. Mass Spectrom. 25 (19), 2697–2703. https://doi.org/10.1002/rcm.5067.

García-Giménez, J.L., Romá-Mateo, C., Pallardó, F.V., 2019. Oxidative post-translational modifications in histones. Biofactors 45 (5), 641–650. https://doi.org/10.1002/biof.1532.

Gavin, D.P., Kusumo, H., Zhang, H., Guidotti, A., Pandey, S.C., 2016. Role of growth arrest and DNA damage-inducible, beta in alcohol-drinking behaviors. Alcohol Clin. Exp. Res. 40 (2), 263–272. https://doi.org/10.1111/acer.12965.

Gentile, C.L., Weir, T.L., 2018. The gut microbiota at the intersection of diet and human health. Science 362 (6416), 776–780. https://doi.org/10.1126/science.aau5812.

Gerhauser, C., 2018. Impact of dietary gut microbial metabolites on the epigenome. Phil. Trans. R. Soc. B 373 (1748), 20170359. https://doi.org/10.1098/rstb.2017.0359.

Ghaffarzad, A., Amani, R., Mehrzad, S.M., Darabi, M., Cheraghian, B., 2016. Correlation of serum lipoprotein ratios with insulin resistance in infertile women with polycystic ovarian syndrome: a case control study. Int. J. Fertil. Steril. 10 (1), 29–35. https://doi.org/10.22074/ijfs.2016.4765.

Glintborg, D., Jensen, R.C., Schmedes, A.V., Brandslund, I., Kyhl, H.B., Jensen, T.K., Andersen, M.S., 2019. Anogenital distance in children born of mothers with polycystic ovary syndrome: the Odense child cohort. Hum. Reprod. 34 (10), 2061–2070. https://doi.org/10.1093/humrep/dez122.

Gorsic, L.K., Dapas, M., Legro, R.S., Hayes, M.G., Urbanek, M., 2019. Functional genetic variation in the anti-müllerian hormone pathway in women with polycystic ovary syndrome. J. Clin. Endocrinol. Metabol. 104 (7), 2855–2874. https://doi.org/10.1210/jc.2018-02178.

Guastella, E., Longo, R.A., Carmina, E., 2010. Clinical and endocrine characteristics of the main polycystic ovary syndrome phenotypes. Fertil. Steril. 94 (6), 2197–2201. https://doi.org/10.1016/j.fertnstert.2010.02.014.

Guo, X., Puttabyatappa, M., Domino, S.E., Padmanabhan, V., 2020. Developmental programming: prenatal testosterone-induced changes in epigenetic modulators and gene expression in metabolic tissues of female sheep. Mol. Cell. Endocrinol. 514, 110913. https://doi.org/10.1016/j.mce.2020.110913.

Guo, Y., Qi, Y., Yang, X., Zhao, L., Wen, S., Liu, Y., Tang, L., 2016. Association between polycystic ovary syndrome and gut microbiota. PLoS One 11 (4), e0153196. https://doi.org/10.1371/journal.pone.0153196.

Gurwara, S., Ajami, N., Jang, A., Hessel, F., Chen, L., Plew, S., Wang, Z., Graham, D., Hair, C., White, D., 2019. Dietary nutrients involved in one-carbon metabolism and colonic mucosa-associated gut microbiome in individuals with an endoscopically normal colon. Nutrients 11 (3), 613. https://doi.org/10.3390/nu11030613.

Handa, H., Murakami, Y., Ishihara, R., Kimura-Masuda, K., Masuda, Y., 2019. The role and function of microRNA in the pathogenesis of multiple myeloma. Cancers 11 (11), 1738. https://doi.org/10.3390/cancers11111738.

Hiam, D., Simar, D., Laker, R., Altıntaş, A., Gibson-Helm, M., Fletcher, E., Moreno-Asso, A., Trewin, A.J., Barres, R., Stepto, N.K., 2019. Epigenetic reprogramming of immune cells in women with PCOS impact genes controlling reproductive function. J. Clin. Endocrinol. Metabol. 104 (12), 6155–6170. https://doi.org/10.1210/jc.2019-01015.

Ho, J., Nicolucci, A.C., Virtanen, H., Schick, A., Meddings, J., Reimer, R.A., Huang, C., 2019. Effect of prebiotic on microbiota, intestinal permeability, and glycemic control in children with type 1 diabetes. J. Clin. Endocrinol. Metabol. 104 (10), 4427—4440. https://doi.org/10.1210/jc.2019-00481.

Hollister, E.B., Riehle, K., Luna, R.A., Weidler, E.M., Rubio-Gonzales, M., Mistretta, T.A., Raza, S., Doddapaneni, H.V., Metcalf, G.A., Muzny, D.M., Gibbs, R.A., Petrosino, J.F., Shulman, R.J., Versalovic, J., 2015. Structure and function of the healthy pre-adolescent pediatric gut microbiome. Microbiome 3 (36). https://doi.org/10.1186/s40168-015-0101-x.

Hosseini, E., Shahhoseini, M., Afsharian, P., Karimian, L., Ashrafi, M., Mehraein, F., Afatoonian, R., 2019. Role of epigenetic modifications in the aberrant CYP19A1 gene expression in polycystic ovary syndrome. Aoms 15 (4), 887—895. https://doi.org/10.5114/aoms.2019.86060.

Hu, M., Zheng, S., Yin, H., Zhu, X., Lu, F., Tong, X.-H., Liu, Y.-S., Zhang, Y., Xu, B., 2020. Identification of microRNAs that regulate the MAPK pathway in human cumulus cells from PCOS women with insulin resistance. Reprod. Sci. 27 (3), 833—844. https://doi.org/10.1007/s43032-019-00086-5.

Huang, A., Brennan, K., Azziz, R., 2010. Prevalence of hyperandrogenemia in the polycystic ovary syndrome diagnosed by the National Institutes of Health 1990 criteria. Fertil. Steril. 93 (6), 1938—1941. https://doi.org/10.1016/j.fertnstert.2008.12.138.

Huang, C.-C., Chen, M.-J., Lan, C.-W., Wu, C.-E., Huang, M.-C., Kuo, H.-C., Ho, H.-N., 2019. Hyperactive CREB signaling pathway involved in the pathogenesis of polycystic ovarian syndrome revealed by patient-specific induced pluripotent stem cell modeling. Fertil. Steril. 112 (3), 594—607.e12. https://doi.org/10.1016/j.fertnstert.2019.05.004.

Huang, Q., Mo, M., Zhong, Y., Yang, Q., Zhang, J., Ye, X., Zhang, L., Cai, C., 2019b. The anticancer role of omega-3 polyunsaturated fatty acids was closely associated with the increase in genomic DNA hydroxymethylation. Anti-Cancer Agent Med. Chem. 19 (3), 330—336. https://doi.org/10.2174/1871520618666181018143026.

Hullar, M.A.J., Fu, B.C., 2014. Diet, the gut microbiome, and epigenetics. Cancer J. 20 (3), 170—175. https://doi.org/10.1097/PPO.0000000000000053.

Igarashi, K., Kashiwagi, K., 2019. The functional role of polyamines in eukaryotic cells. Int. J. Biochem. Cell Biol. 107, 104—115. https://doi.org/10.1016/j.biocel.2018.12.012.

Imran, A., Butt, M.S., Arshad, M.S., Arshad, M.U., Saeed, F., Sohaib, M., Munir, R., 2018. Exploring the potential of black tea based flavonoids against hyperlipidemia related disorders. Lipids Health Dis. 17 (1). https://doi.org/10.1186/s12944-018-0688-6.

Insenser, M., Murri, M., del Campo, R., Martínez-García, M.Á., Fernández-Durán, E., Escobar-Morreale, H.F., 2018. Gut microbiota and the polycystic ovary syndrome: influence of sex, sex hormones, and obesity. J. Clin. Endocrinol. Metabol. 103 (7), 2552—2562. https://doi.org/10.1210/jc.2017-02799.

Jacobsen, V.M., Li, S., Wang, A., Zhu, D., Liu, M., Thomassen, M., Kruse, T., Tan, Q., 2019. Epigenetic association analysis of clinical sub-phenotypes in patients with polycystic ovary syndrome (PCOS). Gynecol. Endocrinol. 35 (8), 691—694. https://doi.org/10.1080/09513590.2019.1576617.

Jambhekar, A., Dhall, A., Shi, Y., 2019. Roles and regulation of histone methylation in animal development. Nat. Rev. Mol. Cell Biol. 20 (10), 625—641. https://doi.org/10.1038/s41580-019-0151-1.

Jia, W., Xie, G., Jia, W., 2018. Bile acid—microbiota crosstalk in gastrointestinal inflammation and carcinogenesis. Nat. Rev. Gastroenterol. Hepatol. 15 (2), 111—128. https://doi.org/10.1038/nrgastro.2017.119.

Jiang, B., Xue, M., Xu, D., Song, Y., Zhu, S., 2020b. Upregulation of microRNA-204 improves insulin resistance of polycystic ovarian syndrome via inhibition of HMGB1 and the inactivation of the TLR4/NF-κB pathway. Cell Cycle 19 (6), 697—710. https://doi.org/10.1080/15384101.2020.1724601.

Jiang, L., Huang, H., Qian, Y., Li, Y., Chen, X., Di, N., Yang, D., 2020a. miR-130b regulates gap junctional intercellular communication through connexin 43 in granulosa cells from patients with polycystic ovary syndrome. Mol. Hum. Reprod. 26 (8), 576—584. https://doi.org/10.1093/molehr/gaaa044.

Jiao, J., Sagnelli, M., Shi, B., Fang, Y., Shen, Z., Tang, T., Dong, B., Li, D., Wang, X., 2019. Genetic and epigenetic characteristics in ovarian tissues from polycystic ovary syndrome patients with irregular menstruation resemble those of ovarian cancer. BMC Endocr. Disord. 19 (1). https://doi.org/10.1186/s12902-019-0356-5.

Jones, H., Sprung, V.S., Pugh, C.J.A., Daousi, C., Irwin, A., Aziz, N., Adams, V.L., Thomas, E.L., Bell, J.D., Kemp, G.J., 2012. Polycystic ovary syndrome with hyperandrogenism is characterized by an increased risk of hepatic steatosis compared to nonhyperandrogenic PCOS phenotypes and healthy controls, independent of obesity and insulin resistance. J. Clin. Endocrinol. Metabol. 97 (10), 3709—3716. https://doi.org/10.1210/jc.2012-1382.

Jung, G., Hernández-Illán, E., Moreira, L., Balaguer, F., Goel, A., 2020. Epigenetics of colorectal cancer: biomarker and therapeutic potential. Nat. Rev. Gastroenterol. Hepatol. 17 (2), 111—130. https://doi.org/10.1038/s41575-019-0230-y.

Kasubuchi, M., Hasegawa, S., Hiramatsu, T., Ichimura, A., Kimura, I., 2015. Dietary gut microbial metabolites, short-chain fatty acids, and host metabolic regulation. Nutrients 7 (4), 2839—2849. https://doi.org/10.3390/nu7042839.

Kelley, S.T., Skarra, D.V., Rivera, A.J., Thackray, V.G., 2016. The gut microbiome is altered in a letrozole-induced mouse model of polycystic ovary syndrome. PLoS One 11 (1), e0146509. https://doi.org/10.1371/journal.pone.0146509.

Kersten, S., Stienstra, R., 2017. The role and regulation of the peroxisome proliferator activated receptor alpha in human liver. Biochimie 136, 75—84. https://doi.org/10.1016/j.biochi.2016.12.019.

Kim, M.-J., Lim, N.-K., Choi, Y.-M., Kim, J.-J., Hwang, K.-R., Chae, S.-J., Park, C.-W., Choi, D.-S., Kang, B.-M., Lee, B.-S., 2014. Prevalence of metabolic syndrome is higher among non-obese PCOS women with hyperandrogenism and menstrual irregularity in Korea. PLoS One 9 (6), e99252. https://doi.org/10.1371/journal.pone.0099252.

Koh, A., De Vadder, F., Kovatcheva-Datchary, P., Bäckhed, F., 2016. From dietary fiber to host physiology: short-chain fatty acids as key bacterial metabolites. Cell 165 (6), 1332—1345. https://doi.org/10.1016/j.cell.2016.05.041.

Kokosar, M., Benrick, A., Perfilyev, A., Fornes, R., Nilsson, E., Maliqueo, M., Behre, C.J., Sazonova, A., Ohlsson, C., Ling, C., 2016. Epigenetic and transcriptional alterations in human adipose tissue of polycystic ovary syndrome. Sci. Rep. 6 (1). https://doi.org/10.1038/srep22883.

Krautkramer, K.A., Kreznar, J.H., Romano, K.A., Vivas, E.I., Barrett-Wilt, G.A., Rabaglia, M.E., Keller, M.P., Attie, A.D., Rey, F.E., Denu, J.M., 2016. Diet-microbiota interactions mediate global epigenetic programming in multiple host tissues. Mol. Cell 64 (5), 982—992. https://doi.org/10.1016/j.molcel.2016.10.025.

Krautkramer, K.A., Dhillon, R.S., Denu, J.M., Carey, H.V., 2017. Metabolic programming of the epigenome: host and gut microbial metabolite interactions with host chromatin. Transl. Res. 189, 30—50. https://doi.org/10.1016/j.trsl.2017.08.005.

Kübeck, R., Bonet-Ripoll, C., Hoffmann, C., Walker, A., Müller, V.M., Schüppel, V.L., Lagkouvardos, I., Scholz, B., Engel, K.-H., Daniel, H., 2016. Dietary fat and gut microbiota interactions determine diet-induced obesity in mice. Mol. Metabol. 5 (12), 1162—1174. https://doi.org/10.1016/j.molmet.2016.10.001.

Kulkarni, S.D., Patil, A.N., Gudi, A., Homburg, R., Conway, G.S., 2019. Changes in diet composition with urbanization and its effect on the polycystic ovarian syndrome phenotype in a Western Indian population. Fertil. Steril. 112 (4), 758—763.

Kumar, H., Lund, R., Laiho, A., Lundelin, K., Ley, R.E., Isolauri, E., Salminen, S., 2014. Gut microbiota as an epigenetic regulator: pilot study based on whole-genome methylation analysis. mBio 5 (6). https://doi.org/10.1128/mBio.02113-14.

Kumar, D.R.N., Seshadri, K.G., Pandurangi, M., 2017. Effect of metformin-sustained release therapy on low-density lipoprotein size and adiponectin in the South Indian women with polycystic ovary syndrome. Indian J. Endocrinol. Metab. 21 (5), 679—683.

Kura, B., Parikh, M., Slezak, J., Pierce, G.N., 2019. The influence of diet on microRNAs that impact cardiovascular disease. Molecules 24 (8), 1509. https://doi.org/10.3390/molecules24081509.

Lacagnina, S., 2020. The developmental origins of health and disease (DOHaD). Am. J. Lifestyle Med. 14 (1), 47—50. https://doi.org/10.1177/1559827619879694.

Leeming, E.R., Johnson, A.J., Spector, T.D., Le Roy, C.I., 2019. Effect of diet on the gut microbiota: rethinking intervention duration. Nutrients 11 (12), 2862. https://doi.org/10.3390/nu11122862.

Lernoux, M., Schnekenburger, M., Dicato, M., Diederich, M., 2018. Anti-cancer effects of naturally derived compounds targeting histone deacetylase 6-related pathways. Pharmacol. Res. 129, 337–356. https://doi.org/10.1016/j.phrs.2017.11.004.

Lewis, B.P., Burge, C.B., Bartel, D.P., 2005. Conserved seed pairing, often flanked by adenosines, indicates that thousands of human genes are microRNA targets. Cell 120 (1), 15–20. https://doi.org/10.1016/j.cell.2004.12.035.

Lim, M.Y., You, H.J., Yoon, H.S., Kwon, B., Lee, J.Y., Lee, S., Song, Y.-M., Lee, K., Sung, J., Ko, G., 2017. The effect of heritability and host genetics on the gut microbiota and metabolic syndrome. Gut 66 (6), 1031–1038. https://doi.org/10.1136/gutjnl-2015-311326.

Lindheim, L., Bashir, M., Münzker, J., Trummer, C., Zachhuber, V., Leber, B., Horvath, A., Pieber, T.R., Gorkiewicz, G., Stadlbauer, V., 2017. Alterations in gut microbiome composition and barrier function are associated with reproductive and metabolic defects in women with polycystic ovary syndrome (PCOS): a pilot study. PLoS One 12 (1), e0168390. https://doi.org/10.1371/journal.pone.0168390.

Liu, H., Wang, J., He, T., Becker, S., Zhang, G., Li, D., Ma, X., 2018. Butyrate: a double-edged sword for health? Adv. Nutr. 9 (1), 21–29. https://doi.org/10.1093/advances/nmx009.

Liu, L., He, D., Wang, Y., Sheng, M., 2020. Integrated analysis of DNA methylation and transcriptome profiling of polycystic ovary syndrome. Mol. Med. Rep. 21 (5), 2138–2150. https://doi.org/10.3892/mmr.2020.11005.

Liu, R., Zhang, C., Shi, Y., Zhang, F., Li, L., Wang, X., Ling, Y., Fu, H., Dong, W., Shen, J., 2017. Dysbiosis of gut microbiota associated with clinical parameters in polycystic ovary syndrome. Front. Microbiol. 8. https://doi.org/10.3389/fmicb.2017.00324.

Lizneva, D., Suturina, L., Walker, W., Brakta, S., Gavrilova-Jordan, L., Azziz, R., 2016. Criteria, prevalence, and phenotypes of polycystic ovary syndrome. Fertil. Steril. 106 (1), 6–15. https://doi.org/10.1016/j.fertnstert.2016.05.003.

Louis, P., Duncan, S.H., McCrae, S.I., Millar, J., Jackson, M.S., Flint, H.J., 2004. Restricted distribution of the butyrate kinase pathway among butyrate-producing bacteria from the human colon. J. Biochem. 186 (7), 2099–2106. https://doi.org/10.1128/jb.186.7.2099-2106.2004.

Louis, P., Hold, G.L., Flint, H.J., 2014. The gut microbiota, bacterial metabolites and colorectal cancer. Nat. Rev. Microbiol. 12 (10), 661–672. https://doi.org/10.1038/nrmicro3344.

Louis, P., Young, P., Holtrop, G., Flint, H.J., 2010. Diversity of human colonic butyrate-producing bacteria revealed by analysis of the butyryl-CoA:acetate CoA-transferase gene. Environ. Microbiol. 12 (2), 304–314. https://doi.org/10.1111/j.1462-2920.2009.02066.x.

Lu, Y., Fan, C., Liang, A., Fan, X., Wang, R., Li, P., Qi, K., 2018. Effects of SCFA on the DNA methylation pattern of adiponectin and resistin in high-fat-diet-induced obese male mice. Br. J. Nutr. 120 (4), 385–392. https://doi.org/10.1017/S0007114518001526.

Macfarlane, S., Macfarlane, G.T., 2003. Regulation of short-chain fatty acid production. Proc. Nutr. Soc. 62 (1), 67–72. https://doi.org/10.1079/pns2002207.

Macut, D., Bjekic-Macut, J., Savic-Radojevic, A., 2013. Dyslipidemia and Oxidative Stress in PCOS, pp. 51–63. https://doi.org/10.1159/000341683.

Makrinou, E., Drong, A.W., Christopoulos, G., Lerner, A., Chapa-Chorda, I., Karaderi, T., Lavery, S., Hardy, K., Lindgren, C.M., Franks, S., 2020. Genome-wide methylation profiling in granulosa lutein cells of women with polycystic ovary syndrome (PCOS). Mol. Cell. Endocrinol. 500, 110611. https://doi.org/10.1016/j.mce.2019.110611.

Maliqueo, M., Sundstrom Poromaa, I., Vanky, E., Fornes, R., Benrick, A., Akerud, H., Stridsklev, S., Labrie, F., Jansson, T., Stener-Victorin, E., 2015. Placental STAT3 signaling is activated in women with polycystic ovary syndrome. Hum. Reprod. 30 (3), 692–700. https://doi.org/10.1093/humrep/deu351.

Mandaviya, P.R., Joehanes, R., Brody, J., Castillo-Fernandez, J.E., Dekkers, K.F., Do, A.N., Graff, M., Hänninen, I.K., Tanaka, T., de Jonge, E.A.L., 2019. Association of dietary folate and vitamin B-12 intake with genome-wide DNA methylation in blood: a large-scale epigenome-wide association analysis in 5841 individuals. Am. J. Clin. Nutr. 110 (2), 437–450. https://doi.org/10.1093/ajcn/nqz031.

March, W.A., Moore, V.M., Willson, K.J., Phillips, D.I.W., Norman, R.J., Davies, M.J., 2010. The prevalence of polycystic ovary syndrome in a community sample assessed under contrasting diagnostic criteria. Hum. Reprod. 25 (2), 544—551. https://doi.org/10.1093/humrep/dep399.

Mason, H.D., Willis, D.S., Beard, R.W., Winston, R.M., Margara, R., Franks, S., 1994. Oestradiol production by granulosa cells of normal and polycystic ovaries: relationship to menstrual cycle history and concentrations of gonadotropins and sex steroids in follicular fluid. J. Clin. Endocrinol. Metabol. 79 (5), 1355—1360. https://doi.org/10.1210/jcem.79.5.7962330.

Mayer, E.A., Tillisch, K., Gupta, A., 2015. Gut/brain axis and the microbiota. J. Clin. Invest. 125 (3), 926—938. https://doi.org/10.1172/jci76304.

McAllister, J.M., Modi, B., Miller, B.A., Biegler, J., Bruggeman, R., Legro, R.S., Strauss, J.F., 2014. Overexpression of a DENND1A isoform produces a polycystic ovary syndrome theca phenotype. Proc. Natl. Acad. Sci. USA 111 (15), E1519—E1527. https://doi.org/10.1073/pnas.1400574111.

McAllister, J.M., Han, A.X., Modi, B.P., Teves, M.E., Mavodza, G.R., Anderson, Z.L., Shen, T., Christenson, L.K., Archer, K.J., Strauss, J.F., 2019. miRNA profiling reveals miRNA-130b-3p mediates DENND1A variant 2 expression and androgen biosynthesis. Endocrinology 160 (8), 1964—1981. https://doi.org/10.1210/en.2019-00013.

Miro-Blanch, J., Yanes, O., 2019. Epigenetic regulation at the interplay between gut microbiota and host metabolism. Front. Genet. 10. https://doi.org/10.3389/fgene.2019.00638.

Moret, M., Stettler, R., Rodieux, F., Gaillard, R.C., Waeber, G., Wirthner, D., Giusti, V., Tappy, L., Pralong, F.P., 2009. Insulin modulation of luteinizing hormone secretion in normal female volunteers and lean polycystic ovary syndrome patients. Neuroendocrinology 89 (2), 131—139. https://doi.org/10.1159/000160911.

Moutinho, C., Esteller, M., 2017. MicroRNAs and Epigenetics, pp. 189—220. https://doi.org/10.1016/bs.acr.2017.06.003.

Munford, R.S., 2016. Endotoxemia-menace, marker, or mistake? J. Leukoc. Biol. 100 (4), 687—698. https://doi.org/10.1189/jlb.3RU0316-151R.

Murray-Stewart, T., Woster, P.M., Casero, R.A., 2014. The re-expression of the epigenetically silenced e-cadherin gene by a polyamine analogue lysine-specific demethylase-1 (LSD1) inhibitor in human acute myeloid leukemia cell lines. Amino Acids 46 (3), 585—594. https://doi.org/10.1007/s00726-013-1485-1.

Muse, M.E., Titus, A.J., Salas, L.A., Wilkins, O.M., Mullen, C., Gregory, K.J., Schneider, S.S., Crisi, G.M., Jawale, R.M., Otis, C.N., 2020. Enrichment of CpG island shore region hypermethylation in epigenetic breast field cancerization. Epigenetics 1—14. https://doi.org/10.1080/15592294.2020.1747748.

Nanda, D., Chandrasekaran, S.P., Ramachandran, V., Kalaivanan, K., Carani Venkatraman, A., 2020. Evaluation of serum miRNA-24, miRNA-29a and miRNA-502-3p expression in PCOS subjects: correlation with biochemical parameters related to PCOS and insulin resistance. Ind. J. Clin. Biochem. 35 (2), 169—178. https://doi.org/10.1007/s12291-018-0808-0.

Narita, T., Weinert, B.T., Choudhary, C., 2019. Functions and mechanisms of non-histone protein acetylation. Nat. Rev. Mol. Cell Biol. 20 (3), 156—174. https://doi.org/10.1038/s41580-018-0081-3.

Neven, A., Laven, J., Teede, H., Boyle, J., 2018. A summary on polycystic ovary syndrome: diagnostic criteria, prevalence, clinical manifestations, and management according to the latest international guidelines. Semin. Reprod. Med. 36 (01), 005—012. https://doi.org/10.1055/s-0038-1668085.

Nilsson, E., Benrick, A., Kokosar, M., Krook, A., Lindgren, E., Källman, T., Martis, M.M., Højlund, K., Ling, C., Stener-Victorin, E., 2018. Transcriptional and epigenetic changes influencing skeletal muscle metabolism in women with polycystic ovary syndrome. J. Clin. Endocrinol. Microbiol. 103 (12), 4465—4477. https://doi.org/10.1210/jc.2018-00935.

Noble, D., 2015. Conrad Waddington and the origin of epigenetics. J. Exp. Biol. 218 (6), 816—818. https://doi.org/10.1242/jeb.120071.

O'Reilly, M., Gathercole, L., Capper, F., Arlt, W., Tomlinson, J., 2015. Effect of insulin on AKR1C3 expression in female adipose tissue: in-vivo and in-vitro study of adipose androgen generation in polycystic ovary syndrome. Lancet 385, S16. https://doi.org/10.1016/S0140-6736(15)60331-2.

Olaniyi, K.S., Amusa, O.A., Areola, E.D., Olatunji, L.A., 2020. Suppression of HDAC by sodium acetate rectifies cardiac metabolic disturbance in streptozotocin—nicotinamide-induced diabetic rats. Exp. Biol. Med. 245 (7), 667—676. https://doi.org/10.1177/1535370220913847.

Ottman, N., Smidt, H., de Vos, W.M., Belzer, C., 2012. The function of our microbiota: who is out there and what do they do? Front. Cell. Infect. Microbiol. 2. https://doi.org/10.3389/fcimb.2012.00104.

Pan, J.-X., Tan, Y.-J., Wang, F.-F., Hou, N.-N., Xiang, Y.-Q., Zhang, J.-Y., Liu, Y., Qu, F., Meng, Q., Xu, J., 2018. Aberrant expression and DNA methylation of lipid metabolism genes in PCOS: a new insight into its pathogenesis. Clin. Epigenet. 10 (1). https://doi.org/10.1186/s13148-018-0442-y.

Pascale, A., Marchesi, N., Marelli, C., Coppola, A., Luzi, L., Govoni, S., Giustina, A., Gazzaruso, C., 2018. Microbiota and metabolic diseases. Endocrine 61 (3), 357—371. https://doi.org/10.1007/s12020-018-1605-5.

Patel, R., Shah, G., 2018. High-fat diet exposure from pre-pubertal age induces polycystic ovary syndrome (PCOS) in rats. Reproduction 155 (2), 139—149. https://doi.org/10.1530/rep-17-0584.

Patel, S., 2018. Polycystic ovary syndrome (PCOS), an inflammatory, systemic, lifestyle endocrinopathy. J. Steroid Biochem. Mol. Biol. 182, 27—36. https://doi.org/10.1016/j.jsbmb.2018.04.008.

Perez, M.F., Lehner, B., 2019. Intergenerational and transgenerational epigenetic inheritance in animals. Nat. Cell Biol. 21 (2), 143—151. https://doi.org/10.1038/s41556-018-0242-9.

Perez, N.B., Dorsen, C., Squires, A., 2020. Dysbiosis of the gut microbiome: a concept analysis. J. Holist. Nurs. 38 (2), 223—232. https://doi.org/10.1177/0898010119879527.

Pergialiotis, V., Trakakis, E., Chrelias, C., Papantoniou, N., Hatziagelaki, E., 2018. The Impact of Mild Hypercholesterolemia on Glycemic and Hormonal Profiles, Menstrual Characteristics and the Ovarian Morphology of Women with Polycystic Ovarian Syndrome. https://doi.org/10.1515/hmbci-2018-0002.

Pervin, M., Unno, K., Takagaki, A., Isemura, M., Nakamura, Y., 2019. Function of green tea catechins in the brain: epigallocatechin gallate and its metabolites. Int. J. Mol. Sci. 20 (15), 3630. https://doi.org/10.3390/ijms20153630.

Poulin, M.B., Schneck, J.L., Matico, R.E., McDevitt, P.J., Huddleston, M.J., Hou, W., Johnson, N.W., Thrall, S.H., Meek, T.D., Schramm, V.L., 2016. Transition state for the NSD2-catalyzed methylation of histone H3 lysine 36. Proc. Natl. Acad. Sci. U.S.A. 113 (5), 1197—1201. https://doi.org/10.1073/pnas.1521036113.

Qin, Y., Wade, P.A., 2018. Crosstalk between the microbiome and epigenome: messages from bugs. J. Biochem. 163 (2), 105—112. https://doi.org/10.1093/jb/mvx080.

Quigley, E.M.M., 2017. Basic definitions and concepts: organization of the gut microbiome. Gastroenterol. Clin. N. Am. 46 (1), 1—8. https://doi.org/10.1016/j.gtc.2016.09.002.

Rabot, S., Membrez, M., Bruneau, A., Gérard, P., Harach, T., Moser, M., Raymond, F., Mansourian, R., Chou, C.J., 2010. Germ-free C57BL/6J mice are resistant to high-fat-diet-induced insulin resistance and have altered cholesterol metabolism. Faseb. J. 24 (12), 4948—4959. https://doi.org/10.1096/fj.10-164921.

Ragsdale, S.W., Pierce, E., 2008. Acetogenesis and the wood—ljungdahl pathway of CO_2 fixation. Biochim. Biophys. Acta Protein Proteonomics 1784 (12), 1873—1898. https://doi.org/10.1016/j.bbapap.2008.08.012.

Ramos-Molina, B., Sánchez-Alcoholado, L., Cabrera-Mulero, A., Lopez-Dominguez, R., Carmona-Saez, P., Garcia-Fuentes, E., Moreno-Indias, I., Tinahones, F.J., 2019. Gut microbiota composition is associated with the global DNA methylation pattern in obesity. Front. Genet. 10. https://doi.org/10.3389/fgene.2019.00613.

Ramos-Romero, S., Hereu, M., Atienza, L., Casas, J., Jáuregui, O., Amézqueta, S., Dasilva, G., Medina, I., Nogués, M.R., Romeu, M., 2017. Mechanistically different effects of fat and sugar on insulin resistance, hypertension, and gut microbiota in rats. Am. J. Physiol. Endocrinol. Metabol. 314 (6), E552—E563. https://doi.org/10.1152/ajpendo.00323.2017.

Randeva, H.S., Tan, B.K., Weickert, M.O., Lois, K., Nestler, J.E., Sattar, N., Lehnert, H., 2012. Cardio-metabolic aspects of the polycystic ovary syndrome. Endocr. Rev. 33 (5), 812—841. https://doi.org/10.1210/er.2012-1003.

Rasines-Perea, Z., Teissedre, P.-L., 2017. Grape polyphenols' effects in human cardiovascular diseases and diabetes. Molecules 22 (1), 68. https://doi.org/10.3390/molecules22010068.

Reichardt, N., Duncan, S.H., Young, P., Belenguer, A., McWilliam Leitch, C., Scott, K.P., Flint, H.J., Louis, P., 2014. Phylogenetic distribution of three pathways for propionate production within the human gut microbiota. ISME J. 8 (6), 1323–1335. https://doi.org/10.1038/ismej.2014.14.

Risal, S., Pei, Y., Lu, H., Manti, M., Fornes, R., Pui, H.-P., Zhao, Z., Massart, J., Ohlsson, C., Lindgren, E., 2019. Prenatal androgen exposure and transgenerational susceptibility to polycystic ovary syndrome. Nat. Med. 25 (12), 1894–1904. https://doi.org/10.1038/s41591-019-0666-1.

Rizk, M.G., Thackray, V.G., 2020. Intersection of polycystic ovary syndrome and the gut microbiome. J. Endocr. Soc. 5 (2), bvaa177. https://doi.org/10.1210/jendso/bvaa177.

Roager, H.M., Dragsted, L.O., 2019. Diet-derived microbial metabolites in health and disease. Nutr. Bull. 44 (3), 216–227. https://doi.org/10.1111/nbu.12396.

Rodriguez, D.M., Benninghoff, A.D., Aardema, N.D.J., Phatak, S., Hintze, K.J., 2019. Basal diet determined long-term composition of the gut microbiome and mouse phenotype to a greater extent than fecal microbiome transfer from lean or obese human donors. Nutrients 11 (7), 1630. https://doi.org/10.3390/nu11071630.

Saben, J.L., Boudoures, A.L., Asghar, Z., Thompson, A., Drury, A., Zhang, W., Chi, M., Cusumano, A., Scheaffer, S., Moley, K.H., 2016. Maternal metabolic syndrome programs mitochondrial dysfunction via germline changes across three generations. Cell Rep. 16 (1), 1–8. https://doi.org/10.1016/j.celrep.2016.05.065.

Sagvekar, P., Kumar, P., Mangoli, V., Desai, S., Mukherjee, S., 2019. DNA methylome profiling of granulosa cells reveals altered methylation in genes regulating vital ovarian functions in polycystic ovary syndrome. Clin. Epigenet. 11 (1). https://doi.org/10.1186/s13148-019-0657-6.

Said, H.M., Nexo, E., 2018. Gastrointestinal Handling of Water-Soluble Vitamins, pp. 1291–1311. https://doi.org/10.1002/cphy.c170054.

Sakamoto, A., Terui, Y., Uemura, T., Igarashi, K., Kashiwagi, K., 2020. Polyamines regulate gene expression by stimulating translation of histone acetyltransferase mRNAs. J. Biol. Chem. 295 (26), 8736–8745. https://doi.org/10.1074/jbc.RA120.013833.

Schnorr, S.L., Candela, M., Rampelli, S., Centanni, M., Consolandi, C., Basaglia, G., Turroni, S., Biagi, E., Peano, C., Severgnini, M., 2014. Gut microbiome of the Hadza hunter-gatherers. Nat. Commun. 5 (1). https://doi.org/10.1038/ncomms4654.

Shamasbi, S.G., Ghanbari-Homayi, S., Mirghafourvand, M., 2020. The effect of probiotics, prebiotics, and synbiotics on hormonal and inflammatory indices in women with polycystic ovary syndrome: a systematic review and meta-analysis. Eur. J. Nutr. 59 (2), 433–450. https://doi.org/10.1007/s00394-019-02033-1.

Sherman, S.B., Sarsour, N., Salehi, M., Schroering, A., Mell, B., Joe, B., Hill, J.W., 2018. Prenatal androgen exposure causes hypertension and gut microbiota dysbiosis. Gut Microb. 1–22. https://doi.org/10.1080/19490976.2018.1441664.

Sinha, N., Roy, S., Huang, B., Wang, J., Padmanabhan, V., Sen, A., 2020. Developmental programming: prenatal testosterone-induced epigenetic modulation and its effect on gene expression in sheep ovary. Biol. Reprod. 102 (5), 1045–1054. https://doi.org/10.1093/biolre/ioaa007.

Soda, K., 2020. Spermine and gene methylation: a mechanism of lifespan extension induced by polyamine-rich diet. Amino Acids 52 (2), 213–224. https://doi.org/10.1007/s00726-019-02733-2.

Soliman, M.L., Rosenberger, T.A., 2011. Acetate supplementation increases brain histone acetylation and inhibits histone deacetylase activity and expression. Mol. Cell. Biochem. 352 (1–2), 173–180. https://doi.org/10.1007/s11010-011-0751-3.

Sonnenburg, E.D., Smits, S.A., Tikhonov, M., Higginbottom, S.K., Wingreen, N.S., Sonnenburg, J.L., 2016. Diet-induced extinctions in the gut microbiota compound over generations. Nature 529 (7585), 212–215. https://doi.org/10.1038/nature16504.

Spałkowska, M., Mrozińska, S., Gałuszka-Bednarczyk, A., Gosztyła, K., Przywara, A., Guzik, J., Janeczko, M., Milewicz, T., Wojas-Pelc, A., 2018. The PCOS patients differ in lipid profile according to their phenotypes. Exp. Clin. Endocrinol. Diabetes 126 (07), 437–444. https://doi.org/10.1055/s-0043-121264.

Stafford, J.M., Raybuck, J.D., Ryabinin, A.E., Lattal, K.M., 2012. Increasing histone acetylation in the hippocampus-infralimbic network enhances fear extinction. Biol. Psychiatr. 72 (1), 25–33. https://doi.org/10.1016/j.biopsych.2011.12.012.

Stoccoro, A., Coppedè, F., 2018. Role of epigenetics in Alzheimer's disease pathogenesis. Neurodegener. Dis. Manag. 8 (3), 181–193. https://doi.org/10.2217/nmt-2018-0004.

Surana, N.K., Kasper, D.L., 2014. Deciphering the tête-à-tête between the microbiota and the immune system. J. Clin. Invest. 124 (10), 4197–4203. https://doi.org/10.1172/jci72332.

Tata, B., Mimouni, N.E.H., Barbotin, A.-L., Malone, S.A., Loyens, A., Pigny, P., Dewailly, D., Catteau-Jonard, S., Sundström-Poromaa, I., Piltonen, T.T., 2018. Elevated prenatal anti-Müllerian hormone re-programs the fetus and induces polycystic ovary syndrome in adulthood. Nat. Med. 24 (6), 834–846. https://doi.org/10.1038/s41591-018-0035-5.

Teede, H.J., Misso, M.L., Costello, M.F., Dokras, A., Laven, J., Moran, L., Piltonen, T., Norman, R.J., Andersen, M., Azziz, R., 2018. Recommendations from the international evidence-based guideline for the assessment and management of polycystic ovary syndrome. Fertil. Steril. 110 (3), 364–379. https://doi.org/10.1016/j.fertnstert.2018.05.004.

Thackray, V.G., 2019. Sex, microbes, and polycystic ovary syndrome. Trends Endocrinol. Metabol. 30 (1), 54–65. https://doi.org/10.1016/j.tem.2018.11.001.

Tirronen, A., Hokkanen, K., Vuorio, T., Ylä-Herttuala, S., 2019. Recent advances in novel therapies for lipid disorders. Hum. Mol. Genet. 28 (R1), R49–R54. https://doi.org/10.1093/hmg/ddz132.

Tofalo, R., Cocchi, S., Suzzi, G., 2019. Polyamines and gut microbiota. Front. Nutr. 6. https://doi.org/10.3389/fnut.2019.00016.

Torchen, L.C., Legro, R.S., Dunaif, A., 2019. Distinctive reproductive phenotypes in peripubertal girls at risk for polycystic ovary syndrome. J. Clin. Endocrinol. Metabol. 104 (8), 3355–3361. https://doi.org/10.1210/jc.2018-02313.

Torres, P.J., Siakowska, M., Banaszewska, B., Pawelczyk, L., Duleba, A.J., Kelley, S.T., Thackray, V.G., 2018. Gut microbial diversity in women with polycystic ovary syndrome correlates with hyperandrogenism. J. Clin. Endocrinol. Metabol. 103 (4), 1502–1511. https://doi.org/10.1210/jc.2017-02153.

Tremblay, B.L., Guénard, F., Rudkowska, I., Lemieux, S., Couture, P., Vohl, M.-C., 2017. Epigenetic changes in blood leukocytes following an omega-3 fatty acid supplementation. Clin. Epigenet. 9 (1). https://doi.org/10.1186/s13148-017-0345-3.

Tremellen, K., Pearce, K., 2012. Dysbiosis of gut microbiota (DOGMA)—a novel theory for the development of polycystic ovarian syndrome. Med. Hypotheses 79 (1), 104–112. https://doi.org/10.1016/j.mehy.2012.04.016.

Valinezhad Orang, A., Safaralizadeh, R., Kazemzadeh-Bavili, M., 2014. Mechanisms of miRNA-mediated gene regulation from common downregulation to mRNA-specific upregulation. Int. J. Genomics 1–15. https://doi.org/10.1155/2014/970607, 2014.

Vázquez-Martínez, E.R., Gómez-Viais, Y.I., García-Gómez, E., Reyes-Mayoral, C., Reyes-Muñoz, E., Camacho-Arroyo, I., Cerbón, M., 2019. DNA methylation in the pathogenesis of polycystic ovary syndrome. Reproduction 158 (1), R27–R40. https://doi.org/10.1530/REP-18-0449.

Volmar, C.-H., Wahlestedt, C., 2015. Histone deacetylases (HDACs) and brain function. Neuroepigenetics 1, 20–27. https://doi.org/10.1016/j.nepig.2014.10.002.

Walters, W.A., Xu, Z., Knight, R., 2014. Meta-analyses of human gut microbes associated with obesity and IBD. FEBS Lett. 588 (22), 4223–4233. https://doi.org/10.1016/j.febslet.2014.09.039.

Wang, Z., Zhao, Y., 2018. Gut microbiota derived metabolites in cardiovascular health and disease. Protein Cell 9 (5), 416–431. https://doi.org/10.1007/s13238-018-0549-0.

Wei, J.-W., Huang, K., Yang, C., Kang, C.-S., 2017. Non-coding RNAs as regulators in epigenetics. Oncol. Rep. 37 (1), 3–9. https://doi.org/10.3892/or.2016.5236.

Willis, D.S., Watson, H., Mason, H.D., Galea, R., Brincat, M., Franks, S., 1998. Premature response to luteinizing hormone of granulosa cells from anovulatory women with polycystic ovary syndrome: relevance to mechanism of Anovulation1. J. Clin. Endocrinol. Metabol. 83 (11), 3984–3991. https://doi.org/10.1210/jcem.83.11.5232.

Xia, H., Zhao, Y., 2020. miR-155 is high-expressed in polycystic ovarian syndrome and promotes cell proliferation and migration through targeting PDCD4 in KGN cells. Artif. Cell Nanomed. Biotechnol. 48 (1), 197–205. https://doi.org/10.1080/21691401.2019.1699826.

Xu, N., Kwon, S., Abbott, D.H., Geller, D.H., Dumesic, D.A., Azziz, R., Guo, X., Goodarzi, M.O., 2011. Epigenetic mechanism underlying the development of polycystic ovary syndrome (PCOS)-Like phenotypes in prenatally androgenized rhesus monkeys. PLoS One 6 (11), e27286. https://doi.org/10.1371/journal.pone.0027286.

Yamada, K., Gherasim, C., Banerjee, R., Koutmos, M., 2015. Structure of human B12Trafficking protein CblD reveals molecular mimicry and identifies a new subfamily of nitro-FMN reductases. J. Biol. Chem. 290 (49), 29155—29166. https://doi.org/10.1074/jbc.m115.682435.

Yao, Q., Chen, Y., Zhou, X., 2019. The roles of microRNAs in epigenetic regulation. Curr. Opin. Chem. Biol. 51, 11—17. https://doi.org/10.1016/j.cbpa.2019.01.024.

Yatsunenko, T., Rey, F.E., Manary, M.J., Trehan, I., Dominguez-Bello, M.G., Contreras, M., Magris, M., Hidalgo, G., Baldassano, R.N., Anokhin, A.P., 2012. Human gut microbiome viewed across age and geography. Nature 486 (7402), 222—227. https://doi.org/10.1038/nature11053.

Ye, J., Wu, W., Li, Y., Li, L., 2017. Influences of the gut microbiota on DNA methylation and histone modification. Dig. Dis. Sci. 62 (5), 1155—1164. https://doi.org/10.1007/s10620-017-4538-6.

Yildiz, B.O., Bozdag, G., Yapici, Z., Esinler, I., Yarali, H., 2012. Prevalence, phenotype and cardiometabolic risk of polycystic ovary syndrome under different diagnostic criteria. Hum. Reprod. 27 (10), 3067—3073. https://doi.org/10.1093/humrep/des232.

Zarzour, A., Kim, H.W., Weintraub, N.L., 2019. Epigenetic regulation of vascular diseases. Arterioscler. Thromb. Vasc. Biol. 39 (6), 984—990. https://doi.org/10.1161/ATVBAHA.119.312193.

Zeng, B., Lai, Z., Sun, L., Zhang, Z., Yang, J., Li, Z., Lin, J., Zhang, Z., 2019. Structural and functional profiles of the gut microbial community in polycystic ovary syndrome with insulin resistance (IR-PCOS): a pilot study. Res. Microbiol. 170 (1), 43—52. https://doi.org/10.1016/j.resmic.2018.09.002.

Zhang, Z., Shi, L., Dawany, N., Kelsen, J., Petri, M.A., Sullivan, K.E., 2016. H3K4 tri-methylation breadth at transcription start sites impacts the transcriptome of systemic lupus erythematosus. Clin. Epigenet. 8 (1). https://doi.org/10.1186/s13148-016-0179-4.

Zhao, J., Li, G., Bo, W., Zhou, Y., Dang, S., Wei, J., Li, X., Liu, M., 2017. Multiple effects of ellagic acid on human colorectal carcinoma cells identified by gene expression profile analysis. Int. J. Oncol. 50 (2), 613—621. https://doi.org/10.3892/ijo.2017.3843.

Zhao, Z., Shilatifard, A., 2019. Epigenetic modifications of histones in cancer. Genome Biol. 20 (1). https://doi.org/10.1186/s13059-019-1870-5.

Introduction to epigenetic programming by gut microbiota

Olugbenga Samuel Michael[1,5,6], Oluwaseun Aremu Adeyanju[2,7], Charles Oluwaseun Adetunji[3], Kehinde Samuel Olaniyi[2], Juliana Bunmi Adetunji[4] and Ayodele Olufemi Soladoye[1]

[1]Cardiometabolic, Microbiome and Applied Physiology Laboratory, Department of Physiology, College of Health Sciences, Bowen University, Iwo, Osun State, Nigeria; [2]Department of Physiology, College of Medicine and Health Sciences, Afe Babalola University, Ado-Ekiti, Nigeria; [3]Applied Microbiology, Biotechnology and Nanotechnology Laboratory, Department of Microbiology, and Directorate of Research and Innovation, Edo State University Uzairue, Iyamho, Auchi, Edo State, Nigeria; [4]Nutritional and Toxicological Research Laboratory, Department of Biochemistry Sciences, Osun State University, Osogbo, Osun State, Nigeria; [5]Department of Physiology, University of Tennessee Health Science Center, Memphis, TN, United States; [6]Department of Medical Pharmacology and Physiology, University of Missouri, Columbia, MO, United States; [7]Department of Cellular and Molecular Biology, The University of Texas Health Science Center at Tyler, Tyler, TX, United States

Introduction

The microbiota—host interactions caused by numerous microorganisms, which may be commensal or pathogenic, contribute to disease pathogenesis or health preservation through their influences on the cellular responses to environment, which have capacity to alter genomic expression without changing the sequence of the DNA. This process is termed epigenetic. Therefore, it is not an understatement that the humans live in a microbial world, which has amazing influences on our physiology. There are accumulating reports that the basis for the integration and responses of host to microbiome signals is through epigenetic changes. Microbiome metabolites including propionate, acetate, and butyrate have been recognized as one of the ways through which the intestinal microbes interact with the host epigenetic mechanisms (Vinolo et al., 2009; Olaniyi et al., 2020).

There are several factors apart from the crucial involvement of the human genes in development of diseases, such as environmental contributions to the development of numerous debilitating illness such as cardiac disease, cancer, diabetes mellitus, and inflammatory bowel disease (Renz et al., 2011). Thus, elucidating or getting a clear perspective on the mechanistic involvement of the epigenomic regulations especially the microbiota—host interactions represents the link between the host and its immediate environment or surroundings that have immense capacity to regulate host physiology and disease pathogenesis through their influences on the genetic makeup of the host (Kamada et al., 2013). The gut microbes are able to cause DNA methylation and overall transcriptional activity because the folate derived from the intestinal microbiota is a crucial methyl contributor for the methylation of DNA (Mischke et al., 2013).

An Introduction to the Microbiome in Health and Diseases
ISBN 978-0-323-91190-0,
https://doi.org/10.1016/B978-0-323-91190-0.00010-2

The human body is constantly in close proximity with microbes in the environment. These body parts such as the skin and gut and respiratory surfaces or pathways are inhabited by these microorganisms forming a resident microbial ecology or community inside the host (Kamada et al., 2013). Skin and intestinal microbial estimation is by far more than human cells. Actually, bacteria genetic contribute about 99% the human genome (Backhed et al., 2005). Therefore, the host—microbiome cross-talk exceeds cellular levels but goes deeper into the genetic composition of the host where they cause significant genetic modifications that influence the host physiology and heredity showing that this has transgenerational consequences. Furthermore, the intestinal microbiome has been reported to have immunomodulatory functions (Gallo and Hooper, 2012; Hooper et al., 2012), which is very crucial for the enhancement and growth of the innate and adaptive immunity (Khosravi et al., 2014), resulting in protection against pathogenic bacteria and infections. These processes ensure that the host is able to cope through internally built defense mechanism against pathogenic microbial infection (Khosravi et al., 2014).

Intestinal microbial-derived metabolites from dietary fiber fermentation possess potent epigenetic modulation capacity through their inhibitory effects on the histone deacetylases (HDACs) (Macfarlane and Macfarlane, 2003). HDACs are said to involve the pathobiology of various chronic ailments such as diabetes, hypertension, cancer, obesity, ulcerative colitis, etc. (de Zoeten et al., 2010; Olaniyi et al., 2020). Likewise, short-chain fatty acids (SCFAs) inhibit inflammatory processes, which further support their capability to mitigate the development of inflammatory-mediated chronic diseases. Chang and coworkers have revealed that butyrate elevated histone acetylation and reduced circulating levels or activities of inflammatory biomarkers (Chang et al., 2014). Some pathogenic microbes contribute to the pathobiology of cancer through the alteration or reprogramming of the host's epigenome (Esteller, 2007; Flanagan, 2007). Acetate administration and acetate-producing bacteria have been documented to confer renoprotection against ischemia—reperfusion injury through the suppression of HDAC enzyme activities and elevated methylation (Andrade-Oliveira et al., 2015).

In addition, nutrition is a modulator of microbial activity and host physiology via epigenetic alterations due to microbial-derived metabolic products that are active signaling molecules and epigenetic modifiers (Krautkramer et al., 2016). Therefore, diet interacts with the intestinal microbiome to initiate broad epigenetic changes, which results in transcriptional regulation of the genes in the host. Epigenetic modifications influence fetal and neonatal-associated growth and development of chronic diseases later in the life of offspring of pregnant women who have been exposed to metabolic stress, high-fat diet, gut dysbiosis, impaired vaginal microbiota homeostasis, exposure to xenobiotics, etc. All these are environmental factors capable of causing alteration in the microbiome—genomic interface, resulting in epigenetic modulations that are transcribed to the offsprings through transcriptional processes. Therefore, the interactions between the microbiome and host have far-reaching epigenetic consequences that influence the host physiology and pathology.

Epigenetics of bacterial infections

Environmental factors influence histone alteration and necessary contributions of DNA in regulating the interpretations that involved epigenetic mode as well as in shaping immune rejoinder that is associated with microbial infections. Past research studies associated with gastric cancer that could be linked to bacterial infection showed that epigenetic changes are caused by methylation of DNA (Maekita et al., 2006). However, it is unclear if methylation of DNA could be as a result of pathogenic bacteria or is subordinate to ongoing infection. All these indicated affirmed the fact that the strategy involved in the epigenetic alteration could be associated to bacterial inflammation.

Some past research work have also established through the process involved in DNA methyltransferase expression and regulation of DNA methylation patterns in their host cells, which might be linked to the action of some microorganisms such as *Campylobacter rectus, Mycobacterium leprae*, and *Escherichia coli* (Masaki et al., 2013). Fascinatingly, it has been established that *M. leprae* portends the capability of reconfigured neuronal cells mainly into cells similar to stem cell. Most of the modifications are very crucial especially toward enhancing bacillary dissemination and changes in methylation (Masaki et al., 2013). Therefore, interpretations increase likelihood that microorganisms could melodramatically cause epigenetic changes that affect the cellular responses in the host, and these epigenetic transcriptional modifications can be transferred across generations as inheritable phenotypic traits (Pereira et al. 2016).

Furthermore, numerous enzymatic activities, which involve significant bacterial-derived metabolites with unswerving capacity to regulate histone acetylation, methylation, or deacetylation, have been reported. Typical examples include Trithorax, variegation suppressor, zeste enhancer as well as generation of proteins BtSET and BaSET, which are transferred into host cellular machinery mainly through the action of *Burkholderia thailandensis* (Mujtaba et al., 2013).

Moreover, BaSET methylates histone H1 when it distributes to the nucleus, which could stimulate the prevention NF-jB, but BtSET is restricted to the nucleolus and enhances the process involved in H3K4 methylation resulting in rRNA transcription. Therefore, with the available facts, it has been stated that the process of histone methylation through the action of bacterial effectors could be projected as a new virulence stratagem and, certainly, SET domain proteins have been established to have a relationship in the other bacterial pathogen which includes *Chlamydia* spp. and *Legionella* spp. (Rolando et al., 2015).

It has been established that some bacteria possess the capability to regulate epigenetic marks available in the host cells incidentally by signaling of mitogen-activated protein kinase (MAPK). MAPKs could stimulate the process involved in the downstream kinases most especially with the phosphorylate histone H3S10 that is related to the stimulation of transcription and H3 acetylation (Sawicka and Seiser, 2012). Also, *Mycobacterium tuberculosis* and *B. anthracis* regulate MAPK-dependent alteration, acetylation, and phosphorylation of histone associated with momentous alterations in inflammatory stimulation of macrophages and epithelial cells (Raymond et al., 2009).

Furthermore, some pathogens possess the capability to stimulate epigenetic alteration in the host cells mainly by influencing the signaling pathways stimulation during inflammation, emphasizing an intricate cross-talk mainly among signaling activities, and epigenetics contribute significantly to disease pathogenesis (Grabiec and Reedquist 2013). Several scientists have done extensive review on the pathoepigenetics of microbial inflammation (Bierne et al., 2012; Rolando et al., 2015).

Moreover, Grabiec and Potempa (2018) discussed the newly discovered methodology that could be employed by pathogenic microorganisms, which is used in penetrating the immune action through acetylation of histone and regulation of the process involved in the protein acetylation most especially HDACi on the eventual effect of bacterial inflammation.

The process involved in the host acetylation system by bacteria

Over the year, several processes have been proven toward the establishment of the proper knowledge involved in the process through which bacteria could capture the acetylation system used in the stimulation of the transcriptional control most especially in the host cells so as to evade its prevention by the immunological mechanistic processes (Bierne et al., 2012). Several methods employed through the action of these microorganism targeting alterations in the acetylation of nonhistone and histone proteins were documented. These involve indirect and direct influence on HDAC expression, epigenetic histone marks, generation of metabolites, and regulation of HAT that stimulates the action of acetylation system components (Hamon and Cossart, 2008; Handy et al., 2011).

Regulation of histone acetylation

Numerous microorganisms have been identified to possess the capability to enhance the defense mechanism of host through improved immunological activity by upsetting acetylation of histone. Typical examples of these microorganisms include *Helicobacter pylori, Listeria monocytogenes,* and *M. tuberculosis.* Some facts have been gathered that some of these bacteria could stimulate the process involved in epigenetic alteration and expression through histone acetylation. Furthermore, it was observed that the infection caused by *M. tuberculosis* didn't have any influence on HDAC2 and HDAC1 manifestation, which could result into development of HDAC-containing complex, upregulation of the corepressor protein, and deacetylation of histone. Reduction in the level of histone acetylation might be related to the rate of inflammation caused by bacterial infection such as *H. pylori* (Wang et al., 2005; Ding et al., 2010).

Moreover, the importance of epigenetic histone modifications is yet to be well established. Therefore, modes of action by the regulation by bacteria on acetylation of histone in the host have been described for microbial pathogenic organisms. It has been identified that *L. monocytogenes* possesses that capability to induce illness in newborns and pregnant women, which has established a sophisticated stratagems that could enhance their survival and entry into infected cells; most especially the management of the host transcriptional machinery performs a significant function (Niller and Minarovits, 2016). It has been stated that inflammation of HeLa cells most especially in the *L. monocytogenes* could led to fast total deacetylation or specific deacetylation of histone H4 (Eskandarian et al., 2013). Also, histone H4 deacetylation could also be related to decrease phosphorylation of H3S10 (Hamon et al., 2007).

The histone alteration by *L. monocytogenes* involves changes in listeriolysin O (LLO) causing a drastic dephosphorylation of Ser^{10} on histone H3 and deacetylation of histone H4. This decreases manifestation of the genetic composition of the host, entailing significant control of the immunoregulatory reaction most especially phosphatase DUSP4. Also, it has been stated that some bacterial toxins are derived from some bacterial such as *Streptococcus pneumonia, Clostridium perfringens*, and *Clostridium pneumolysin*. Moreover, perfringolysin derived from *Streptococcus pneumonia* indicated a comparable influence on the phosphorylation of H3S10 (Hamon et al., 2007). Deacetylation of histone H3K18 through the downregulation of Sirt2 manifestation of subclass of genes entails the immune responses alongside transcriptional regulation, thereby enhancing the survival and invasion of *L. monocytogenes*. The function of Sirt2-dependent transcriptional reprograming during the process involved in the inflammation by the bacterial infection has been documented by several assays (Eskandarian et al., 2013).

Certain acetylation of histone codes is decreased inside the host of *Legionella pneumophila* in relationship to the disease pathology that involved several factors of severe pneumonia called Legionnaires' disease. This could be linked to the effector molecule RomA of *L. pneumophila* that possesses the potential to induce trimethylation of histone H3K14 that could mitigate against acetylation of this residue as well as inhibit innate immune genes, which includes interleukin-6 (IL6) and receptor 5 (TLR5) (Rolando et al., 2013). Moreover, the ability of microbial pathogens to regulate epigenetic modulations such as histone acetylation, which improves their adaptation inside the host, has been established through experimental facts.

Gut microbiome and epigenetic regulations

The human gut is made up of several bacterial with metabolic potentials. They are involved in various complex processes such as digestion (Hooper et al., 2002), synthesis of vitamins (Hill, 1997), metabolism, and degradation (Hill, 1997; Hylemon and Harder, 1998) and support the immune system (Braun-Fahrlander et al., 2002). Each individual has his/her own unique microbiome and may vary with advance in age. Epigenetics involves different developments that result in gene expression, which can be inherited without compromising the DNA (Egger et al., 2004). Histone modification, DNA methylation, and noncoding RNA are the main machineries for epigenetic alteration, and intestinal microbiome is involved in these modifications through diet (Jirtle, 2007; Li and Tollefsbol, 2010).

Epigenetic machineries and the microbiome

How does epigenetic mechanism establish a cross-talk with the microbiome? These will be explained with the various mechanisms involved in epigenetic modification.

Noncoding RNAs are RNA transcription products not converted into proteins (Dempsey et al., 2018). These have been linked to various disease states (Devaux and Raoult, 2018). Liang et al. (2015) show the important role it plays in the immune process and proves that noncoding RNA is influenced by gut microbiome, which consequently initiates metabolic disorders such as insulin resistance and obesity. Another study also evaluated the expression of these RNAs in different tissues such as the adipose tissue amid others and found that in the absence of gut microbiome, the noncoding RNAs (nc-RNAs) were differentially regulated in the concerned organs, further implicating a possible involvement of the gut microbiota in the expression of nc-RNA (Dempsey et al., 2018). Microbiome makes use of nc-RNA in the pathogenesis or regulation of metabolic homeostasis, suggesting involvement of gut microbiota in nc-RNA regulation and, by extension, epigenetic regulations (Davelos et al., 2011; Virtue et al., 2019).

DNA methylation is an enzyme-catalyzed process of adding a methyl group to carbon-5 of cysteine ring (Romano and Rey, 2018). This enzyme is highly sensitive to nutrients, which activate the microbiomes residing in the gastrointestinal tract (Romano and Rey, 2018). Some metabolites such as folate, vitamin B12, etc. also contribute to methylation of DNA (Kok et al., 2015) due to the fact that they can be regulated by gut microbiome (Rossi et al., 2011), hence showing pivotal relationship that gut microbiome has with DNA methylation, which causes regulation of the body metabolic homeostasis. Metabolites produced by intestinal microbiota also play a cardinal role in DNA methylation (Luu et al., 2019).

Histone modification is simply the alteration of histones and includes processes such as acetylation, methylation, ubiquitination, etc. (Bernstein et al., 2007). These modulation processes, histone deacetylation, methylation, and acetylation processes stand out. Microbial end-products, SCFAs, control acetylation of histone (Krautkramer et al., 2016) as an experimental study shows that supplementation with SCFAs induced acetylation of the histones (Soliman and Rosenberger, 2011) and another study proves its beneficial role on inflammation (Wang et al., 2019). Histone deacetylate (HDAC) inhibitors are used therapeutically for management of metabolic derangement. Intestinal microbiome can control activities of histone deacetylates through production of epigenetic metabolic products—SCFA that maintains metabolic homeostasis (Marlicz et al., 2018).

Gut microbiome: Before or after birth?

The influence of intestinal microbiome is pivotal in the preservation of health; diseases pathogenesis and microbiome alterations have been connected to various disorder (Paul et al., 2015; Bloomfield et al., 2016). Funnily, it has been shown that when new babies are given birth to, the microbial flora that reside in their gut are actually gotten from the maternal gut, skin, or vaginal duct (Rook et al., 2014). Reports have shown that the fetuses begin to obtain intestinal microbiota while they are inside the womb as these might go through the mothers' gastrointestinal tract to the growing fetus digestive tract—hence, maternal diet contributes to this (Collado et al., 2016). An experimental study corroborated this fact by showing that mothers who consume high-fat diet cause fetal programming with gut microbiome imbalance (Ma et al., 2014). There was further confirmation by the study that makes use of rodent with totally absent microbiota at birth (Chu et al., 2016). Modern lifestyles have, however, brought its disadvantages on intestinal microbiome. Bacterial infections are treated with antibiotics, but regardless of their benefits, they can effectively destroy the gut microbiome as shown by clinical studies in which usage of antibiotics showed decrease microbiome activity (Lankelma et al., 2016). Intestinal microbiome assumes its original state but in most cases not the same again after antibiotics administration. The antibiotics-associated microbiome disruption affects basic immune homeostasis especially in early life (Francino, 2014).

Environmental microbiome influences fetal development

Fetal microbial constitution is pivotal to the eventual well-being in the later life (Perry et al., 2016) as placenta formally believed to be sterile is now known to contain some of these microbes (DiGiulio et al., 2008). Maternal environment (both internal and external) influences intestinal microflora growth. In fact, delivery type affects microbiome community. Improved intestinal microbiome constitution has been associated with vaginal delivery, which tends to have increased microbiome constitution (Milani et al., 2017) than those through caesarean section who are also susceptible to various disorders (Huh et al., 2012; Almgren et al., 2014; Kristensen and Henriksen, 2016). Another influencer on the gut microbiome development and/or composition is the gestational age (Korpela et al., 2018). Infants given birth to prematurely or not to term tend to have lower composition of gut microbiome and are at risk of gut dysbiosis (Arboleya et al., 2012). Breastfeeding is another factor that has effect on neonatal gut microbiome development (Rossi et al., 2011). Hence, the microbiome—epigenetic interaction is interesting and may serve as therapeutic targets for later life disease (Indrio et al., 2017).

Maternal nutrients effect on gut microbiome and epigenetic regulation

The mother's eating lifestyle during pregnancy also goes a long way to affect the fetal and neonatal microflora and epigenetic regulation. Microbiome metabolites such as SCFAs amid others modulate epigenetic processes (Hullar and Fu, 2014; Gerhauser, 2018); hence, maternal gut microbial metabolites can influence these processes (Krautkramer et al., 2017).

Some vitamins are not produced by the body during pregnancy and need to be taken in through food. The gut microbiota is a vital source of vitamin B during pregnancy (Yoshii et al., 2019). These vitamin B family are needed in the folate cycle and can affect epigenetic regulation (Krautkramer et al., 2017). SCFAs improve immunological regulatory system via epigenetic modulation (Dalile et al., 2019). Maternal acetate can prevent asthma through HDAC inhibition (Thorburn et al., 2015); the same way butyrate can induce histone acetylation (Shimazu et al., 2013). Propionate also prevents histone acetylation and, by extension, inflammatory processes by inhibiting interleukin production (Luu et al., 2019). Additionally, maternal supplementations with polyphenols (which have immune-modulatory, antioxidant, and antiinflammatory properties) help prevent the risk of intrauterine growth restriction (Vazquez-Gomez et al., 2017). Other types of polyphenols help prevent metabolic disorder (Wu and Tian, 2017) and lipid dysmetabolism, prevent histone acetylation activity (Kang, 2023), and reduce cardiovascular disease (CVD) risk factors in later life of the newborn baby (Thakur et al., 2012). Omega-3 polyunsaturated fats (PUFAs), which can only be obtained by diet enhance new born immunity via epigenetic modulation (Lee et al., 2013; Costantini et al., 2017). Omega-3 PUFA from the mother causes methylation in different organs (Van Dijk et al., 2016) and also controls or prevents obesity in infants via microbiome-dependent mechanism (Robertson et al., 2018). Maternal high-fat diet promotes development of CVD in offspring by altering intestinal microbiome homeostasis associated with reduced SCFAs production (Hsu et al., 2019) of which supplementation with SCFAs prevents this through reduction of HDAC activity (Whitt et al., 2018). Lastly, maternal overnutrition or undernutrition can also influence epigenetic changes through the gut microbiome increasing the risk of inflammatory diseases and obesity in infants (Stanislawski et al., 2017; Connor et al., 2018). High-fat diet intake by pregnant mothers disrupts the gut microbiomes of the offspring as well (Wankhade et al., 2017)—this through histone acetylation.

Therefore, alteration of microbiome or gut dysbiosis during developmental periods has been revealed to regulate multifaceted phenotypic characteristics in the adulthood (Bai et al., 2019; Sanna et al., 2019). These epigenetic programming may be either favorable or deleterious. Improved metabolic responses by the host subsequently in adulthood or later life are due to enhanced growth of favorable intestinal microbes during fetal developmental stages (Backhed, 2011; Zhong et al., 2019). Caesarean section delivery altered fetal microbiome constitution, which is connected with metabolic disorders (Decker et al., 2011). Maternal intestinal microbiome dysregulation in pregnancy and lactation causes epigenetic manipulations that influence the physiological/pathological responses in the offspring (Priyadarshini et al., 2014; Paul et al., 2018). Early fetal life-associated microbiota alteration induced by administration of antibiotics (amoxicillin) causes cardiac protective effects by reducing the blood pressure, signifying the involvement of beneficial epigenetic imprinting during the fetal life (Galla et al., 2020).

Epigenetic programming during fetal life could be from either parents, i.e., mother or father. These environmental-mediated heritable phenotypic modifications are expressed in future generations as such it is referred to as transgenerational epigenomic traits transmission. This type of transmission allows subsequent generations to cope successfully or become vulnerable to environmental challenges or insults influencing their physiological responses (Dunn and Bale, 2011; Aiken and Ozanne, 2014). Nutrition—microbiome—host interactions have been shown as a potent transgenerational epigenetic modifier responsible for the observed phenotype (Kaminsky et al., 2006). Diet—microbiota-induced epigenetic modifications can be passed down multiple generations (i.e., transgenerational inheritance) causing considerable evolutionary changes (Miller, 2016). It is noteworthy that not all transgenerational phenotypic transmission involves epigenetic manipulations. There are some factors that induced nongenetic developmental programming, such as inadequate reproductive tract environment or impaired maternal pregnancy adaptations.

Furthermore, aging in mothers is a possible factor for programmed changes in pregnancy being transmitted to subsequent generations (Aiken and Ozanne, 2014). Therefore, proper elucidation of the mechanisms responsible for the observed epigenetic modifications and transgenerational transmissions in developing fetuses is necessary for the formulation of interventional approaches to mitigate the epigenetic modulations. Epigenetic modifications are persistent but can be reversed with appropriate and targeted mechanistic inventions aimed at the root cause of the phenotypic changes. But there are several considerations to be put in place such as fflexibility of the treatments, and timing of mitigation during development, how many generations will it take to undo the epigenetic signatures (Milagro et al., 2013). The gut microbes are key modulators of epigenetic transmission across different generations, and appropriate attention must be given to the gut microbiome balance especially during critical windows of development. How the gut microbiota—induced epigenetic imprinting must be properly documented so as to develop therapeutic strategies and approaches for proper mitigation.

Microbiota—immune regulation—epigenetics axis

The microbiota—immune interaction alliance permits induction of protective response, and regulatory pathways involve the maintenance of pathogenic microbial tolerance. Early-life environmental factor has been reported to modulate microbiome, which in turn regulates epigenetic changes that influence human health (Wu and Wu, 2012; Lee, 2019). Recent studies reported that microbiota—host interactions contribute to immune regulation through epigenetics control of host tissues by the microbial metabolic response to diet or environmental factor (Hullar and Fu, 2014; Miro-Blanch and Yanes, 2019).

Diet, gut microbiota, and epigenetic regulation

It has been discovered that the type and amount of macromolecules present in a diet could stimulate the synthesis of some active microbiome metabolites. The basic metabolites synthesized are mainly SCFAs provide energy via microbial degradation of macromolecules. However, the most examined active metabolites entail butyrate, propionate, acetate, and SCFAs. Moreover, the SCFAs derived from the action of microbiome have been reported to possess valuable metabolites with great physiological potential on human health (Macfarlane and Macfarlane, 2012). Hence, the microbiomes are responsible for the maintenance of sound immune response for good health (Alasmar et al., 2019). Animals maintained on protein-derived diet increased the count of *Bifidobacteria* spp. and *Lactobacilli* spp. significantly in their feces (Elwood et al., 2008; Sprong et al., 2010). Sprong et al. (2010) concluded that the gut microbiome could be reshaped through the isolation of protein in accordance with the dosages administered.

Also, fiber derived from carbohydrate is routed through the small intestine to the colon where it has been acted upon by the microbiota to generate the needed energy before removing the remaining fibers via the feces. The example of carbohydrates could easily be degraded by the gut microbiomes including starch, oligosaccharides, and insulin, respectively. Moreover, there are several reports on the beneficial effect of the microbiome on health. Microbial flora often produces lots of bioactive compounds, mostly vitamins necessary for health; however, some are toxic. Moreover, the host immune defense such as the mucus barrier assists in preventing detrimental bacteria from harming the tissues. Therefore, diet such as fibers helps in the maintenance of a healthy and stable intestinal microflora.

Intestinal microbiome environment has been observed that it could be altered within a day through the process of diet modification (Rasnik et al., 2017; Valdes et al., 2018). O'Keefe et al. (2015) also claimed that a change in diet can make a tangible change to the microbiome of the gut. For instance, the process of alteration could result in the activation of butyrate synthesis by gut microbiomes while decreasing bile acid production. It was reported also that an excess inflammation and stress could result from injury and can also alter the gut microbiomes to induce an insult on the immune system (Earley et al., 2015). Hence, excessive of immune system response to gut microbiome can induce the inflammation of the bowel, which can cause stunted growth in children, cancer, and even lead to some other complications such as ulcerative colitis (Kim and Chang, 2014; Ronald et al., 2019).

Furthermore, the gut microbiomes often get their nutrition from the undigested carbohydrates during fermentation and those that boycott the peak digestion stage (Jandhyala et al., 2015). However, some fermented fibers could provide nourishment in the bowel for some microbes (Deehan et al., 2017). But, the electrons generated are serving as a waste product of respiration from response to inflammation, which could become a stressor that might stimulate the growth of bacterial (Winter and Baumjer, 2014). Moreover, some undigested fibers present in food serve as prebiotics (Bindels et al., 2015) and thus enhance the beneficial potential of some microbiomes present in the bowel. However, inadequate fiber intake could cause drastic reduction in beneficial metabolites synthesis, leading to synthesis of metabolites that are harmful (Cummings and Macfarlane, 1991; Russell et al., 2011). Interestingly, low fibers present in diet could degrade colonic mucus layer causing obstruction of the microbiome to pathogen vulnerability and inflammation, which might eventually lead to chronic disease.

Studies have shown that fiber portends that capability to maintain and support the bowel by improving the digesta mass through toxins dilution which decreases intracolonic pressure and increases defecation turnout, respectively. However, it has been established that dietary fibers play a vital role in enlarging fecal mass especially through the process of fermentation in the presence of bacteria. This is a benefit derived from fermentation of the carbohydrate by gut microbiota and the bioactive component produced. The fermented fibers produce organic acids such as small-chain fatty acids serving as a metabolic fuel source in peripheral tissues and epithelial lining of bowel. One of the organic acids, butyrate, is a unique colonic metabolic energy source (Roediger, 1980), regulates colonocytes differentiation and apoptosis, removes dysfunctional cells, maintains integrity of the mucosal lining, controls inflammatory processes in the intestine, and stimulates genomic constancy (Fung et al., 2012).

The involvement of microbiota in immune regulation

Microbiome drives early and postnatal immune development in neonates (Belkaid and Hand, 2014). The mechanism has been partly attributable to factors contained in maternal milk such as live microbes, microbial dendritic cells, metabolites such as oligosaccharides, cytokines, and IgA immune cells. Many of these factors synergize and influence the neonatal microbiome and the host response. Take, for example, IgA of the mother binds nutritional and innocuous antigen prevents immune activation and microbial, and likewise the existence of oligosaccharides in maternal milk promotes the expansion of gut microbiome containment like *Bifidobacterium* (Perez et al., 2007; Belkaid and Hand, 2014). Take for example, the Toll Like Receptors (TLR) ligands expressed in neonate innate cells respond distinctively to microbial ligands with notable impaired inflammatory mediators production and increased production of regulatory cytokines (IL-10) compared with TLR in adult cells (Kollmann et al., 2012).

Microbiome facilitates induction of regulatory T cells, which restricts mucosal inflammation and tolerance promotion. This forms physiological rationale for the identification of future microbes possessing better potential to stimulate regulatory T cells (Belkaid and Hand, 2014). Nevertheless, the factors underlying innate immune-microbial integration is under elucidation, but recent findings support the involvement of epigenetics (the expression of epigenome-modifying enzymes) in microbiota-immune regulation (Alenghat et al., 2013; Belkaid and Hand, 2014; Miro-Blanch and Yanes, 2019). Butyrate is reported to epigenetically regulate gene expression by suppression of HDAC activities (Davie, 2003; Furusawa et al., 2013), and this property is currently suggested as a mechanistic link to enhancement of Treg production in the gut environment (Belkaid and Hand, 2014).

Microbial metabolites, including SCFAs, have been demonstrated to induce T reg generation through modification of epigenetics (Canani et al., 2012; Hullar and Fu, 2014). For example, at physiological condition, there is an experience of higher levels of butyrate by the normal cells near the lumen; this subsequently accumulates in the nucleus and suppresses the activity of HDAC.

Therefore, the use of HDAC inhibitors, especially butyrate, was previously demonstrated to enhance cathelicidin expression and strengthened innate immunity (Canani et al., 2012). Hence, the activity of microbial metabolites resulting in epigenetic modulation of numerous genes contributes to the induction of regulatory pathways, particularly T cell and dendritic cell, revealing the crucial involvement of microbiome in regulating immunity via the machinery of epigenetics.

Conclusion

The microbial environmental influences have been reported to be of significant effects on fetal development in utero and during lactation because of the reshaping of gut microbiome by nutrition, delivery mode, vaginal microbial community, etc. These microbiome alterations have ability to modify the microbial genome resulting in phenotypic alterations due to epigenetics. Epigenetic alterations can be transferred across multiple generations termed transgenerational transference of trait. Therefore, the gut microbiome is a major epigenetic modifier with lasting influence on health of the host. Furthermore, metabolic products of microbial fermentation have been found to be a major epigenetic modifier because of their ability to inhibit histone modification.

References

Aiken, C.E., Ozanne, S.E., 2014. Transgenerational developmental programming. Hum. Reprod. Update 20, 63–75.

Alasmar, R.M., Varadharajan, K., Shanmugakonar, M., Hamda, A., Al-Naemi1, 2019. Gut microboiota and health: understanding the role of diet. Food Nutr. Sci. 10, 1344–1373, 2019.

Alenghat, T., Osborne, L.C., Saenz, S.A., Kobuley, D., Ziegler, C.G., Mullican, S.E., Choi, I., Grunberg, S., Sinha, R., Wynosky-Dolfi, M., et al., 2013. Histone deacetylase 3 coordinates commensal-bacteria-dependent intestinal homeostasis. Nature 504, 153–157.

Almgren, M., Schlinzig, T., Gomez-Cabrero, D., Gunnar, A., Sundin, M., Johansson, S., et al., 2014. Cesarean delivery and hematopoietic stem cell epigenetics in the newborn infant: implications for future health? Am. J. Obstet. Gynecol. 211, 502.

Andrade-Oliveira, V., Amano, M.T., Correa-Costa, M., et al., 2015. Gut bacteria products prevent AKI induced by ischemia–reperfusion. J. Am. Soc. Nephrol. 26, 1877–1888.

Arboleya, S., Binetti, A., Salazar, N., Fernandez, N., Solis, G., Hernandez-Barranco, A., et al., 2012. Establishment and development of intestinal microbiota in preterm neonates. FEMS Microbiol. Ecol. 79, 763–772.

Backhed, F., 2011. Programming of host metabolism by the gut microbiota. Ann. Nutr. Metab. 58 (Suppl. 2), 44–52.

Backhed, F., Ley, R.E., Sonnenburg, J.L., Peterson, D.A., Gordon, J.I., 2005. Host-bacterial mutualism in the human intestine. Science 307, 1915–1920.

Bai, J., Hu, Y., Bruner, D.W., 2019. Composition of gut microbiota and its association with body mass index and lifestyle factors in a cohort of 7–18 years old children from the American Gut Project. Pediatr. Obes. 14, e12480.

Belkaid, Y., Hand, T.W., 2014. Role of the microbiota in immunity and inflammation. Cell 157 (1), 121–141.

Bernstein, B.E., Meissner, A., Lander, E.S., 2007. The mammalian epigenome. Cell 128 (4), 669–681.

Bierne, H., Hamon, M., Cossart, P., 2012. Epigenetics and bacterial infections. Cold Spring Harb Perspect. Med. 2 (12), a010272.

Bindels, L.B., Delzenne, N.M., Cani, P.D., Walter, J., 2015. Towards a more comprehensive concept for prebiotics. Nat. Rev. Gastroenterol. Hepatol. 12, 303–310. https://doi.org/10.1038/nrgastro.2015.47.

Bloomfield, P.S., Selvaraj, S., Veronese, M., Rizzo, G., Bertoldo, A., Owen, D.R., … Howes, O.D., 2016. Microglial activity in people at ultra high risk of psychosis and in schizophrenia: an [(11)C]PBR28 PET brain imaging study. Am. J. Psychiatr. 173, 44–52.

Braun-Fahrlander, C., Riedler, J., Herz, U., et al., 2002. Environmental exposure to endotoxin and its relation to asthma in school age children. N. Engl. J. Med. 347 (12), 869–877.

Canani, R.B., Di Costanzo, M., Leone, L., 2012. The epigenetic effects of butyrate: potential therapeutic implications for clinical practice. Clin. Epigenet. 4 (1), 4.

Chang, P.V., Hao, L., Offermanns, S., Medzhitov, R., 2014. The microbial metabolite butyrate regulates intestinal macrophage function via histone deacetylase inhibition. Proc. Natl. Acad. Sci. U.S.A. 111, 2247–2252.

Chu, D.M., Antony, K.M., Ma, J., Prince, A.L., Showalter, L., Moller, M., Aagaard, K.M., 2016. The early infant gut microbiome varies in association with a maternal high-fat diet. Genome Med. 8 (1), 77.

Collado, M.C., Rautava, S., Aakko, J., Isolauri, E., Salminen, S., 2016. Human gut colo-nisation may be initiated in utero by distinct microbial communities in the placenta and amniotic fluid. Sci. Rep. 6, 23129. https://doi.org/10.1038/srep23129.

Connor, K.L., Chehoud, C., Altrichter, A., Chan, L., DeSantis, T.Z., Lye, S.J., 2018. Maternal metabolic, immune, and microbial systems in late pregnancy vary with malnutrition in mice. Biol. Reprod. 98, 579–592.

Costantini, L., Molinari, R., Farinon, B., Merendino, N., 2017. Impact of omega-3 fatty acids on the gut microbiota. Int. J. Mol. Sci. 18, E2645.

Cummings, J.H., Macfarlane, G.T., 1991. The control and consequences of bacterial fermentation in the human colon. J. Appl. Bacteriol. 70, 443–459. https://doi.org/10.1111/j.1365-2672.1991.tb02739.x.

Dalile, B., Van Oudenhove, L., Vervliet, B., Verbeke, K., 2019. The role of short-chain fatty acids in microbiota-gut-brain communication. Nat. Rev. Gastroenterol. Hepatol. 16, 461–478.

Davalos, A., Goedeke, L., Smibert, P., Ramirez, C.M., Warrier, N.P., Andreo, U., et al., 2011. miR-33a/b contribute to the regulation of fatty acid metabolism and insulin signaling. Proc. Natl. Acad. Sci. U.S.A. 108 (22), 9232–9237.

Davie, J.R., 2003. Inhibition of histone deacetylase activity by butyrate. J. Nutr. 133, 2485S–2493S.

de Zoeten, E.F., Wang, L., Sai, H., Dillmann, W.H., Hancock, W.W., 2010. Inhibition of HDAC9 increases T regulatory cell function and prevents colitis in mice. Gastroenterology 138, 583–594.

Decker, E., Hornef, M., Stockinger, S., 2011. Cesarean delivery is associated with celiac disease but not inflammatory bowel disease in children. Gut Microb. 2, 91–98.

Deehan, E.C., Duar, R.M., Armet, A.M., Perez-Munoz, M.E., Jin, M., Walter, J., 2017. Modulation of the gastrointestinal microbiome with nondigestible fermentable carbohydrates to Improve human health. Microbiol. Spectr. 5, 5. https://doi.org/10.1128/microbiolspec.BAD-0019-2017.

Dempsey, J., Zhang, A., Cui, J.Y., 2018. Coordinate regulation of long noncoding RNAs and protein-coding genes in germ-free mice. BMC Genom. 19 (1), 834.

Devaux, C.A., Raoult, D., 2018. The microbiological memory, an epigenetic regulator governing the balance between good health and metabolic disorders. Front. Microbiol. 9, 1379.

DiGiulio, D.B., Romero, R., Amogan, H.P., Kusanovic, J.P., Bik, E.M., Gotsch, F., et al., 2008. Microbial prevalence, diversity and abundance in amniotic fluid during preterm labor: a molecular and culture-based investigation. PLoS One 3, e3056.

Ding, S.Z., Fischer, W., Kaparakis-Liaskos, M., Liechti, G., Merrell, D.S., Grant, P.A., Ferrero, R.L., Crowe, S.E., Haas, R., Hatakeyama, M., Goldberg, J.B., 2010. *Helicobacter pylori*-induced histone modification, associated gene expression in gastric epithelial cells, and its implication in pathogenesis. PLoS One 5 (4), e9875. https://doi.org/10.1371/journal.pone.0009875.

Dunn, G.A., Bale, T.L., 2011. Maternal high-fat diet effects on third-generation female body size via the paternal lineage. Endocrinology 152, 2228–2236.

Earley, Z.M., Akhtar, S., Green, S.J., Naqib, A., Khan, O., Cannon, A.R., et al., 2015. Burn Injury alters the intestinal microbiome and increases gut permeability and bacterial translocation. PLoS One 10, e0129996. http://dx.plos.org/10.1371/journal.pone.0129996.

Egger, G., Liang, G., Aparicio, A., Jones, P.A., 2004. Epigenetics in human disease and prospects for epigenetic therapy. Nature 429 (6990), 457–463. https://doi.org/10.1038/nature02625.

Elwood, P.C., Givens, D.I., Beswick, A.D., Fehily, A.M., Pickering, J.E., Gallacher, J., 2008. The survival advantage of milk and dairy consumption: anoverview of evidence from cohort studies of vascular diseases diabetesand cancer. J. Am. Coll. Nutr. 27, 723S–734S.

Eskandarian, H.A., Impens, F., Nahori, M.A., Soubigou, G., Coppée, J.Y., Cossart, P., Hamon, M.A., 2013. A role for SIRT2-dependent histone H3K18 deacetylation in bacterial infection. Science 341 (6145), 1238858.

Esteller, M., 2007. Cancer epigenomics: DNA methylomes and histone modification maps. Nat. Rev. Genet. 8, 286–298.

Flanagan, J.M., 2007. Host epigenetic modifications by oncogenic viruses. Br. J. Cancer 96, 183–188.

Francino, M.P., 2014. Early development of the gut microbiota and immune health. Pathogens 3 (3), 769–790.

Fung, K.Y.C., Cosgrove, L., Lockett, T., Head, R., Topping, D.L., 2012. A review of the potential mechanisms for the lowering of colorectal oncogenesis by butyrate. Br. J. Nutr. 108, 820–831.

Furusawa, Y., Obata, Y., Fukuda, S., Endo, T.A., Nakato, G., Takahashi, D., Nakanishi, Y., Uetake, C., Kato, K., Kato, T., et al., 2013. Commensal microbe-derived butyrate induces the differentiation of colonic regulatory T cells. Nature.

Galla, S., Chakraborty, S., Cheng, X., Yeo, J.Y., Mell, B., Chiu, N., Wenceslau, C.F., Vijay-Kumar, M., Joe, B., 2020. Exposure to amoxicillin in early life is associated with changes in gut microbiota and reduction in blood pressure: findings from a study on rat dams and offspring. J. Am. Heart Assoc. 9, e014373. https://doi.org/10.1161/JAHA.119.014373.

Gallo, R.L., Hooper, L.V., 2012. Epithelial antimicrobial defence of the skin and intestine. Nat. Rev. Immunol. 12, 503–516.

Gerhauser, C., 2018. Impact of dietary gut microbial metabolites on the epigenome. Philos. Trans. R. Soc. Lond. B Biol. Sci. 373, 20170359.

Grabiec, A.M., Potempa, J., 2018. Epigenetic regulation in bacterial infections: targeting histone deacetylases. Crit. Rev. Microbiol. 44 (3), 336–350.

Grabiec, A.M., Reedquist, K.A., 2013. The ascent of acetylation in the epigenetics of rheumatoid arthritis. Nat. Rev. Rheumatol. 9 (5), 311–318.

Hamon, M.A., Cossart, P., 2008. Histone modifications and chromatin remodeling during bacterial infections. Cell Host Microbe 4 (2), 100–109.

Hamon, M.A., Batsché, E., Régnault, B., Tham, T.N., Seveau, S., Muchardt, C., Cossart, P., 2007. Histone modifications induced by a family of bacterial toxins. Proc. Natl. Acad. Sci. U.S.A. 104 (33), 13467–13472.

Handy, D.E., Castro, R., Loscalzo, J., 2011. Epigenetic modifications: basic mechanisms and role in cardiovascular disease. Circulation 123 (19), 2145–2156.

Hill, M.J., 1997. Intestinal flora and endogenous vitamin synthesis. Eur. J. Cancer Prev. 6 (Suppl. 1), S43–S45.

Hooper, L.V., Midtvedt, T., Gordon, J.I., 2002. How host-microbial interactions shape the nutrient environment of the mammalian intestine. Annu. Rev. Nutr. 22, 283–307.

Hooper, L.V., Littman, D.R., Macpherson, A.J., 2012. Interactions between the microbiota and the immune system. Science 336, 1268–1273.

Hsu, C.N., Hou, C.Y., Lee, C.T., Chan, J.Y., Tain, Y.L., 2019. The interplay between maternal and post-weaning high-fat diet and gut microbiota in the developmental programming of hypertension. Nutrients 11, E1982.

Huh, S.Y., Rifas-Shiman, S.L., Zera, C.A., Edwards, J.W., Oken, E., Weiss, S.T., et al., 2012. Delivery by caesarean section and risk of obesity in preschool age children: a prospective cohort study. Arch. Dis. Child. 97, 610–616.

Hullar, M.A., Fu, B.C., 2014. Diet, the gut microbiome, and epigenetics. Cancer J. (Sudbury, MA) 20 (3), 170–175.

Hylemon, P.B., Harder, J., 1998. Biotransformation of monoterpenes, bile acids, and other isoprenoids in anaerobic ecosystems. FEMS Microbiol. Rev. 22 (5), 475–488.

Indrio, F., Martini, S., Francavilla, R., Corvaglia, L., Cristofori, F., Mastrolia, S.A., … Loverro, G., 2017. Epigenetic matters: the link between early nutrition, microbiome, and long-term health development. Front. Pediatr. 5, 178.

Jandhyala, S.M., Talukdar, R., Subramanyam, C., Vuyyuru, H., Sasikala, M., Nageshwar Reddy, D., 2015. Role of the normal gut microbiota. World J. Gastroenterol. 21, 8787–8803.

Jirtle, R.L., Skinner, M.K., 2007. Environmental epigenomics and disease susceptibility. Nat. Rev. Genet. 8, 253–262.

Kamada, N., Seo, S.U., Chen, G.Y., Nunez, G., 2013. Role of the gut microbiota in immunity and inflammatory disease. Nat. Rev. Immunol. 13, 321–335.

Kaminsky, Z., Wang, S.C., Petronis, A., 2006. Complex disease, gender and epigenetics. Ann. Med. 38, 530–544.

Kang, H., 2023. Regulation of acetylation states by nutrients in the inhibition of vascular inflammation and atherosclerosis. Int. J. Mol. Sci. 24 (11), 9338.

Khosravi, A., Yanez, A., Price, J.G., Chow, A., Merad, M., Goodridge, H.S., Mazmanian, S.K., 2014. Gut microbiota promote hematopoiesis to control bacterial infection. Cell Host Microbe 15, 374–381.

Kim, E.R., Chang, D.K., 2014. Colorectal cancer in inflammatory bowel disease: the risk, pathogenesis, prevention and diagnosis. World J. Gastroenterol. 20, 9872–9881.

Kok, D.E., Dhonukshe-Rutten, R.A., Lute, C., Heil, S.G., Uitterlinden, A.G., van der Velde, N., et al., 2015. The effects of long-term daily folic acid and vitamin B12 supplementation on genome-wide DNA methylation in elderly subjects. Clin. Epigenet. 7, 121.

Kollmann, T.R., Levy, O., Montgomery, R.R., Goriely, S., 2012. Innate immune function by Toll-like receptors: distinct responses in newborns and the elderly. Immunity 37, 771–783.

Korpela, K., Blakstad, E.W., Moltu, S.J., Strommen, K., Nakstad, B., Ronnestad, A.E., et al., 2018. Intestinal microbiota development and gestational age in preterm neonates. Sci. Rep. 8, 2453.

Krautkramer, K.A., Kreznar, J.H., Romano, K.A., Vivas, E.I., Barrett-Wilt, G.A., Rabaglia, M.E., Keller, M.P., Attie, A.D., Rey, F.E., Denu, J.M., 2016. Diet-microbiota interactions mediate global epigenetic programming in multiple host tissues. Mol. Cell. 64 (5), 982–992. https://doi.org/10.1016/j.molcel.2016.10.025.

Krautkramer, K.A., Dhillon, R.S., Denu, J.M., Carey, H.V., 2017. Metabolic programming of the epigenome: host and gut microbial metabolite interactions with host chromatin. Transl. Res. 189, 30–50.

Kristensen, K., Henriksen, L., 2016. Cesarean section and disease associated with immune function. J. Allergy Clin. Immunol. 137, 587–590.

Lankelma, J.M., Belzer, C., Hoogendijk, A.J., de Vos, A.F., de Vos, W.M., van der Poll, T., Wiersinga, W.J., 2016. Antibiotic-induced gut microbiota disruption decreases TNF-α release by mononuclear cells in healthy adults. Clin. Transl. Gastroenterol. 7, e186.

Lee, H.S., 2019. The interaction between gut microbiome and nutrients on development of human disease through epigenetic mechanisms. Genomics Inform. 17 (3).

Lee, H.S., Barraza-Villarreal, A., Hernandez-Vargas, H., Sly, P.D., Biessy, C., Ramakrishnan, U., et al., 2013. Modulation of DNA methylation states and infant immune system by dietary supplementation with omega-3 PUFA during pregnancy in an intervention study. Am. J. Clin. Nutr. 98, 480–487.

Li, Y., Tollefsbol, T.O., 2010. Impact on DNA methylation in cancer prevention and therapy by bioactive dietary components. Curr. Med. Chem. 17, 2141—2151.

Liang, L., Ai, L., Qian, J., Fang, J.Y., Xu, J., 2015. Long noncoding RNA expression profiles in gut tissues constitute molecular signatures that reflect the types of microbes. Sci. Rep. 5, 11763.

Luu, M., Pautz, S., Kohl, V., Singh, R., Romero, R., Lucas, S., et al., 2019. The short-chain fatty acid pentanoate suppresses autoimmunity by modulating the metabolic-epigenetic crosstalk in lymphocytes. Nat. Commun. 10, 760.

Ma, J., Prince, A.L., Bader, D., Hu, M., Ganu, R., Baquero, K., … Aagaard, K.M., 2014. High-fat maternal diet during pregnancy persistently alters the offspring microbiome in a primate model. Nat. Commun. 5, 3889.

Macfarlane, S., Macfarlane, G.T., 2003. Regulation of short-chain fatty acid production. Proc. Nutr. Soc. 62, 67—72.

Macfarlane, G.T., Macfarlane, S., 2012. Bacteria colonic fermentation, and gas-trointestinal health. J. AOAC Int. 95, 50—60.

Maekita, T., Nakazawa, K., Mihara, M., Nakajima, T., Yanaoka, K., Iguchi, M., Arii, K., Kaneda, A., Tsukamoto, T., Tatematsu, M., Tamura, G., Saito, D., Sugimura, T., Ichinose, M., Ushijima, T., 2006. High levels of aberrant DNA methylation in *Helicobacter pylori*-infected gastric mucosae and its possible association with gastric cancer risk. Clin. Cancer Res. 12 (3 Pt 1), 989—995.

Marlicz, W., Skonieczna-Zydecka, K., Dabos, K.J., Loniewski, I., Koulaouzidis, A., 2018. Emerging concepts in non-invasive monitoring of Crohn's disease. Ther. Adv. Gastroenterol. 11, 1756284818769.

Masaki, T., Qu, J., Cholewa-Waclaw, J., Burr, K., Raaum, R., Rambukkana, A., 2013. Reprogramming adult Schwann cells to stem cell-like cells by leprosy bacilli promotes dissemination of infection. Cell 152 (1—2), 51—67.

Milagro, F.I., Mansego, M.L., de Miguel, C., Martinez, J.A., 2013. Dietary factors, epigenetic modifications and obesity outcomes: progresses and perspectives. Mol. Aspects Med. 34, 782—812.

Milani, C., Duranti, S., Bottacini, F., Casey, E., Turroni, F., Mahony, J., et al., 2017. The first microbial colonizers of the human gut: composition, activities, and health implications of the infant gut microbiota. Microbiol. Mol. Biol. Rev. 81, e00036-17.

Miller, W.B., 2016. The eukaryotic microbiome: origins and implications for fetal and neonatal life. Front. Pediatr. 4, 96. https://doi.org/10.3389/fped.2016.00096.

Miro-Blanch, J., YanesO, 2019. Epigenetic regulation at the interplay between gut microbiota and host metabolism. Front. Genet. 10.

Mischke, M., Plosch, T., 2013. More than just a gut instinct-the potential interplay between a baby's nutrition, its gut microbiome, and the epigenome. Am. J. Physiol. Regul. Integr. Comp. Physiol. 304, R1065—R1069.

Mujtaba, S., Winer, B.Y., Jaganathan, A., Patel, J., Sgobba, M., Schuch, R., Gupta, Y.K., Haider, S., Wang, R., Fischetti, V.A., 2013. Anthrax SET protein: a potential virulence determinant that epigenetically represses NF-κB activation in infected macrophages. J. Biol. Chem. 288 (32), 23458—23472.

Niller, H.H., Minarovits, J., 2016. Patho-epigenetics of infectious diseases caused by intracellular bacteria. Adv. Exp. Med. Biol. 879, 107—130.

Olaniyi, K.S., Amusa, O.A., Areola, E.D., Olatunji, L.A., 2020. Suppression of HDAC by sodium acetate rectifies cardiac metabolic disturbance in streptozotocin—nicotinamide-induced diabetic rats. Exp. Biol. Med. 1535370220913847.

O'Keefe, S.J., Li, J.V., Lahti, L., 2015. Fat, fibre and cancer risk in. Afr. Am. Rural Afr. 6, 6342.

Paul, B., Barnes, S., Demark-Wahnefried, W., Morrow, C., Salvador, C., Skibola, C., Tollefsbol, T.O., 2015. Influences of diet and the gut microbiome on epigenetic modulation in cancer and other diseases. Clin. Epigenet. 7, 112.

Paul, H.A., Collins, K.H., Bomhof, M.R., Vogel, H.J., Reimer, R.A., 2018. Potential impact of metabolic and gut microbial response to pregnancy and lactation in lean and diet-induced obese rats on offspring obesity risk. Mol. Nutr. Food Res. 62, 1—11. https://doi.org/10.1002/mnfr.201700820.

Pereira, J.M., Hamon, M.A., Cossart, P., 2016. A lasting impression: epigenetic memory of bacterial infections? Cell Host Microbe 19 (5), 579—582. https://doi.org/10.1016/j.chom.2016.04.012.

Perez, P.F., Dore, J., Leclerc, M., Levenez, F., Benyacoub, J., Serrant, P., Segura-Roggero, I., Schiffrin, E.J., Donnet-Hughes, A., 2007. Bacterial imprinting of the neonatal immune system: lessons from maternal cells? Pediatrics 119, e724—e732.

Perry, R.J., Peng, L., Barry, N.A., Cline, G.W., Zhang, D., Cardone, R.L., et al., 2016. Acetate mediates a microbiome-brain-beta-cell axis to promote metabolic syndrome. Nature 534, 213—217.

Priyadarshini, M., Thomas, A., Reisetter, A.C., Scholtens, D.M., Wolever, T.M., Josefson, J.L., Layden, B.T., 2014. Maternal short-chain fatty acids are associated with metabolic parameters in mothers and newborns. Transl. Res. 164, 153—157.

Rasnik, K.S., Chang, H.-W., Yan, D., Lee, K.M., Ucmak, D., Wong, K., Abrouk, M., Farahnik, B., Nakamura, M., Zhu, T.H., Bhutani, T., Liao, W., 2017. Influence of diet on the gut microbiome and implications for human health. J. Transl. Med. 15 (73), 1—17.

Raymond, B., Batsche, E., Boutillon, F., Wu, Y.Z., Leduc, D., Balloy, V., Raoust, E., Muchardt, C., Goossens, P.L., Touqui, L., 2009. Anthrax lethal toxin impairs IL-8 expression in epithelial cells through inhibition of histone H3 modification. PLoS Pathog. 5 (4), e1000359.

Renz, H., von Mutius, E., Brandtzaeg, P., Cookson, W.O., Autenrieth, I.B., Haller, D., 2011. Gene-environment interactions in chronic inflammatory disease. Nat. Immunol. 12, 273—277.

Robertson, R.C., Kaliannan, K., Strain, C.R., Ross, R.P., Stanton, C., Kang, J.X., 2018. Maternal omega-3 fatty acids regulate offspring obesity through persistent modulation of gut microbiota. Microbiome 6, 95.

Roediger, W.E.W., 1980. Role of anaerobic-bacteria in the metabolic welfare of the colonic mucosa in man. Gut 21, 793—798.

Rolando, M., Sanulli, S., Rusniok, C., Gomez-Valero, L., Bertholet, C., Sahr, T., Margueron, R., Buchrieser, C., 2013. Legionella pneumophila effector RomA uniquely modifies host chromatin to repress gene expression and promote intracellular bacterial replication. Cell Host Microbe 13 (4), 395—405.

Rolando, M., Gomez-Valero, L., Buchrieser, C., 2015. Bacterial remodelling of the host epigenome: functional role and evolution of effectors methylating host histones. Cell Microbiol. 17 (8), 1098—1107.

Romano, K.A., Rey, F.E., 2018. Is maternal microbial metabolism an early life determinant of health? Lab. Anim. (NY) 47 (9), 239—243.

Ronald, D.H., Pontefract, B.A., Mishcon, H.R., Black, C.A., Sutton, S.C., Theberge, C.R., 2019. Microbiome: profound implications for diet and disease. Nutrients 11, 1613. https://doi.org/10.3390/nu11071613.

Rook, G.A., Lowry, C.A., Raison, C.L., 2014. Hygiene and other early childhood influences on the subsequent function of the immune system. Brain Res. 1617, 47—62.

Rossi, M., Amaretti, A., Raimondi, S., 2011. Folate production by probiotic bacteria. Nutrients 3, 118—134.

Russell, W.R., Gratz, S.W., Duncan, S.H., et al., 2011. High-protein, reduced-carbohydrate weight-loss diets promote metabolite profiles likely to be detrimental to co lonic health. Am. J. Clin. Nutr. 93, 1062—1072. https://doi.org/10.3945/ajcn.110.002188.

Sanna, S., van Zuydam, N.R., Mahajan, A., Kurilshikov, A., Vich Vila, A., Vosa, U., Mujagic, Z., Masclee, A.A.M., Jonkers, D., Oosting, M., Joosten, L.A.B., Netea, M.G., Franke, L., Zhernakova, A., Fu, J., Wijmenga, C., McCarthy, M.I., 2019. Causal relationships among the gut microbiome, short-chain fatty acids and metabolic diseases. Nat. Genet. 51, 600—605.

Sawicka, A., Seiser, C., 2012. Histone H3 phosphorylation - a versatile chromatin modification for different occasions. Biochimie 94 (11), 2193—2201.

Shimazu, T., Hirschey, M.D., Newman, J., He, W., Shirakawa, K., Le Moan, N., et al., 2013. Suppression of oxidative stress by beta-hydroxybutyrate, an endogenous histone deacetylase inhibitor. Science 339, 211—214.

Soliman, M.L., Rosenberger, T.A., 2011. Acetate supplementation increases brain histone acetylation and inhibits histone deacetylase activity and expression. Mol. Cell. Biochem. 352 (1—2), 173—180.

Sprong, R., Schonewille, A., Van der Meer, R., 2010. Dietary cheese whey proteinprotects rats against mild dextran sulfate sodium?induced colitis: role ofmucin and microbiota. J. Dairy Sci. 93, 1364—1371.

Stanislawski, M.A., Dabelea, D., Wagner, B.D., Sontag, M.K., Lozupone, C.A., Eggesbo, M., 2017. Prepregnancy weight, gestational weight gain, and the gut microbiota of mothers and their infants. Microbiome 5, 113.

Thakur, V.S., Gupta, K., Gupta, S., 2012. Green tea polyphenols increase p53 transcriptional activity and acetylation by suppressing class I histone deacetylases. Int. J. Oncol. 41, 353—361.

Thorburn, A.N., McKenzie, C.I., Shen, S., Stanley, D., Macia, L., Mason, L.J., Roberts, L.K., Wong, C.H.Y., Shim, R., Robert, R., et al., 2015. Evidence that asthma is a developmental origin disease influenced by maternal diet and bacterial metabolites. Nat. Commun. 6, 7320.

Valdes, A.M., Walter, J., Segal, E., Spector, T.D., 2018. Role of the gut microbiota in nutrition and health. BMJ 361. http://www.bmj.com/. Accessed 15 January 2019.

van Dijk, S.J., Zhou, J., Peters, T.J., Buckley, M., Sutcliffe, B., Oytam, Y., et al., 2016. Effect of prenatal DHA supplementation on the infant epigenome: results from a randomized controlled trial. Clin. Epigenetics 8, 114.

Vazquez-Gomez, M., Garcia-Contreras, C., Torres-Rovira, L., Pesantez, J.L., Gonzalez-Anover, P., Gomez-Fidalgo, E., et al., 2017. Polyphenols and IUGR pregnancies: maternal hydroxytyrosol supplementation improves prenatal and early-postnatal growth and metabolism of the offspring. PLoS One 12, e0177593.

Vinolo, M.A., Rodrigues, H.G., Hatanaka, E., Hebeda, C.B., Farsky, S.H., Curi, R., 2009. Short-chain fatty acids stimulate the migration of neutrophils to inflammatory sites. Clin. Sci. (Lond.) 117, 331—338.

Virtue, A.T., McCright, S.J., Wright, J.M., Jimenez, M.T., Mowel, W.K., Kotzin, J.J., Joannas, L., Basavappa, M.G., Spencer, S.P., Clark, M.L., Eisennagel, S.H., Williams, A., Levy, M., Manne, S., Henrickson, S.E., Wherry, E.J., Thaiss, C.A., Elinav, E., Henao-Mejia, J., 2019. The gut microbiota regulates white adipose tissue inflammation and obesity via a family of microRNAs. Sci. Transl. Med. 11 (496), eaav1892. https://doi.org/10.1126/scitranslmed.aav1892.

Wang, Y., Curry, H.M., Zwilling, B.S., Lafuse, W.P., 2005. Mycobacteria inhibition of IFN-gamma induced HLA-DR gene expression by up-regulating histone deacetylation at the promoter region in human THP-1 monocytic cells. J. Immunol. 174 (9), 5687—5694.

Wang, X., Wang, W., Wang, L., Yu, C., Zhang, G., Zhu, H., et al., 2019. Lentinan modulates intestinal microbiota and enhances barrier integrity in a piglet model challenged with lipopolysaccharide. Food Funct. 10 (1), 479—489.

Wankhade, U.D., Zhong, Y., Kang, P., Alfaro, M., Chintapalli, S.V., Thakali, K.M., et al., 2017. Enhanced offspring predisposition to steatohepatitis with maternal high-fat diet is associated with epigenetic and microbiome alterations. PLoS One 12, e0175675.

Whitt, J., Woo, V., Lee, P., Moncivaiz, J., Haberman, Y., Denson, L., et al., 2018. Disruption of epithelial HDAC3 in intestine prevents diet-induced obesity in mice. Gastroenterology 155, 501—513.

Winter, S.E., Baumler, A.J., 2014. Dysbiosis in the inflamed intestine: chance favors the prepared microbe. Gut Microb. 5, 71—73.

Wu, S., Tian, L., 2017. Diverse phytochemicals and bioactivities in the ancient fruit and modern functional food pomegranate (*Punica granatum*). Molecules 22, E1606.

Wu, H.J., Wu, E., 2012. The role of gut microbiota in immune homeostasis and autoimmunity. Gut Microb. 3 (1), 4—14.

Yoshii, K., Hosomi, K., Sawane, K., Kunisawa, J., 2019. Metabolism of dietary and microbial vitamin B family in the regulation of host immunity. Front. Nutr. 6, 48.

Zhong, H., Penders, J., Shi, Z., Ren, H., Cai, K., Fang, C., Ding, Q., Thijs, C., Blaak, E.E., Stehouwer, C.D.A., Xu, X., Yang, H., Wang, J., Wang, J., Jonkers, D., Masclee, A.A.M., Brix, S., Li, J., Arts, I.C.W., Kristiansen, K., 2019. Impact of early events and lifestyle on the gut microbiota and metabolic phenotypes in young school-age children. Microbiome 7, 2.

CHAPTER 11

Gut microbiota, nutrition, and health: Fundamental and basic principle

Olugbenga Samuel Michael[1,6,7], Juliana Bunmi Adetunji[2], Oluwaseun Aremu Adeyanju[3,9], Charles Oluwaseun Adetunji[4], Olufunto Olayinka Badmus[5,8], Kehinde Samuel Olaniyi[2] and Ayodele Olufemi Soladoye[1]

[1]Cardiometabolic, Microbiome and Applied Physiology Laboratory, Department of Physiology, College of Health Sciences, Bowen University, Iwo, Osun State, Nigeria; [2]Nutritional and Toxicological Research Laboratory, Department of Biochemistry Sciences, Osun State University, Osogbo, Osun State, Nigeria; [3]Department of Physiology, College of Medicine and Health Sciences, Afe Babalola University, Ado–Ekiti, Ekiti, Nigeria; [4]Applied Microbiology, Biotechnology and Nanotechnology Laboratory, Department of Microbiology, and Directorate of Research and Innovation, Edo State University Uzairue, Iyamho, Auchi, Edo State, Nigeria; [5]Department of Public Health, Kwara State University, Malete, Kwara, Nigeria; [6]Department of Physiology, University of Tennessee Health Science Center, Memphis, TN, United States; [7]Department of Medical Pharmacology and Physiology, University of Missouri, Columbia, MO, United States; [8]Department of Physiology and Biophysics, Cardiorenal, and Metabolic Diseases Research Center, University of Mississippi Medical Center, Jackson, MS, United States; [9]Department of Cellular and Molecular Biology, The University of Texas Health Science Center at Tyler, Tyler, TX, United States

Introduction

Gut microbiome has gained enormous recognition in recent times for many reasons, part of which include regulation of physiological processes in the host and contribution to pathogenesis of some diseases; the microbiome homeostasis is altered. It weighs around 2 kg, and microorganisms in the gut are very enormous numerically (Eckburg et al., 2005). The gut microorganisms carry out regulatory roles in human health and diseases spanning from metabolism, immune response, cardiovascular to neurological effects (Patterson et al., 2016). Intestinal microbiome consists of constellation microbes. The interplay and balance of the intestinal microbes is very important for preservation of physiological homeostasis (Victor and Quigley, 2016; Dominguez-Bello et al., 2019). Recently, there is an astounding growth and technological advances in the field of microbiome research, which enhances the identification of microorganisms, their genome, and metabolic functions (Claesson et al., 2017).

Microbiome—host cross-talks have been examined using diverse experimental models. Germ-free mice have been for determination of detrimental effects of lacking microorganisms on the growth and effectiveness of the immunological system (Bhattarai and Kashyap, 2016; Dominguez-Bello et al., 2019; Uzbay, 2019). Microbiome develops alongside the host meaning that the developmental progress made by the host also affects the intestinal microbes. At birth, the gut is germ-free (Rehbinder et al., 2018; Theis et al., 2019) but gained its microbial flora during the birthing process via the vagina, subsequently oral ingestion then surrounding environment (Ryan et al., 2019; Zhuang

An Introduction to the Microbiome in Health and Diseases
ISBN 978-0-323-91190-0,
https://doi.org/10.1016/B978-0-323-91190-0.00011-4

et al., 2019). Interestingly, diverse intestinal microbes swiftly enlarge in the space of the foremost 3 years of life and subsequently become stable with microbial constellation similar to adult's intestinal microbes (Ryan et al., 2019; Zhuang et al., 2019). This premature and decisive developmental period of the intestinal microbiota is exceptionally susceptible to either favorable or harmful modulation, suggesting that diseases that manifest in adults frequently have developmental origins (Codagnone et al., 2019).

The cross-talk between colonizing microorganisms and the host is quite interesting because the host offers the intestinal microbial flora steady environmental conditions to thrive while the microbes supply the host with many interesting functions including fermentation of dietary fibers, proteins, and carbohydrates, generation of metabolic products such as vitamins, nutrients, protection against pathogenic microorganisms, and immunoregulatory function. There are also negative consequences of intestinal microbes on the health of the host involving generation of detrimental products linked with pathogenesis of diseases, whereas favorable or positive products that offer protection against diseases are also generated by metabolic processes of the microbes (Bhattarai and Kashyap, 2016; Dominguez-Bello et al., 2019; Uzbay, 2019).

In addition, it is very likely that many factors reshape or modulate the gut microbiome dynamics in adults; such factors include age, environment, genetics, delivery mode at birth, stress, drugs, and diet that has emerged as the sole stimulator of microbiome dynamics and differences in individuals (Claesson et al., 2012; Falony et al., 2016). Dietary manipulation drives intestinal microbiome diversity and metabolic functions, making the intestinal microorganisms a connection between nutrition and health (Makki et al., 2018). Therefore, human nutritional consequences are enormous and cannot be overemphasized because it has capacity to reshape the microbiome composition and functions in mediating diet—microbiota—host interactions are very clear from infancy when milk from lactating mothers enhances microbiome growth in infants (Charbonneau et al., 2016), after which gut flora's bacteria community is enlarged due to the commencement of adult's solid food (Laursen et al., 2017), and finally, there is reduction in the gut flora commensal bacteria community as a result of advanced age in elderly likely because of decreased food variety (Claesson et al., 2012). A healthy neonatal microbiome is enhanced by the following factors; vaginal-type delivery, full-term birth, breastfeeding, and contact with diverse microbes. However, a caesarean section, preterm birth, formulated milk, and antibiotics administration have adverse effects on the variety and constellation of the microbiome in the newborn (Gritz and Bhandari, 2015; Mueller et al., 2015).

Colonic microbiota fermentation of indigestible fibers generates bioactive products called short-chain fatty acids (SCFAs) such as propionate, acetate, and butyrate (Macfarlane and Macfarlane, 2012). When these dietary fibers are lacking, the gut microbiota change to less energy-rich foods such as amino acids or fats for development and metabolism (Cummings and Macfarlane, 1997), causing decreased gut microbiome fermentation function resulting in low SCFAs metabolic production (Russell et al., 2011). Fermentation of protein produces mainly branched-chain fatty acids and little SCFAs (Smith and Macfarlane, 1997). The metabolites from protein fermentation have been associated with metabolic disorder (Newgard et al., 2009). However, addition of fiber-rich diets to high-fat diets (HFDs) will reinstate or increase the number of commensal bacteria, leading to reduction in the harmful microbiota metabolic products and heightened SCFAs production (Sanchez et al., 2009). SCFAs provide protection by increasing intestinal integrity, through enhancement of immune response, increase energy expenditure, and elevated insulin sensitivity (Koh et al., 2016; Omolekulo et al., 2019). They are also important signaling molecules that play epigenetic roles through histone deacetylases (HDACs) inhibition or by activating G protein—coupled receptors (GPCRs) (Olaniyi et al., 2020).

Therefore, gut dysbiosis that occurs as a result of dietary manipulations results in disruption of intestinal integrity causing inflammation, which has detrimental impacts on the host (Amabebe et al., 2020; Singh et al., 2021). Experimental studies have demonstrated that consumption of Western diets or fast foods, fat-rich diets, ans diets low in fiber can cause to wear down the mucosal epithelial layer of the host, resulting in inflammation and development of cardiometabolic disorders (Yao et al., 2021; Aziz et al., 2024; O'Donnell et al., 2023). Interestingly, intake of dietary fibers mitigates HFD or Western diets—associated metabolic dysregulation (Yao et al., 2021; Aziz et al., 2024; O'Donnell et al., 2023). Butyrate, a metabolite from microbial fermentation, protects against cardiometabolic syndrome such as obesity, diabetes, glucose dysregulation, and hypertension (Adeyanju et al., 2021; Badejogbin et al., 2019). Diet—microbiota—host interactions and their implication in health and diseases will be explained in this chapter. Furthermore, microbiome has gained therapeutic application in some disease condition through fecal or vagina microbiota transplantation. Therefore, health application of gut microbiome will also be explored.

Diet and gut microbiota

Diets have the ability to interfere with gut microbiota directly by stimulating or inhibiting some of their developmental and metabolic processes such as growth, and energy generation from particular dietary composition provides a unique benefit to this preferred microorganism of the intestinal microbiota. Therefore, dietary supplementation has the capability to alter the amount, viability, and development of gut microbes (Korem et al., 2015). Diet-mediated reshaping of the gut microbiome activity could cause favorable or harmful health effects, which may be as a result of the immunoregulatory functions due to impaired microbial homeostasis, epigenetic alterations, or varying types of fermentation products or metabolites with the capacity to induce localized effects in the intestine or system-wide or interorgan effects. Therefore, the diet—microbiota—host interaction health consequences may not mandatorily need modification of the entire microbial composition, but specific microbes in the gut may be altered by the dietary manipulations (Smith et al., 2013a).

Diet has emerged as very important modulators of the gut microbiome activity (Graf et al., 2015). Food constituents such as carbohydrate, protein and fat, probiotics, prebiotics, food additives, and polyphenols can reshape microbial constitution as well as activity (Roca-Saavedra et al., 2018). Researches revealed that diets rich in fiber are beneficial in ameliorating disease conditions (Kovatcheva-Datchary et al., 2015). Nutrition influences intestinal microbial function, constituents as well as host health through diverse mechanistic approach as will be discussed in the following.

Dietary fiber

Microbial fermentation of dietary fibers generates end products such as hydrogen, carbon dioxide, and SCFAs, which have enormous roles in reshaping the gut ecology (Chassard and Lacroix, 2013). The intake of plant fibers improved microbiome community with abundant *Roseburia, Lachnospira,* and *Prevotella* with elevated SCFA generation (De Filippis et al., 2015). However, obese humans on low dietary fibers resulted in altered intestinal microbial flora with reduced quantity of gut microbes such as *Roseburia* spp., *Eubacterium rectale,* and *Bifidobacteria* alongside proportionate reduction in SCFAs (butyrate) levels in the feces of the subjects (Duncan et al., 2007). Therefore, dietary fiber-induced alteration in gut microbiome constituents and increased metabolic end products have the potential to autonomously control the end results of diseases.

Interestingly, dietary fibers have been shown to positively modify or reshape intestinal microbial flora across generations as demonstrated by improved variety and abundance of commensal bacteria in the gut of overweight pregnant women following consumption of fibers (Roytio et al., 2017). However, consumption of diets low in fibers resulted in persistence of negatively altered intestinal microbial community across multiple generations of rodents. The fascinating thing here is that the observed effect cannot be reversed in the offspring even with improved consumption of fiber-rich diets. This clearly emphasizes the developmental origin of diseases in adulthood due to unfavorable conditions during pregnancy on the fetus (Sonnenburg et al., 2016).

Fat

HFD supplementation plays a significant contributory role in the development of obesity and cardiovascular diseases. HFD-associated dysregulation is connected to alteration of the intestinal microbial flora. Butyrate, a fermentation product, was reported to reverse cardiac damage induced by HFD through uric acid—dependent mechanistic pathway (Badejogbin et al., 2019). HFD possess the capability to reshape gut microbiome negatively (Collins et al., 2016) resulting in inflammation. Butyrate production by intestinal microbiome is reduced by HFD probably due to alteration of intestinal microbiome constitution (Cheng et al., 2016).

Fecal microbiome transplantation alleviates HFD-induced liver damage through the enhanced modulation gut microbial flora (Zhou et al., 2017). HFD can stimulate the development of a proinflammatory gut microbiota, which results in elevated permeability of the intestine causing increased plasma levels of lipopolysaccharides (LPS) (Kim et al., 2012; Araújo et al., 2017). Neurodegenerative diseases may develop due to prolonged HFD intake, which may be through impaired microbiome homeostasis. This interaction is called microbiota—gut—brain interaction (Proctor et al., 2017).

Obesity induced by HFD has been reported to be alleviated by sodium butyrate an SCFA through the reversal of gut dysbiosis, decreased circulating LPS levels, and reshaping of intestinal microflora (Fang et al., 2019). Therefore, HFD impairs intestinal microflora constitution and functionality resulting in development of chronic diseases such as obesity, cognitive decline, cardiac damage, and hepatic steatosis and even fatty pancreas (You et al., 2022; Liang et al., 2023). This further underscores dietary potentials to reshape intestinal microbiota to initiate disease development.

Protein

Protein-rich red meat is continuously connected to cardiovascular diseases (CVDs) (Rohrmann and Linseisen, 2016; Wolk, 2017). L-carnitine is a major amino acid constituent of red meat, and it undergoes metabolic degradation by the gut microbes to form trimethylamine (TMA) and is futher oxidized to trimethylamine N-oxide (TMAO) by flavin monooxygenase (Koeth et al., 2013).

TMAO is connected with CVD pathogenesis especially atherothrombotic CVD. Alteration of intestinal microbiome induced by L-carnitine supplementation in a rodent model and human subjects result in elevated concentration of TMA and TMAO in the circulation and augmented risk of developing atherothrombotic CVD. Interestingly, these observed changes were mitigated by antibiotics. During this study, it was revealed that L-carnitine supplementation favored reshaping of the gut microbiome with abundant *Prevotella*, which are largely responsible for the formation of TMAO in human subjects and rodent model (Koeth et al., 2013). Furthermore, excessive generation of TMAO by the gut microbiota has been connected with platelet hyperreactivity and thrombotic diseases (Zhu et al., 2016).

Food additives

Food additives such as emulsifiers carboxymethylcellulose and polysorbate-80 reshape the gut microbial flora resulting in gut dysbiosis and associated proinflammatory status and cardiometabolic dysfunction (Chassaing et al., 2015). Microbiome transplantation from a mice fed with food additive (emulsifier) into germ-free mice stimulates development of classical signs and gene expression such as inflammation and altered gut microbial flora, which were associated with the mice fed with emulsifier (Chassaing et al., 2017). Phophatidylcholine (a lecithin) is another emulsifier that has been shown to also alter the intestinal microbiome (in a similar fashion as L-carnitine) resulting in detrimental health consequences in humans due to increased circulating levels of TMAO, which has been known to be connected with CVD pathogenesis (Tang et al., 2013).

Noncaloric artificial sweeteners (NAS) are a class of food additive frequently used because of its associated weight-reducing potential. However, this claim has been contradicted by a study that revealed that NAS induces weight gain and metabolic dysregulation in both rodents and humans via impairment of microbiome composition and function, thereby causing metabolic derangement (Suez et al., 2015).

NAS-induced metabolic dysregulation and microbiota alteration have been noticed following examples of NAS administered to rats such as saccharin, aspartame, sucralose, etc. (Palmnäs et al., 2014). Glucose dysregulation is a major metabolic disturbance induced by saccharin in rodent. This was further established in study where fecal microbiota of a saccharin-treated rodent was injected into germ-free mice through fecal microbiota transplantation (Michael et al., 2022). The germ-free mice then developed impaired glucose tolerance such as saccharin-treated rodent (Suez et al., 2014). Another study demonstrated that NAS (saccharin and aspartame) led to increased generation of SCFAs specifically acetate and propionate, which may be associated with elevated energy production by the intestinal microbes of the NAS-treated mice (Palmnäs et al., 2014). Saccharin has been shown to also cause liver damage through impaired microbiome composition and metabolic activities (Bian et al., 2017). Therefore, food additives are not just metabolically inert component of our diets, but they possess unique potential to reshape our intestinal microflora.

Polyphenols

Most fruits and plants contain phytochemicals and phenols, which are known to have medicinal effects. They are sometimes referred to as flavonoids (Santhakumar et al., 2018). Flavonoids or polyphenols have been reported to possess antioxidative, cardiovascular protective, and neuronal protective attributes (Selma et al., 2009; Cardona et al., 2013). Microbiome activates the phenols into a bioactive compound that is compatible and beneficial to the body (Selma et al., 2009; Cardona et al., 2013).

Sun and coworkers (2018) demonstrated that phenolic compounds present in tea have potential to modulate the intestinal microbial flora through elevated quantity of *Bifidobacterium, Enterococcus,* and *Lactobacillus* and simultaneously caused elevated SCFA generation. In addition, resveratrol, a polyphenol, present in wine has been reported to have positive health benefits such as cardioprotective, neuroprotective, and antioxidative effects largely in part by altering gut microbiota composition and activity. Larrosa et al. (2009) demonstrated that resveratrol ameliorates colitis by increasing *Bifidobacteria* and *Lactobacilli* and decreasing the proliferation of *E. coli* as well as *Enterobacteria* in the gut. Therefore, phenolic compounds can reshape the gut microbiome through the enhancement of the commensal bacteria community, which causes positive health activities (Espín et al., 2017). Furthermore, phenolic compounds in red wine have obesity lowering activities and confer protection against cardiometabolic disorders by causing the proliferation of some intestinal microbes such as *Faecalibacterium* spp. These microbes are mainly butyrate-producing bacteria, and butyrate is an SCFA that has been demonstrated to possess metabolic protective effects (Moreno-Indias et al., 2016).

Interestingly, dieting rodents study revealed the link between microbiome and polyphenols (Thaiss et al., 2016). HFD led to reduced levels of flavonoids such as naringenin and apigenin due to gut dysbiosis resulting in abundance of microbes with the ability to breakdown the flavonoids (Tan et al., 2018). However, when the diet was changed to polysaccharide, the impaired metabolic homeostasis was normalized, but the gut dysbiosis was not reversed meaning that there are still low polyphenols such as naringenin and apigenin. The metabolic dysregulation and gut dysbiosis were exaggerated due to "microbiome memory" upon reintroduction of HFD to the mice. But when the mice were given diets supplemented with naringenin and apigenin, the metabolic impairment, excess weight gain, adiposity, and gut dysbiosis were abrogated as a result of improved energy utilization. Therefore, a fascinating relationship exists between the gut microbiome and low polyphenol diet or polyphenol rich diets, which can either have harmful or beneficial health implications (Thaiss et al., 2016).

Prebiotics

Prebiotics are dietary substrates, which are carefully used by gut microbiota to enhance or improve host metabolic processes resulting in improved microbiome homeostasis (Gibson et al., 2017). Prebiotics sources are fruit, vegetables, tomatoes, onions, wheat, garlic, bananas, etc. (Crittenden et al., 2008). Carbohydrates, like fiber, are possible prebiotics. Prebiotics could be a fiber-rich diet, but fiber-rich diet is not always a prebiotic (Slavin, 2013; Holscher, 2017; Gill et al., 2021).

Prebiotics are for food constituents that are indigested in the gut. Reports exist about the connection between the gut and prebiotics. Interestingly, prebiotics are able to increase the microbes in the gut of hosts, thereby resulting in increased fermentation of the indigested carbohydrates or fibers producing SCFAs that have the capacity to impact the host's health positively (Van Den Abbeele et al., 2013).

Prebiotics function by enhancing proliferation of gut *Bifidobacterium* and *Lactobacillus* species, which causes the inhibition of growth of some gut pathogens through its antimicrobial activity; also it leads to decrease in the pH of the intestine, which may assist in increasing the rate of digestion (Gibson et al., 1994), decrease constipation, improve glucose regulation, and improve inflammatory bowel disease (Gibson et al., 2010; Barengolts, 2013).

Probiotics

Probiotics are microorganisms with the ability to promote good health upon administration in specific quantity to the host. Most well-known probiotics belong to the human microbiome group of microorganisms and are likely to exert similar effects as the symbionts (Druart et al., 2014). Probiotics found in species *Lactobacillus* and *Bifidobacterium*, *Lactococcus*, *Streptococcus*, *Enterococcus*, *Bacillus* spp., and *Saccharomyces* are frequently used (Markowiak and Slizewska, 2017).

Live strains cultures of probiotics are found in fermented foods including yogurt, fermented bean paste, fermented vegetables, etc. (Rezac et al., 2018). Comparable probiotics are found in fermented milk and plants. Experimentally, probiotics are used in the treatment of obesity to induce weight loss, improve insulin-mediated metabolic functions, decreased hepatic steatosis, inflammation, and dyslipidemia (Kang et al., 2013; Stenman et al., 2014).

In addition, probiotics enhance gut microbiome proliferation by stimulating adequate equilibrium in the pathogenic and commensal microorganisms needed by the body for its activities (Oelschlaeger, 2010). Interestingly, many living microbes that serve as probiotics are used for the manufacturing of food with health benefits and food protection. Likewise, the useful effects of probiotics are engaged for the rejuvenation of the microbiome following antibiotic assaults (Johnston et al., 2006). Probiotics also inhibit the potency of harmful gut bacteria by preventing their growth and expansion, e.g., *Escherichia coli* (Markowiak and Śliżewska, 2018) and *Staphylococcus* (Sikorska and Smoragiewicz, 2013), consequently averting poisonous food content. Furthermore, probiotics enhance gut motility for proper breakdown and absorption of food, management of food-induced hypersensitivity reaction, and tooth problem (Thomas and Greer, 2010). They also serve as immune booster, anticancer, and sometimes antibiotics (Ishikawa et al., 2005).

Intestinal microbiomes actively maintain and enhance health of the host through the regulation of metabolism, and immunological system cannot be overlooked (Upadrasta and Madempudi, 2016). Diets have ability to reshape microbiome constitution and activities; this has emerged as a research hotspot in recent times due to their involvement in the treatment of metabolic derangement and associated complications such as micro- and macrovascular diseases. There is need to further increase interest in determining probiotics—microbiome—metabolic health mechanistic involvement or relationships (Upadrasta and Madempudi, 2016).

SCFAs, gut-derived and exogenous

Diet is needed for everyday sustenance of all living things and determines their well-being as a result of the nutrient components of such diets. Increase in science and technological knowledge has coincided with the intake of processed foods and decreased in ingestion of natural foods that contain high amount of fibers. Surprisingly, this modern era and change in lifestyle have come with its negative effect in various pathological conditions as well (Bhatti et al., 2020; Przybyłowicz and Danielewicz, 2022). Hence, there is emergence of significant contributions of dietary fibers to health as there are microorganisms living in the intestine that breakdown dietary fibers into SCFAs (Hammer et al., 2008). Indigestible dietary fibers are not broken down by the mammalian digestive system due to absence of enzymatic agents needed for this process, thereby needing fermentation (Eswaran et al., 2013). Importance of SCFAs in normal life is further reiterated in studies that show decrease in particular microbiome composition involved in the generation of SCFAs from dietary fibers (Frank et al., 2007). Human's intestinal lumen contains several microorganisms and is generally called the gut microbiota—involved with the metabolic breakdown of the fiber-rich diets to SCFAs (Gibson et al., 2010).

Origin, production, and basic functions

SCFAs are the main product of metabolism of dietary fibers in the gut lumen. Butyrate, acetate, and propionate are mostly the abundant of this product of fermentation (den Besten et al., 2013). Over some years now, SCFAs have been shown to play different pathophysiological roles (Donohoe et al., 2011). Over 90% of SCFAs in the colon are acetate, butyrate as well as propionate of which their production in the colon depends mainly on the substrate used rather than microbiome constitution (Cummings et al., 1987). Previous experimental studies have hypothesized that glucose dysmetabolism results in increased endogenous of SCFAs as noticed in diabetic states (Akanji et al., 1989) and metabolism of proteins (Wolever et al., 1997). SCFAs are absorbed by colonic epithelial cells (Ritzhaupt et al., 1998) and high-affinity transport mediated by sodium-coupled MCT1 (Gopal et al., 2007). Once in the colonic tissue, a significant proportion of the SCFAs are metabolized (Zambell et al., 2003), thereby resulting in smaller amount of SCFAs leaving for the portal circulation, making their studies a bit difficult (Ganapathy et al., 2013) although they have largely been studied by their interactions with histone deacetylases (HDACs; Thangaraju et al., 2009). Butyrate serves as a bioenergetics source for cells of colon for normal functioning of colonic processes such as proliferation and differentiation amid others (Zheng et al., 2013; Lu et al., 2015). Propionate is involved in hepatic gluconeogenesis by ensuring that acetate reaches the peripheral tissues as well (Huang et al., 2017). Generally, the composition of the SCFAs changes over a wide range, exhibiting tissue-specific roles in change in concentration gradients. Furthermore, their circulating and intestinal concentration may promote various diseases conditions (Kaczmarczyk et al., 2012).

Receptors

Receptors for the SCFAs (commonly referred to as GPCRs) are nutrient receptors and depend on the gut—kidney axis response to SCFAs. Four of them have been discovered to respond to SCFAs (Miyamoto et al., 2016) via free fatty acid receptor 3 FFAR3, FFAR2, GPR109A as well as Olfr78 (an olfactory receptor). FFAR3 and FFAR2 are involved in insulin homeostasis, thereby involving in the activation of the satiety-hunger center and control of energy metabolism (Psichas et al., 2015). FFAR3 and

FFAR2 are found in the kidney as both are expressed in ischemia—reperfusion injury and acetate treatment, respectively (Andrade-Oliveira et al., 2015), and by extension have beneficial metabolic effects (Nilsson et al., 2003; Vinolo et al., 2011a; Pluznick et al., 2013). Positive properties notwithstanding the therapeutic benefits of these FFARs remain largely inconclusive (Ang and Ding, 2016) and need further investigation (Huang et al., 2017). GPR109A is activated by ingestion of high-fiber diet and help to promote gut homeostasis (Macia et al., 2015). Olfr78 mediates sense of smell. It has been shown to be receptor for acetate and propionate (Pluznick, 2013) and colocalized with FFAR3 and FFAR2 (Pluznick, 2013, 2014). Olfr78 is present in the colon but absent in the GI compartment (Fleischer et al., 2015); hence, this receptor needs further investigation to elucidate its role.

SCFA receptors as therapeutic targets

Recent studies have shown the pivotal roles played by SCFA receptors not only in disease but also the fact that they influence and control other GI functions (Bologini et al., 2016). One of such is niacin, a GPR109A agonist whose importance has been demonstrated in an experimental pathological state (Hegyi et al., 2004) although another clinical outcome was conflicting (Bologini et al., 2016). Therefore, the therapeutic benefit of these receptors still needs further investigation so as to know whether other factors influence the SCFA receptors and know whether agonists or antagonists of these receptors are more therapeutically beneficial (Cornall et al., 2013; Bologini et al., 2016).

SCFAs in Health and Disease

1. Glucose metabolism

Dietary fiber helps to maintain normal blood glucose level and body weights as several studies on the role of SCFA-containing diets have shown. Experimental animals fed with diets rich in butyrate showed improved energy metabolism with prevention of obesity and IR (Gao et al., 2009; Badejogbin et al., 2019). Acetate normalizes the body weight and corrects glucose intolerance (Yamashita et al., 2007). Different investigators also show similar benefits with supplementation with butyrate or propionates (Lin et al., 2012; De Vadder et al., 2014). Levels of glucagon-like peptide-1 (GLP-1) and peptide YY (PYY) in the blood are raised by infusion of acetate into the venous system (Freeland and Wolever, 2010) and help to improve glucose tolerance suggesting an association between SCFAs and glucoregulation. The metabolic benefits played by butyrate and propionate on intestinal gluconeogenesis have been described. Propionate is substrate from nonglucose source in the intestine, while butyrate contributes to intestinal gluconeogenesis by raising the levels of cAMP in the colon. Therefore, the favorable SCFAs metabolic action is prevention of glucose intolerance by enhancing insulin sensitivity (De Vadder et al., 2014). Several researches are still needed, however, to ascertain the role of SCFAs on glucose metabolism.

2. Immunity

SCFAs and their receptor are very crucial in gut and/or whole systemic immunity. Immune-suppressive mechanisms are essential for the gut homeostasis. GPR1094, stimulated by butyrate supplementation, can increase IL-18 (Singh et al., 2014). SCFAs activate receptors and promote immunity (Macia et al., 2015). SCFA receptors present in large quantities in immune cells play an important role on T cell functions and differentiation (Furusawa et al., 2013; Smith et al., 2013b; Singh et al., 2014; Park et al., 2015).

3. Inflammation

Oxidative stress and inflammation are inseparable as they have a cross-talk; the effect of one affects the other. SCFA affects the production of ROS (Viera et al., 2015) and has beneficial effect on tissue inflammation (Vinolo et al., 2009). Butyrate inhibits macrophage in response to inflammation (Lopez-Barrera et al., 2016). Additionally, propionate and butyrate attenuate inflammatory biomarkers in experimental studies (Vinolo et al., 2011b; Kim et al., 2014). SCFAs work to prevent inflammation via reduction of endothelial cell adhesion molecules, which prevent movement of white blood cells to sites of inflammation (Miller et al., 2005). However, studies also show that SCFAs stimulate Th1 and Th17 following decreased immunity and possess unique ability to promote inflammation (Park et al., 2015; Corea-Oliveira et al., 2016).

4. Cancer

The gut microbiome has recently gained attention as an interesting site for cancer (de Martel et al., 2012) and persistent inflammatory responses resulting in colorectal cancer (Medzhitov, 2008). Through their beneficial role on inflammation, SCFAs can help alleviate colorectal cancer (Cresci et al., 2010). One limitation, however, is that butyrate has been shown to promote tumorigenesis in a genetically modified experimental animal (Belcheva et al., 2014), and this is independent of inflammation (Liang et al., 2010). Thus, when looking at the broad beneficial effect of SCFAs on cancer, genetical background and environment of the cell need to be put into consideration.

5. Respiratory disease

The airway serves as a connection between the internal and external environment and hence is constantly exposed to pathogens from the external environment. Asthma has been shown to be a chronic respiratory disease affecting quite a great number of people globally (Bruselle et al., 2013). The intestinal microbiome was reported to exert a crucial function against this chronic airway disease (Russell et al., 2012). Acetate has been shown to prevent airway disease to a point by enhancing Treg cells through inhibition of HDAC9 (Thorburn et al., 2015). In the same vein, propionate prevents allergy of the airway by promoting hematopoiesis and improving the lung ventilation (Zaiss et al., 2015).

6. Nervous system

Butyrate controls the functions of the nervous cells of the gastrointestinal system (Soret et al., 2010); as one of SCFA receptors, FFAR3 is highly present in the gastrointestinal nervous cells (Nøhr et al., 2015). Furthermore, FFAR3 is highly present in the peripheral nervous system (PNS) such as the vasal and dorsal root amid others (Kimura et al., 2011; Nøhr et al., 2015). Concomitant stimulation of the FFAR3 by SCFAs activates sympathetic nervous system, thereby increasing energy use (Kimura et al., 2011), thereby implicating SCFAs in nervous system transmission. SCFAs have also been linked to the brain (Frost et al., 2014; Erny et al., 2015). More so, SCFAs modulate blood—brain barrier (BBB) penetrability. Butyrate decreases BBB penetrability (Braniste et al., 2014) and helps prevent breakdown of the BBB while promoting blood vessels and nerve cell formation (Kim et al., 2009).

Therefore, SCFAs action is not restricted to the gut but plays a role in other systemic functions of the organism as a whole (Nogal et al., 2021). Effects of exogenous administration of SCFAs might depend on the route and, from the studies shown before, might quite be different from the effect of SCFAs produced through the fermentation. For propionate supplementation, the gluconeogenesis it causes in the intestine is beneficial metabolically, but the same gluconeogenesis in the hepatic tissue is dangerous (Yoshida et al., 2019). The exact role of SCFAs in health and disease still needs further research as evidence by its continuous interest from the scientific world.

Gut dysbiosis

Gut microbiota dysbiosis is a situation in which there is a persistent disproportion in the microorganisms within the intestines (Thursby and Juge, 2017). This condition is also referred to as "intestinal dysbiosis" or "gastrointestinal dysbiosis." Within the gut, there are microorganisms, collectively known as gut flora or gut's microbial community that needs to be in consistent balance. The gut's microbial community consists mainly of bacteria than protozoa, fungi, as well as archaea (Belizário et al., 2018). The gut microbiome in an adult is predominantly colonized by three bacterial phyla, *Actinobacteria* spp., *Bacteroidetes* spp., and *Firmicutes* spp. (Gill et al., 2006). The population of colonic microbiome is mostly anaerobic bacteria much more than aerobic bacteria at a ratio of 100—1000:1 (Belizário and Faintuch, 2018). The microbes in the gut are usually found attached to its mucus lining forming a physical barrier for the gut (Hawrelak and Myers, 2004). In early life of an infant, the microbes that form the gut microbiome are transmitted from the breast milk to babies as the human breast milk contains large quantities of microbes such as Bifidobacteria (Gregory et al., 2016). However, bacteria such as Clostridium, Parabacteroides, and Bacteroides form the major part of the gut microbia (Eckburg et al., 2005; Lepage et al., 2013). Interestingly, in gastrointestinal tract, each site has its own distinctive gut flora and their bacteria density is highest in the colon, large intestine, and jejunum compared with its density in the stomach and duodenum (Belizário and Faintuch, 2018).

An imbalance in the gut's microbiota, otherwise known as gut dysbiosis, in this case, beneficial microorganisms exhibit decreased ability and are overpowered by unhelpful microbes. Unfortunately, this condition has a vicious effect because once the beneficial microbes are decreased, they do not have the ability to check each other's growth and are unable to keep the unhelpful microbes from rapidly multiplying. As more beneficial microbes are being damaged in the gut, they are outnumbered by the unhelpful microbes resulting in disturbance of balance of the gut flora. Hence, when this condition goes unconstrained for a while, a persistence imbalance between the beneficial and unhelpful microbes sets in. The gut microbiome is very essential to the functions of the body; hence, its disruption can affect many areas of host health (de Vos et al., 2022; Hou et al., 2022) Fig. 11.1. Gut microbiome carries out immunoregulatory functions, enhancement of the gut integrity, and functions including motility, digestion, absorption, synthesis, protection against pathogenic bacteria, and fermentation of dietary fibers (Gibson and Roberfroid, 1995; Noack et al., 1998; Holzapfel et al., 1998). The intestinal microflora are not only essential for digestion, but it is also relevant in immune functioning (Hooper et al., 2012) and metabolic activity (Cani and Delzenne, 2009).

Factors causing gut dysbiosis

1. Antibiotics

The use of antibiotics is the commonest cause of gut flora disruption (Gismondo, 1998). Recent study indicated that infection and antibiotic treatment led to dysbiosis that can affect host systemic energy metabolism and elicit phenotypic and health modifications (Le Roy et al., 2019). The ability of antibiotics to influence gut flora, however, depends on its dosage, length of administration, pharmacokinetics, and spectrum of activity (Nord, 1990; Gismondo, 1998). Hence, the gut flora is greatly influenced by antibiotics. Also increased length of administration and dosage has increased negative effect on the gut flora.

Use of antibiotics during infancy elicits imbalances in gut microbiota leading to dysbiosis (Biedermann and Rogler, 2015; Vangay et al., 2015). An earlier animal study suggests that antibiotic use alters the gut microbiome, which contributes immensely to inflammatory responses through the disruption of metabolic status in host (Sun et al., 2019).

2. Unhealthy diet

Several studies have proposed the existence of the diet—microbiota—host relationship (Afzaal et al., 2022). Microbiome homeostasis in neonate has been reported to be modified by nutrition resulting in dysbiosis, which can lead to altered immune responses (Nash et al., 2017). Diets can either stimulate proliferation of commensal microbes or pathogenic microbes with deleterious health consequences in the host (Makki et al., 2018; Perler et al., 2023). Western diets rich in fats and sugar while lacking in fiber greatly

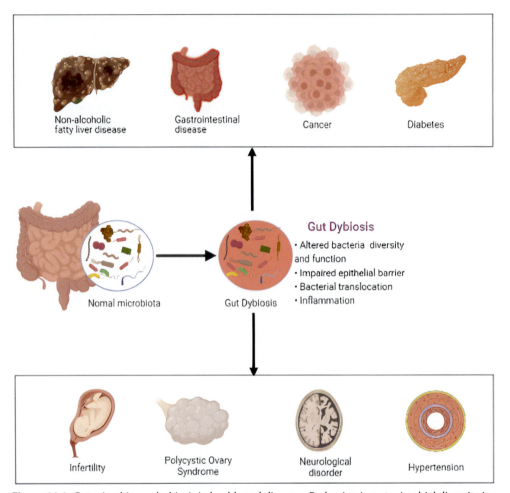

Figure 11.1 Gut microbiome dysbiosis in health and diseases. Reduction in gut microbial diversity is a crucial pathogenic mechanism responsible for many diseases such as obesity, hypertension, cancer, diabetes, infertility, PCOS, gastrointestinal disorders, neurological disorders, chronic liver disease, etc. Created with biorender.com.

influence the composition of gut flora (Martinez et al., 2017; Malesza et al., 2021). Therefore, diets can reshape the gut microbiome community resulting metabolic dysregulations due to the alterations of the gut microbial homeostasis (Bootz-Maoz et al., 2022; Cox et al., 2022; Vijay and Valdes, 2022). Dietary intake of human milk or infant formula in the neonate plays a vital role in their microbiota composition. Breast milk influences beneficial microbial growth in infant's intestine unlike infant formula (Li et al., 2022). In breastfed babies, the predominant gut flora are the *Bifidobacteria* and *Lactobacilli*, whereas *Enterococci* and *Enterobacteria* were largely predominant in formula-fed infants (Palmer et al., 2007).

3. High-fat diet

HFD consumption decreases *Bacteroidetes* but increases *Firmicutes, a shift connected to impaired metabolism and obesity. Possible ways by which HFD alters* gut microbiome associated with obesity include an elevated energy storing capabilities, proliferation of lipopolysaccharide-producing bacteria in the gut, and stimulatory effects of Toll-like receptors on immune cells (Kim et al., 2012), Also, HFD can change intestinal integrity through the decrease in structural proteins (Shen et al., 2014) and enhanced gut permeability (Zhang et al., 2012; Murphy et al., 2015).

4. Simple sugar/carbohydrate diet

The bowel motility and transport time is lowered by simple sugar-rich diet. However, the microbial fermentation activity and colonic bile composition is elevated by simple sugars (Kruis et al., 1991). The downside to the slow bowel movement is the elevated contact with toxic contents of the colon (Lewis and Heaton, 1999).

5. High-protein diet

Production of harmful microbial metabolites has been associated with the increased intake of protein-rich diet. Usually, Western diet contains increased dietary protein of approximately 100 g protein/day, and some of these proteins reach the colon, escaping digestive activity of the higher part of the gut (Linder, 1991). Undigested protein that escaped into the colon is metabolized by colonic microbes generating SCFAs, BCFAs, sulfide, ammonia, polyphenols, and indoles (Macfarlane and Macfarlane, 1995; Smith and Macfarlane, 1997). Production of these dangerous compounds is directly related to intake of dietary protein and can be attenuated by increased intake of high-fiber diet (Birkett et al., 1996). Ammonia, a by-product of protein is harmful to the gut mucosal epithelia and can shorten the existence of mucosal epithelial cells (Macfarlane and Gibson, 1994; Murawaki et al., 2000). Likewise, indoles and phenols act as cocarcinogens and are involved in cancer of the bladder and bowel (Birkett et al., 1996).

6. Sulfates

Sulfate and sulfite, which are derivatives of food additives, have been associated with elevated generation of harmful bacteria metabolites in the intestine (Levitt et al., 1995). Sulfate-reducing bacteria, a specific class of gram-negative anaerobes, are present in the human colon. Sulfate-reducing bacteria reduce sulfate and sulfite to sulfide, leading to the production of hydrogen sulfide, which is a harmful substance (Cummings and Macfarlane, 1991). Hydrogen sulfide can cause abdominal gas distension, inhibit butyric oxidation, and damage colonic mucosa (Levitt et al., 1995). However, suppression of butyrate oxidative reaction is a defective symptom of colitis resulting in deficient intracellular energy (Cummings and Macfarlane, 1997).

Health problems linked to gut dysbiosis

Health problems arising from gut dysbiosis are not all digestive in nature but include altered immune responses and metabolic disorders. Dysbiosis-associated diseases include obesity, type 2 diabetes, autoimmune diseases (Kumagai et al., 2018), skin diseases, dementia (Manichanh et al., 2006; Kolho et al., 2015; Miyoshi and Chang, 2017), ankylosing spondylitis (Zhang et al., 2019), and inflammation of the colon and cancer (Sheflin et al., 2014).

Analytical approach for gut microbiota identification

Microbiome relevance in human and animal health has been undermined due to inability to cultivate 400–500 bacterial species that take residence in human and animal bodies. Different approaches to microbial cultivation and analysis have been explored and detailed in the following.

Culture-dependent approach

Culture-dependent approach of gut microbes' identification has been in existence since the 20th century. Consistently, advance isolates characterization requires their profiles of fermentation and in vitro growth features. This method is fairly cheap but quite intensive in terms of labor requirement and does not produce extensive information on gut microbial composition, because to date, <30% of the gut microbes have been cultured (Eckburg et al., 2005). Some microbes that are culturable are not cultured due to unfavorable permissive growth conditions. Microbial culture chips and gel microdroplets are improved culture methods (Ingham et al., 2007). Being a high-throughput technique makes it possible to culture previously uncultured microbes. Nevertheless, the uses of new pure-culture techniques have also been limited, especially in organisms that have a symbiotic relationship and depend on the metabolic activity of each other (Fraher et al., 2012).

16S rRNA approach

This approach is known as culture-independent method and is dependent on small subunit ribosomal RNA (16S rRNA) sequence divergences. This is able to demonstrate the gut microbiome composition, quantitative and qualitative information on microbial types as well as disease-associated microbiome alteration. This approach involves the use of fluorescence in situ hybridization (FISH), DNA microarrays, PCR, terminal restriction fragment length polymorphism (T-RFLP), denaturing gradient gel electrophoresis (DGGE), and next-generation sequencing (NGS) of the 16S rRNA gene. It is important to know that 70S ribosomes, including 30S and 50S, are dispersed in the cytoplasmic portion of microbial cell. The 50S subunit also comprises of 5S and 23S RNA molecules, while 30S subunit, which is the small subunit, contains16S rRNA molecule. Initiation and extension of protein synthesis was mediated by 16S rRNA molecule. The 16S rRNA approach is achieved with the use of any of the following techniques.

PCR technique

The introduction of PCR is a major technical advance in science but limited due to the possible introduction of bias from the steps involved in sample collection (Fraher et al., 2012). Laboratory reports may be altered by differential bacterial cell lysis producing impaired microbiome composition result, for instance, lysing the bacterial cell wall of gram-positive organisms required a rigorous condition, which may subject the gram-negative chromosomal DNA to excessive fragmentation (Carey et al., 2007). In addition, the use of primers to target all phyla may be challenging in the choice of PCR technique. PCR alone may not be reliable for DNA quantification in a given sample and the use of quantitative/real-time PCR, which ensures amplification and quantification of DNA molecule of a given sample is encouraged (Clarridge, 2004).

Moreover, this technique involves the extraction of DNA and formation of reaction mixture, which consists of compounds with fluorescent property when binding to double-stranded DNA molecule, and this is known as the product of PCR (Carey et al., 2007). Candidate primers are needed, and qPCR can be singly used or combined with other techniques when data validation is required. However, the use of qPCR in analyzing gut microbiota can be technically challenging and labor-intensive compared with culture-dependent method.

Fingerprinting technique

This is a community-based analytical tool that allows two or more DNA fragments in a given sample to be compared, especially the DGGE separates complex mixtures of 16S rRNA gene amplicon with the same length but different DNA sequence (Sekirov et al., 2010). This mixture is applied to an increasing gradient of denaturant called polyacrylamide gel-like formamide or urea. Following the application of electrophoretic current, the 16S rRNA gene amplicons begin to migrate down the gel, and the DNA strands begin to denature. Once the DNA strands are completely denatured, then migration stops. The DNA sequence separates into segments and "bands" (Noor et al., 2010), and the gel is stained for band visualization, which can then be excised for sequencing and probe hybridization. The technique permits separation of different species of 16S from a given sample. DGGE is essential for comparison, take for instance, a diseased condition can be differentiated from a healthy state (Fraher et al., 2012). This technique has been demonstrated to be fast and allows the analysis of multiple samples (Muyzer, 1999; Fraher et al., 2012). Because of the visual impressions of band and species intensity and abundance that are associated with the use of this technique, together with FISH, TGGE, and T-RFLP, they are referred to as quantitative because it gives visual impressions of band and species intensity and abundance. Nevertheless, the techniques are semi-quantitative and may include bias and lack of direct phylogenetic detection unless the analyst performs sequencing and probe hybridization. Similarly, TGGE has proven its usefulness for intestinal microbiome analysis and works in a related manner to DGGE.

Fluorescence in situ hybridization technique

This uses fluorescently labeled oligonucleotide probes to hybridize target 16S rRNA sequences in a given sample. FISH is combined with flow cytometry for intestinal microbiome assessment (McCartney, 2002). It can differentiate the gut microbiota in disease condition from healthy state, and it is fast and semiquantitative. Notwithstanding, this technique cannot identify unknown species, but specific phyla or species can be targeted using probe designs.

Terminal restriction fragment length polymorphism

In T–RFLP technique, there is fragmentation of 16S rRNA gene amplicons by restricting endonucleases with band visualization. T–RFLP is mainly used to determine the diversity of microbial community, and compare and profile gut microbiota (Nadal et al., 2007). It is fast and cheap but semiquantitative, which could be challenging when analyzing multiple 16S rRNA. Additionally, T–RFLP does not permit phylogenetic identification, except in combination with 16S rRNA clone library analysis (Matsumoto et al., 2005).

Gene (DNA) microarray technique

This is in other words referred to as phylogenetic microarray, gene array, and DNA chip. Being a high-throughput method identifies phylogenetic containments of intestinal microbiome (Fraher et al., 2012). This technique can detect microbial DNA molecule of a given sample at a level of 0.00025% (Paliy et al., 2009). Microarray chips that target the human microbiome are now available commercially, such as the human PhyloChip and intestinal tract chip microarrays, which are used to distinguish microbiota of different populations (Paliy et al., 2009). This method is also fast, qualitative, and semiquantitative and can detect all species concomitantly. However, it is limited due to possibility of cross-hybridization (Fraher et al., 2012).

Sequencing

This is the taxonomic identification gold standard for species at all levels, and here in, information is required from the whole-length 6S rRNA gene, which could be pragmatically obtained from sequenced clone library insert. Following determination of sequence by cloned 16S rRNA gene amplicons or NGS, comparison is made to a database, which contains nucleotide sequences for more than PCR, for example, GenBank Database (Johnson et al., 2019).

Cloned 16S rRNA sequencing

To sequence a cloned whole-length 16S rRNA gene amplicons, Sanger method is employed. This method involves chain termination by dideoxynucleotides (Clarridge, 2004). The approach has been shown to reveal specific variability of the microbiota in the gut (Cole et al., 2009). It also quantifies/phylogenetically identifies the gut microbiome and can analyze uncultured bacteria. However, it is expensive and labor intensive.

Next-generation sequencing

NGS involves the sequencing of full or complex microbial genomes with barcodes and significantly less labor intensive and reduced cost per base compared with the traditional sequencing. In NGS, the sequence outputs are given in range of hundreds of megabases to gigabases, and it provides informative data in terms of quantity and quality (Ignys et al., 2014). This technique has high impact on basic and clinical researches and is currently used in the analysis of gut microbiota to differentiate between normal and dysbiotic conditions (Ignys et al., 2014; Ai et al., 2019).

The use of 16S rRNA gene-based technologies to characterize the gut microbial composition and its dynamics has become so popular in the past few years. With NGS, more can be done at the level of DNA, such as through the metagenomics approach in which a high representative DNA sample from a whole bacterial community is isolated and sequenced in a random fashion. In addition, NGS has made possible both targeted 16S rRNA gene and random metagenomic sequencing (Said et al., 2014; Yoneda et al., 2016). Multiplexing is also achieved through the addition of a barcode sequence and one or two indexing reads, located outside the primer region. For throughput, there is preparation of libraries for each sample using universal primers targeting different regions of the 16S rRNA gene. Primers are adapted for throughput sequencing with the addition of adapter sequences and barcoded forward and reverse sequences taken from the earlier studies (Zheng et al., 2015).

Microbiota applications in health and disease

Human microbiota influences host's physiology, immunology, and nutrition (Neish, 2009). Previous studies revealed that impaired gut bacterial composition alters the host's immunological functions, obesity, and gastrointestinal dysregulation (Kipanyula et al., 2013; Chen et al., 2016). Likewise, aberrant intestinal microbiome composition causes immune-related diseases, diabetes obesity, hypertension, cancer, etc. (Rea et al., 2018). Recently, gut microbiota is found to influence human physiological and immunological status through soluble or insoluble dietary fibers fermentation mediated by nutrient–gut microbiota interaction (Chen et al., 2016; Zeng et al., 2017). Nevertheless, implications of gut microbiome on host well-being are enormous, and some are furnished in the following.

Gut microbiota and metabolic homeostasis

Accumulating evidence reveals that commensal bacteria are seeded shortly after birth in the host. The simple community of bacteria gradually develops into diverse colony as the host grows (Gill et al., 2006; Wang et al., 2017), with consequent uprising to host-bacterial mutualistic relationship. Gut microbes metabolize indigestible soluble and insoluble compounds with the generation of essential nutrients. They also protect intestinal epithelium against opportunistic pathogenic colonization and involves in the formation of intestinal mucosal architecture (Wang et al., 2017). Similarly, the use of metabolomic and metagenomic sequencing technologies to analyze human fecal sample demonstrate huge involvement in metabolism of amino acid, polysaccharides, xenobiotics, and micronutrients, attributable to gut microbiome, which implies endogenous microorganisms, promotes efficient bioenergetics and effective metabolism (Payne et al., 2012; Kho and Lal, 2018). In addition, microbiome relevance in lipid/protein metabolism and production of essential nutrient vitamins, such as folates, biotin, vitamin K, riboflavin (B2), and cobalamin (B12), have been documented (Kang et al., 2012). The fermented end products of nutrient—gut microbial interactions exist in the form of SCFAs. SCFAs generate ATP for day-to-day biological processes of the host (Gill et al., 2006; Kho and Lal, 2018). SCFAs regulate gastrointestinal motility, inflammation as well as energy harvesting (Wang et al., 2017).

SCFAs are GPCR ligands expressed by a gut epithelial enteroendocrine cell, and they have been demonstrated to control energy homeostasis via stimulation of GPCR-linked leptin synthesis in rodent adipocytes (Samuel et al., 2008). SCFAs' modulatory effects in metabolic pathologies such as cardiometabolic disorders, diabetes mellitus, insulin resistance, and obesity (Koeth et al., 2014; Nie et al., 2015; Olaniyi et al., 2020).

Immunological functions of microbiome

Intricate connections exist between immune system, particularly mucosal immune system and intestinal microbiome (Clemente et al., 2012). Immunological development is apparently influenced by microbial interactions, which are needed for the development of both innate and adaptive immune systems (Chow et al., 2010). IgA alteration corresponds to disruption of immune status because IgA is primarily involved in immunological defense of the mucosal surface (Clemente et al., 2012). Absence and mutation of TLRs lead to the formation of aberrant gut—immune interaction that initiates the activation of inflammatory pathway (Clemente et al., 2012). Development of immunological protection necessary for the maintenance of pregnancy is enhanced by the microbiome via antiinflammatory dependent mechanisms (Round et al., 2011; Belkaid and Hand, 2014). Gut commensals gain an advantage by lowering IgA by modulating immunodominant determinants (Clemente et al., 2012).

Conclusion

Nutrition/diets have very significant modulatory effects on quantity/constellation of intestinal microbial flora, which usually start as early as infancy. Nutrition/diet has effect on the host health and physiology due to the ability to reshape the intestinal microbiome community and function. Applications of gut-derived microbial fermentation products such as SCFAs have emerged as a regulator of metabolic processes and energy utilization.

It is safe to say that diet—microbiota—host interaction influences the host physiology and pathophysiolgy. Microbiota transplantation has gained recent application for the management of disease conditions, and studies are still ongoing to explain the diet—microbiome mechanistic link. Therefore, proper understanding of the intestinal microorganisms, their activities, and molecular products is warranted because microbiome research is gaining center stage and the full capabilities of the gut microbiome as it relates to health and diseases remain to be discovered.

References

Adeyanju, O.A., Badejogbin, O.C., Areola, D.E., Olaniyi, K.S., Dibia, C., Soetan, O.A., Oniyide, A.A., Michael, O.S., Olatunji, L.A., Soladoye, A.O., 2021. Sodium butyrate arrests pancreato-hepatic synchronous uric acid and lipid dysmetabolism in high fat diet fed Wistar rats. Biomed. Pharmacother. 133, 110994.

Afzaal, M., Saeed, F., Shah, Y.A., Hussain, M., Rabail, R., Socol, C.T., Hassoun, A., Pateiro, M., Lorenzo, J.M., Rusu, A.V., Aadil, R.M., September 26, 2022. Human gut microbiota in health and disease: unveiling the relationship. Front. Microbiol. 13, 999001. https://doi.org/10.3389/fmicb.2022.999001.

Ai, D., Pan, H., Li, X., Gao, Y., Liu, G., Xia, L.C., 2019. Identifying gut microbiota associated with colorectal cancer using a zero-inflated lognormal model. Front. Microbiol. 10, 826.

Akanji, A.O., Humphreys, S., Thursfield, V., Hockaday, T.D., 1989. The relationship of plasma acetate with glucose and other blood intermediary metabolites in non-diabetic and diabetic subjects. Clin. Chim. Acta 185, 25—34.

Amabebe, E., Robert, F.O., Agbalalah, T., Orubu, E.S.F., May 28, 2020. Microbial dysbiosis-induced obesity: role of gut microbiota in homoeostasis of energy metabolism. Br. J. Nutr. 123 (10), 1127—1137. https://doi.org/10.1017/S0007114520000380.

Andrade-Oliveira, V., Amano, M.T., Correa-Costa, M., et al., 2015. Gut bacteria products prevent AKI induced by ischemia-reperfusion. J. Am. Soc. Nephrol. 26, 1877—1888.

Ang, Z., Ding, J.L., 2016. GPR41 and GPR43 in obesity and inflammation - protective or causative? Front. Immunol. 7, 28.

Araújo, J.R., Tomas, J., Brenner, C., Sansonetti, P.J., October 2017. Impact of high-fat diet on the intestinal microbiota and small intestinal physiology before and after the onset of obesity. Biochimie 141, 97—106. https://doi.org/10.1016/j.biochi.2017.05.019.

Aziz, T., Hussain, N., Hameed, Z., Lin, L., 2024 January—December. Elucidating the role of diet in maintaining gut health to reduce the risk of obesity, cardiovascular and other age-related inflammatory diseases: recent challenges and future recommendations. Gut Microb. 16 (1), 2297864. https://doi.org/10.1080/19490976.2023.2297864.

Badejogbin, C., Areola, D.E., Olaniyi, K.S., Adeyanju, O.A., Adeosun, I.O., November 2019. Sodium butyrate recovers high-fat diet-fed female Wistar rats from glucose dysmetabolism and uric acid-associated cardiac tissue damage. Naunyn-Schmiedeberg's Arch. Pharmacol. 392 (11), 1411—1419. https://doi.org/10.1007/s00210-019-01679-2.

Barengolts, E., 2013. Vitamin D and prebiotics may benefit the intestinal microbacteria and improve glucose homeostasis in prediabetes and type 2 diabetes. Endocr. Pract. 19, 497—510.

Belcheva, A., Irrazabal, T., Robertson, S.J., Streutker, C., Maughan, H., Rubino, S., Moriyama, E.H., Copeland, J.K., Kumar, S., Green, B., et al., 2014. Gut microbial metabolism drives transformation of MSH2-deficient colon epithelial cells. Cell 158, 288—299.

Belizário, J.E., Faintuch, J., 2018. Microbiome and gut dysbiosis. Exp. Suppl. 109, 459—476.

Belizário, J.E., Faintuch, J., Garay-Malpartida, M., 2018. Gut microbiome dysbiosis and immunometabolism: new frontiers for treatment of metabolic diseases. Mediat. Inflamm. 2018, 2037838.

Belkaid, Y., Hand, T.W., 2014. Role of the microbiota in immunity and inflammation. Cell 157, 121—141.

Bhattarai, Y., Kashyap, P.C., 2016. Germ-free mice model for studying host-microbial interactions. Methods Mol. Biol. 1438, 123–135.

Bhatti, G.K., Reddy, A.P., Reddy, P.H., Bhatti, J.S., January 10, 2020. Lifestyle modifications and nutritional interventions in aging associated cognitive decline and Alzheimer's disease. Front. Aging Neurosci. 11, 369. https://doi.org/10.3389/fnagi.2019.00369.

Bian, X., Tu, P., Chi, L., Gao, B., Ru, H., Lu, K., 2017. Saccharin induced liver inflammation in mice by altering the gut microbiota and its metabolic functions. Food Chem. Toxicol. 107 (Pt B), 530–539. https://doi.org/10.1016/j.fct.2017.04.045.

Biedermann, L., Rogler, G., 2015. The intestinal microbiota: its role in health and disease. Eur. J. Pediatr. 174, 151–167.

Birkett, A., Muir, J., Phillips, J., et al., 1996. Resistant starch lowers fecal concentrations of ammonia and phenols in humans. Am. J. Clin. Nutr. 63, 766–772.

Bolognini, D., Tobin, A.B., Milligan, G., Moss, C.E., 2016. The pharmacology and function of receptors for short-chain fatty acids. Mol. Pharmacol. 89, 388–398.

Bootz-Maoz, H., Pearl, A., Melzer, E., Malnick, S., Sharon, E., Bennet, Y., Tsentsarevsky, R., Abuchatzera, S., Amidror, S., Aretz, E., Azriel, S., Gam Ze Letova, C., Naama, M., Shoval, I., Yaron, O., Karako-Lampert, S., Bel, S., Yissachar, N., November 15, 2022. Diet-induced modifications to human microbiome reshape colonic homeostasis in irritable bowel syndrome. Cell Rep. 41 (7), 111657. https://doi.org/10.1016/j.celrep.2022.111657.

Braniste, V., Al-Asmakh, M., Kowal, C., Anuar, F., Abbaspour, A., Toá th, M., Korecka, A., Bakocevic, N., Ng, L.G., Kundu, P., et al., 2014. The gut microbiota influences blood-brain barrier permeability in mice. Sci. Transl. Med. 6, 263ra158.

Brusselle, G.G., Maes, T., Bracke, K.R., 2013. Eosinophils in the spotlight: eosinophilic airway inflammation in nonallergic asthma. Nat. Med. 19, 977–979.

Cani, P.D., Delzenne, N.M., 2009. The role of the gut microbiota in energy metabolism and metabolic disease. Curr. Pharmaceut. Des. 15, 1546–1558.

Cardona, F., Andrés-Lacueva, C., Tulipani, S., Tinahones, F.J., Queipo-Ortuño, M.I., August 2013. Benefits of polyphenols on gut microbiota and implications in human health. J. Nutr. Biochem. 24 (8), 1415–1422. https://doi.org/10.1016/j.jnutbio.2013.05.001.

Carey, C.M., Kirk, J.L., Ojha, S., Kostrzynska, M., 2007. Current and future uses of realtime polymerase chain reaction and microarrays in the study of intestinal microbiota, and probiotic use and effectiveness. Can. J. Microbiol. 53, 537–550.

Charbonneau, M.R., et al., 2016. Sialylated milk oligosaccharides promote microbiota- dependent growth in models of infant undernutrition. Cell 164, 859–871.

Chassaing, B., et al., 2015. Dietary emulsifiers impact the mouse gut microbiota promoting colitis and metabolic syndrome. Nature 519, 92–96.

Chassaing, B., Van de Wiele, T., De Bodt, J., Marzorati, M., Gewirtz, A.T., 2017. Dietary emulsifiers directly alter human microbiota composition and gene expression ex vivo potentiating intestinal inflammation. Gut 66, 1414–1427.

Chassard, C., Lacroix, C., 2013. Carbohydrates and the human gut microbiota. Curr. Opin. Clin. Nutr. Metab. Care 16 (4), 453–460.

Chen, M.L., Yi, L., Zhang, Y., Zhou, X., Ran, L., Yang, J., Zhu, J.D., Zhang, Q.Y., Mi, M.T., 2016. Resveratrol attenuates trimethylamine-N-oxide (TMAO)-induced atherosclerosis by regulating TMAO synthesis and bile acid metabolism via remodeling of the gut microbiota. mBio 7 (2), e02210–e02215.

Cheng, L., et al., 2016. High fat diet exacerbates dextran sulfate sodium induced colitis through disturbing mucosal dendritic cell homeostasis. Int. Immunopharmacol. 40, 1–10.

Chow, J., Lee, S.M., Shen, Y., Khosravi, A., Mazmanian, S.K., 2010. Host-bacterial symbiosis in health and disease. Adv. Immunol. 107, 243–274.

Clarridge 3rd, J.E., 2004. Impact of 16S rRNA gene sequence analysis for identification of bacteria on clinical microbiology and infectious diseases. Clin. Microbiol. Rev. 17, 840–862.

Claesson, M.J., Jeffery, I.B., Conde, S., et al., 2012. Gut microbiota composition correlates with diet and health in the elderly. Nature 488 (7410), 178–184.

Claesson, M.J., Clooney, A.G., O'Toole, P.W., 2017. A clinician's guide to microbiome analysis. Nat. Rev. Gastroenterol. Hepatol. 14 (10), 585–595.

Clemente, J.C., Ursell, L.K., Parfrey, L.W., Knight, R., 2012. The impact of the gut microbiota on human health: an integrative view. Cell 148 (6), 1258—1270.

Codagnone, M.G., Spichak, S., O'Mahony, S.M., et al., 2019. Programming bugs: microbiota and the developmental origins of brain health and disease. Biol. Psychiatr. 85 (2), 150—163.

Cole, J.R., et al., 2009. The Ribosomal Database Project: improved alignments and new tools for rRNA analysis. Nucleic Acids Res. 37, D141—D145.

Collins, K.H., Paul, H.A., Hart, D.A., Reimer, R.A., Smith, I.C., Rios, J.L., , … Herzog, W., 2016. A high-fat high-sucrose diet rapidly alters muscle integrity, inflammation and gut microbiota in male rats. Sci. Rep. 6, 37278.

Cornall, L.M., Mathai, M.L., Hryciw, D.H., McAinch, A.J., 2013. The therapeutic potential of GPR43: a novel role in modulating metabolic health. Cell. Mol. Life Sci. 70, 4759—4770.

Corrêa-Oliveira, R., Fachi, J.L., Vieira, A., Sato, F.T., Vinolo, M.A., 2016. Regulation of immune cell function by short-chain fatty acids. Clin. Transl. Immunol. 5, e73.

Cox, T.O., Lundgren, P., Nath, K., Thaiss, C.A., July 29, 2022. Metabolic control by the microbiome. Genome Med. 14 (1), 80. https://doi.org/10.1186/s13073-022-01092-0.

Cresci, G.A., Thangaraju, M., Mellinger, J.D., Liu, K., Ganapathy, V., 2010. Colonic gene expression in conventional and germ-free mice with a focus on the butyrate receptor GPR109A and the butyrate transporter SLC5A8. J. Gastrointest. Surg. 14, 449—461.

Crittenden, R., Playne, M.J., 2008. Nutrition News. Facts and functions of prebiotics, probiotics and synbiotics. In: Lee, Y.K., Salminen, S. (Eds.), Handbook of Probiotics and Prebiotics. Wiley-Interscience, Kansas State University, Hoboken, NJ, USA; Manhattan, KS, USA, pp. 535—582.

Cummings, J.H., Macfarlane, G.T., June 1991. The control and consequences of bacterial fermentation in the human colon. J. Appl. Bacteriol. 70 (6), 443—459.

Cummings, J.H., Macfarlane, G.T., 1997. Role of intestinal bacteria in nutrient metabolism. JPEN - J. Parenter. Enter. Nutr. 21, 357—365.

Cummings, J.H., Pomare, E.W., Branch, W.J., Naylor, C.P., Macfarlane, G.T., 1987. Short chain fatty acids in human large intestine, portal, hepatic and venous blood. Gut 28 (10), 1221—1227. https://doi.org/10.1136/gut.28.10.1221.

De Filippis, F., Pellegrini, N., Vannini, L., Jeffery, I.B., La Storia, A., Laghi, L., Ercolini, D., 2015. High-level adherence to a Mediterranean diet beneficially impacts the gut microbiota and associated metabolome. Gut 65 (11), 1812—1821.

de Martel, C., Ferlay, J., Franceschi, S., Vignat, J., Bray, F., Forman, D., Plummer, M., 2012. Global burden of cancers attributable to infections in 2008: a review and synthetic analysis. Lancet Oncol. 13, 607—615.

De Vadder, F., Kovatcheva-Datchary, P., Goncalves, D., Vinera, J., Zitoun, C., Duchampt, A., Backhed, F., Mithieux, G., 2014. Microbiota-generated metabolites promote metabolic benefits via gut-brain neural circuits. Cell 156, 84—96.

de Vos, W.M., Tilg, H., Van Hul, M., Cani, P.D., May 2022. Gut microbiome and health: mechanistic insights. Gut 71 (5), 1020—1032. https://doi.org/10.1136/gutjnl-2021-326789.

den Besten, G., van Eunen, K., Groen, A.K., Venema, K., Reijngoud, D., Bakker, B.M., 2013. The role of short-chain fatty acids in the interplay between diet, gut microbiota and host energy metabolism. J. Lipid Res. 54, 2325—2340.

Dominguez-Bello, M.G., Godoy-Vitorino, F., Knight, R., et al., 2019. Role of the microbiome in human development. Gut 68 (6), 1108—1114.

Donohoe, D.R., Garge, N., Zhang, X., Sun, W., O'Connell, T.M., Bunger, M.K., Bultman, S.J., 2011. The microbiome and butyrate regulate energy metabolism and autophagy in the mammalian colon. Cell Metabol. 13, 517—526.

Druart, C., Alligier, M., Salazar, N., Neyrinck, A.M., Delzenne, N.M., 2014. Modulation of the gut microbiota by nutrients with prebiotic and probiotic properties. Adv. Nutr. 5, S624—S633.

Duncan, S.H., Belenguer, A., Holtrop, G., Johnstone, A.M., Flint, H.J., Lobley, G.E., 2007. Reduced dietary intake of carbohydrates by obese subjects results in decreased concentrations of butyrate and butyrate-producing bacteria in feces. Appl. Environ. Microbiol. 73 (4), 1073—1078.

Eckburg, P.B., Bik, E.M., Bernstein, C.N., Purdom, E., Dethlefsen, L., Sargent, M., Gill, S.R., Nelson, K.E., Relman, D.A., June 10, 2005. Diversity of the human intestinal microbial flora. Science 308 (5728), 1635—1638.

Erny, D., Hrab_e de Angelis, A.L., Jaitin, D., Wieghofer, P., Staszewski, O., David, E., Keren-Shaul, H., Mahlakoiv, T., Jakobshagen, K., Buch, T., et al., 2015. Host microbiota constantly control maturation and function of microglia in the CNS. Nat. Neurosci. 18, 965—977.

Espín, J.C., González-Sarrías, A., Tomás-Barberán, F.A., September 1, 2017. The gut microbiota: a key factor in the therapeutic effects of (poly)phenols. Biochem. Pharmacol. 139, 82—93. https://doi.org/10.1016/j.bcp.2017.04.033.

Eswaran, S., Muir, J., Chey, W.D., 2013. Fiber and functional gastrointestinal disorders. Am. J. Gastroenterol. 108, 718—727.

Falony, G., Joossens, M., Vieira-Silva, S., et al., 2016. Population-level analysis of gut microbiome variation. Science 352 (6285), 560—564.

Fang, W., Xue, H., Chen, X., Chen, K., Ling, W., 2019. Supplementation with sodium butyrate modulates the composition of the gut microbiota and ameliorates high-fat diet-induced obesity in mice. J. Nutr. 149 (5), 747—754.

Fleischer, J., Bumbalo, R., Bautze, V., Strotmann, J., Breer, H., 2015. Expression of odorant receptor Olfr78 in enteroendocrine cells of the colon. Cell Tissue Res. 361, 697—710.

Fraher, M.H., O'toole, P.W., Quigley, E.M., 2012. Techniques used to characterize the gut microbiota: a guide for the clinician. Nat. Rev. Gastroenterol. Hepatol. 9 (6), 312.

Frank, D.N., St Amand, A.L., Feldman, R.A., Boedeker, E.C., Harpaz, N., Pace, N.R., 2007. Molecular-phylogenetic characterization of microbial community imbalances in human inflammatory bowel diseases. Proc. Natl. Acad. Sci. U. S. A. 104, 13780—13785.

Freeland, K.R., Wolever, T.M.S., 2010. Acute effects of intravenous and rectal acetate on glucagon-like peptide-1, peptide YY, ghrelin, adiponectin and tumour necrosis factor-alpha. Br. J. Nutr. 103, 460—466.

Frost, G., Sleeth, M.L., Sahuri-Arisoylu, M., Lizarbe, B., Cerdan, S., Brody, L., Anastasovska, J., Ghourab, S., Hankir, M., Zhang, S., et al., 2014. The short chain fatty acid acetate reduces appetite via a central homeostatic mechanism. Nat. Commun. 5, 3611.

Furusawa, Y., Obata, Y., Fukuda, S., Endo, T.A., Nakato, G., Takahashi, D., Nakanishi, Y., Uetake, C., Kato, K., Kato, T., et al., 2013. Commensal microbe-derived butyrate induces the differentiation of colonic regulatory T cells. Nature 504, 446—450.

Ganapathy, V., Thangaraju, M., Prasad, P.D., Martin, P.M., Singh, N., 2013. Transporters and receptors for short-chain fatty acids as themolecular link between colonic bacteria and the host. Curr. Opin. Pharmacol. 13, 869—874.

Gao, Z., Yin, J., Zhang, J., Ward, R.E., Martin, R.J., Lefevre, M., Cefalu, W.T., Ye, J., 2009. Butyrate improves insulin sensitivity and increases energy expenditure in mice. Diabetes 58, 1509—1517.

Gibson, G.R., Roberfroid, M.B., June 1995. Dietary modulation of the human colonic microbiota: introducing the concept of prebiotics. J. Nutr. 125 (6), 1401—1412. https://doi.org/10.1093/jn/125.6.1401.

Gibson, G.R., Wang, X., 1994. Regulatory effects of bifidobacteria on the growth of other colonic bacteria. J. Appl. Microbiol. 77, 412—420.

Gibson, G.R., Scott, K., A Rastall, R., Tuohy, K., Hotchkiss, A., Dubert-Ferrandon, A., Buddington, R., 2010. Dietary prebiotics: current status and new definition. Food Sci. Technol. Bull. 7 (1), 1—19.

Gibson, G.R., Hutkins, R., Sanders, M.E., Prescott, S.L., Reimer, R.A., Salminen, S.J., Reid, G., 2017. Expert consensus document: the International Scientific Association for Probiotics and Prebiotics (ISAPP) consensus statement on the definition and scope of prebiotics. Nat. Rev. Gastroenterol. Hepatol. 14 (8), 491—502.

Gill, S.R., Pop, M., Deboy, R.T., Eckburg, P.B., Turnbaugh, P.J., Samuel, B.S., et al., 2006. Metagenomic analysis of the human distal gut microbiome. Science 312, 1355—1359.

Gill, S.K., Rossi, M., Bajka, B., Whelan, K., February 2021. Dietary fibre in gastrointestinal health and disease. Nat. Rev. Gastroenterol. Hepatol. 18 (2), 101—116. https://doi.org/10.1038/s41575-020-00375-4.

Gismondo, M.R., 1998. Antibiotic impact on intestinal microflora. Gastroenterol. Int. 11, 29—30.

Gopal, E., Miyauchi, S., Martin, P.M., et al., 2007. Transport of nicotinate and structurally related compounds by human SMCT1 (SLC5A8) and its relevance to drug transport in the mammalian intestinal tract. Pharm. Res. (N. Y.) 24, 575—584.

Graf, D., Di Cagno, R., Fåk, F., Flint, H.J., Nyman, M., Saarela, M., Watzl, B., February 4, 2015. Contribution of diet to the composition of the human gut microbiota. Microb. Ecol. Health Dis. 26, 26164. https://doi.org/10.3402/mehd.v26.26164.

Gregory, K.E., Samuel, B.S., Houghteling, P., Shan, G., Ausubel, F.M., Sadreyev, R.I., Walker, W.A., December 30, 2016. Influence of maternal breast milk ingestion on acquisition of the intestinal microbiome in preterm infants. Microbiome 4 (1), 68. https://doi.org/10.1186/s40168-016-0214-x.

Gritz, E.C., Bhandari, V., 2015. The human neonatal gut microbiome: a brief review. Front. Pediatr. 3, 17. https://doi.org/10.3389/fped.2015.00017.

Hamer, H.M., Jonkers, D., Venema, K., Vanhoutvin, S., Troost, F.J., Brummer, R.J., 2008. Review article: the role of butyrate on colonic function. Aliment. Pharmacol. Ther. 27, 104–119.

Hawrelak, J.A., Myers, S.P., June 2004. The causes of intestinal dysbiosis: a review. Altern. Med. Rev. 9 (2), 180–197.

Hegyi, J., Schwartz, R.A., Hegyi, V., 2004. Pellagra: dermatitis, dementia, and diarrhea. Int. J. Dermatol. 43, 1–5.

Holscher, H.D., March 4, 2017. Dietary fiber and prebiotics and the gastrointestinal microbiota. Gut Microb. 8 (2), 172–184. https://doi.org/10.1080/19490976.2017.1290756.

Holzapfel, W.H., Haberer, P., Snel, J., et al., 1998. Overview of gut flora and probiotics. Int. J. Food Microbiol. 41, 85–101.

Hooper, L.V., Littman, D.R., Macpherson, A.J., 2012. Interactions between the microbiota and the immune system. Science 336 (6086), 1268–1273.

Hou, K., Wu, Z.X., Chen, X.Y., Wang, J.Q., Zhang, D., Xiao, C., Zhu, D., Koya, J.B., Wei, L., Li, J., Chen, Z.S., April 23, 2022. Microbiota in health and diseases. Signal Transduct. Targeted Ther. 7 (1), 135. https://doi.org/10.1038/s41392-022-00974-4.

Huang, W., Zhou, L., Guo, H., Xu, Y., Xu, Y., 2017. The role of short-chain fatty acids in kidney injury induced by gut-derived inflammatory response. Metabolism 68, 20–30.

Ignys, I., Szachta, P., Galecka, M., Schmidt, M., Pazgrat-Patan, M., 2014. Methods of analysis of gut microorganism—actual state of knowledge. Ann. Agric. Environ. Med. 21 (4).

Ingham, C.J., Sprenkels, A., Bomer, J., Molenaar, D., van den Berg, A., van HylckamaVlieg, J.E., de Vos, W.M., 2007. The micro-Petri dish, a million-well growth chip for the culture and high-throughput screening of microorganisms. Proc. Natl. Acad. Sci. USA 104 (46), 18217–18222.

Ishikawa, H., Akedo, I., Otani, T., Suzuki, T., Nakamura, T., Takeyama, I., Ishiguro, S., Miyaoka, E., Sobue, T., Kakizoe, T., 2005. Randomized trial of dietary fiber and Lactobacillus casei administration for prevention of colorectal tumors. Int. J. Cancer 116, 762–767.

Johnson, J.S., Spakowicz, D.J., Hong, B.Y., Petersen, L.M., Demkowicz, P., Chen, L., Leopold, S.R., Hanson, B.M., Agresta, H.O., Gerstein, M., Sodergren, E., Weinstock, G.M., November 6, 2019. Evaluation of 16S rRNA gene sequencing for species and strain-level microbiome analysis. Nat. Commun. 10 (1), 5029. https://doi.org/10.1038/s41467-019-13036-1.

Johnston, B.C., Supina, A.L., Vohra, S., 2006. Probiotics for pediatric antibiotic-associated diarrhea: a meta-analysis of randomized placebo-controlled trials. Can. Med. Assoc. J. 175, 377–383.

Kaczmarczyk, M.M., Miller, M.J., Freund, G.G., 2012. The health benefits of dietary fiber: beyond the usual suspects of type 2 diabetes, cardiovascular disease and colon cancer. Metabolism 61, 1058–1066.

Kang, Z., Zhang, J., Zhou, J., Qi, Q., Du, G., Chen, J., 2012. Recent advances in microbial production of δ-aminolevulinic acid and vitamin B12. Biotechnol. Adv. 30 (6), 1533–1542.

Kang, J.H., Yun, S.I., Park, M.H., Park, J.H., Jeong, S.Y., Park, H.O., 2013. Anti-obesity effect of Lactobacillus gasseri BNR17 in high-sucrose diet-induced obese mice. PLoS One 8, e54617.

Kho, Z.Y., Lal, S.K., August 14, 2018. The human gut microbiome - a potential controller of wellness and disease. Front. Microbiol. 9, 1835. https://doi.org/10.3389/fmicb.2018.01835.

Kim, H.J., Leeds, P., Chuang, D.M., 2009. The HDAC inhibitor, sodium butyrate, stimulates neurogenesis in the ischemic brain. J. Neurochem. 110, 1226–1240.

Kim, K.A., Gu, W., Lee, I.A., Joh, E.H., Kim, D.H., 2012. High fat diet-induced gut microbiota exacerbates inflammation and obesity in mice via the TLR4 signaling pathway. PLoS One 7 (10), e47713. https://doi.org/10.1371/journal.pone.0047713.

Kim, C.H., Park, J., Kim, M., 2014. Gut microbiota-derived short-chain Fatty acids, T cells, and inflammation. Immune Netw. 14, 277−288.

Kimura, I., Inoue, D., Maeda, T., Hara, T., Ichimura, A., Miyauchi, S., Kobayashi, M., Hirasawa, A., Tsujimoto, G., 2011. Short-chain fatty acids and ketones directly regulate sympathetic nervous system via G protein-coupled receptor 41 (GPR41). Proc. Natl. Acad. Sci. USA 108, 8030−8035.

Kipanyula, M.J., Etet, P.F., Vecchio, L., Farahna, M., Nukenine, E.N., Kamdje, A.H., 2013. Signaling pathways bridging microbial-triggered inflammation and cancer. Cell. Signal. 25 (2), 403−416.

Koeth, R.A., Wang, Z., Levison, B.S., Buffa, J.A., Org, E., Sheehy, B.T., Britt, E.B., Fu, X., Wu, Y., Li, L., Smith, J.D., DiDonato, J.A., Chen, J., Li, H., Wu, G.D., Lewis, J.D., Warrier, M., Brown, J.M., Krauss, R.M., Tang, W.H., Bushman, F.D., Lusis, A.J., Hazen, S.L., May 2013. Intestinal microbiota metabolism of L-carnitine, a nutrient in red meat, promotes atherosclerosis. Nat. Med. 19 (5), 576−585. https://doi.org/10.1038/nm.3145.

Koeth, R.A., Levison, B.S., Culley, M.K., Buffa, J.A., Wang, Z., Gregory, J.C., Org, E., Wu, Y., Li, L., Smith, J.D., Tang, W.W., 2014. γ-Butyrobetaine is a proatherogenic intermediate in gut microbial metabolism of L-carnitine to TMAO. Cell Metabol. 20 (5), 799−812.

Koh, A., De Vadder, F., Kovatcheva-Datchary, P., Backhed, F., 2016. From dietary fiber to host physiology: short-chain fatty acids as key bacterial metabolites. Cell 165. https://doi.org/10.1016/j.cell.2016.05.041.

Kolho, K.L., Korpela, K., Jaakkola, T., Pichai, M.V., Zoetendal, E.G., Salonen, A., de Vos, W.M., 2015. Fecal microbiota in pediatric inflammatory bowel disease and its relation to inflammation. Am. J. Gastroenterol. 110 (6), 921−930.

Korem, T., Zeevi, D., Suez, J., Weinberger, A., Avnit-Sagi, T., Pompan-Lotan, M., Matot, E., Jona, G., Harmelin, A., Cohen, N., Sirota-Madi, A., Thaiss, C.A., Pevsner-Fischer, M., Sorek, R., Xavier, R., Elinav, E., Segal, E., September 4, 2015. Growth dynamics of gut microbiota in health and disease inferred from single metagenomic samples. Science 349 (6252), 1101−1106. https://doi.org/10.1126/science.aac4812.

Kovatcheva-Datchary, P., Nilsson, A., Akrami, R., Lee, Y.S., De Vadder, F., Arora, T., Hallen, A., Martens, E., Björck, I., Bäckhed, F., December 1, 2015. Dietary fiber-induced improvement in glucose metabolism is associated with increased abundance of Prevotella. Cell Metabol. 22 (6), 971−982. https://doi.org/10.1016/j.cmet.2015.10.001.

Kruis, W., Forstmaier, G., Scheurlen, C., Stellaard, F., 1991. Effect of diets low and high in refined sugars on gut transit, bile acid metabolism, and bacterial fermentation. Gut 32, 367−371.

Kumagai, T., Rahman, F., Smith, A.M., 2018. The microbiome and radiation induced-bowel injury: evidence for potential mechanistic role in disease pathogenesis. Nutrients 10 (10), 1405.

Larrosa, M., Yañéz-Gascón, M.J., Selma, M.V., González-Sarrías, A., Toti, S., Cerón, J.J., Tomás-Barberán, F., Dolara, P., Espín, J.C., March 25, 2009. Effect of a low dose of dietary resveratrol on colon microbiota, inflammation and tissue damage in a DSS-induced colitis rat model. J. Agric. Food Chem. 57 (6), 2211−2220. https://doi.org/10.1021/jf803638d.

Laursen, M.F., Bahl, M.I., Michaelsen, K.F., Licht, T.R., 2017. First foods and gut microbes. Front. Microbiol. 8, 356.

Le Roy, C.I., Woodward, M.J., Ellis, R.J., La Ragione, R.M., Claus, S.P., 2019. BMC Vet. Res. 15, 37.

Lepage, P., Leclerc, M.C., Joossens, M., Mondot, S., Blottière, H.M., Raes, J., Ehrlich, D., Doré, J., January 2013. A metagenomic insight into our gut's microbiome. Gut 62 (1), 146−158. https://doi.org/10.1136/gutjnl-2011-301805.

Levitt, M.D., Gibson, G.R., Christl, S.U., 1995. Gas metabolism in the large intestine. In: Gibson, G.R., Macfarlane, G.T. (Eds.), Human Colonic Bacteria: Role in Nutrition, Physiology, and Pathology. CRC Press, Boca Raton, FL, pp. 131−149, 1995.

Lewis, S.J., Heaton, K.W., 1999. The metabolic conse- quences of slow colonic transit. Am. J. Gastroenterol. 94, 2010−2016.

Li, Y., Ren, L., Wang, Y., Li, J., Zhou, Q., Peng, C., Li, Y., Cheng, R., He, F., Shen, X., December 19, 2022. The effect of breast milk microbiota on the composition of infant gut microbiota: a cohort study. Nutrients 14 (24), 5397. https://doi.org/10.3390/nu14245397.

Liang, G., Taranova, O., Xia, K., Zhang, Y., 2010. Butyrate promotes induced pluripotent stem cell generation. J. Biol. Chem. 285, 25516−25521.

Liang, L., Saunders, C., Sanossian, N., March 7, 2023. Food, gut barrier dysfunction, and related diseases: a new target for future individualized disease prevention and management. Food Sci. Nutr. 11 (4), 1671–1704. https://doi.org/10.1002/fsn3.3229.

Lin, H.V., Frassetto, A., Kowalik Jr., E.J., Nawrocki, A.R., Lu, M.M., Kosinski, J.R., Hubert, J.A., Szeto, D., Yao, X., Forrest, G., Marsh, D.J., 2012. Butyrate and propionate protect against diet-induced obesity and regulate gut hormones via free fatty acid receptor 3-independent mechanisms. PLoS One 7, e35240.

Linder, M.C., 1991. Nutrition and metabolism of proteins. In: Linder, M.C. (Ed.), Nutritional Biochemistry and Metabolism, second ed. Appleton and Lange, Norwalk, CT, pp. 87–110. 1991.

López-Barrera, D.M., Vázquez-Sánchez, K., Loarca-Piña, M.G., Campos-Vega, R., 2016. Spent coffee grounds, an innovative source of colonic fermentable compounds, inhibit inflammatory mediators in vitro. Food Chem. 212, 282–290.

Lu, Z., Gui, H., Yao, L., Yan, L., Martens, H., Aschenbach, J.R., Shen, Z., 2015. Short-chain fatty acids and acidic pH upregulate UT-B, GPR41, and GPR43 in rumen epithelial cells of goats. Am. J. Physiol. Regul. Integr. Comp. Physiol. 308, R283–R293.

Macfarlane, G.T., Gibson, G.R., 1994. Metabolic activities of the normal colonic flora. In: Gibson, S.A.W. (Ed.), Human Health: The Contribution of Microorganisms. Springer- Verlag, London, pp. 17–53, 1994.

Macfarlane, S., Macfarlane, G.T., 1995. Proteolysis and amino acid fermentation. In: Gibson, G.R., Macfarlane, G.T. (Eds.), Human Colonic Bacteria: Role in Nutrition, Physiology, and Pathology. CRC Press, Boca Raton, FL, pp. 75–100, 1995.

Macfarlane, G.T., Macfarlane, S., 2012 . Bacteria, colonic fermentation, and gastrointestinal health. J. AOAC Int. 95 (1), 50–60. https://doi.org/10.5740/jaoacint.sge_macfarlane.

Macia, L., Tan, J., Vieira, A.T., Leach, K., Stanley, D., Luong, S., Maruya, M., Ian McKenzie, C., Hijikata, A., Wong, C., Binge, L., Thorburn, A.N., Chevalier, N., Ang, C., Marino, E., Robert, R., Offermanns, S., Teixeira, M.M., Moore, R.J., Flavell, R.A., Fagarasan, S., Mackay, C.R., April 1, 2015. Metabolite-sensing receptors GPR43 and GPR109A facilitate dietary fibre-induced gut homeostasis through regulation of the inflammasome. Nat. Commun. 6, 6734. https://doi.org/10.1038/ncomms7734.

Makki, K., Deehan, E.C., Walter, J., Bäckhed, F., June 13, 2018. The impact of dietary fiber on gut microbiota in host health and disease. Cell Host Microbe 23 (6), 705–715. https://doi.org/10.1016/j.chom.2018.05.012.

Malesza, I.J., Malesza, M., Walkowiak, J., Mussin, N., Walkowiak, D., Aringazina, R., Bartkowiak-Wieczorek, J., Mądry, E., November 14, 2021. High-fat, western-style diet, systemic inflammation, and gut microbiota: a narrative review. Cells 10 (11), 3164. https://doi.org/10.3390/cells10113164.

Manichanh, C., Rigottier-Gois, L., Bonnaud, E., Gloux, K., Pelletier, E., Frangeul, L., Nalin, R., Jarrin, C., Chardon, P., Marteau, P., et al., 2006. Reduced diversity of faecal microbiota in Crohn's disease revealed by a metagenomic approach. Gut 55 (2), 205–211.

Markowiak, P., Śliżewska, K., June 6, 2018. The role of probiotics, prebiotics and synbiotics in animal nutrition. Gut Pathog. 10, 21. https://doi.org/10.1186/s13099-018-0250-0.

Markowiak, P., Slizewska, K., 2017. Effects of probiotics, prebiotics, and synbiotics on human health. Nutrients 9, 1021.

Martinez, K.B., Leone, V., Chang, E.B., March 4, 2017. Western diets, gut dysbiosis, and metabolic diseases: are they linked? Gut Microb. 8 (2), 130–142. https://doi.org/10.1080/19490976.2016.1270811.

Matsumoto, M., Sakamoto, M., Hayashi, H., Benno, Y., 2005. Novel phylogenetic assignment database for terminal-restriction fragment length polymorphism analysis of human colonic microbiota. J. Microbiol. Methods 61, 305–319.

McCartney, A.L., 2002. Application of molecular biological methods for studying probiotics and the gut flora. Br. J. Nutr. 88 (Suppl. 1), S29–S37.

Medzhitov, R., 2008. Origin and physiological roles of inflammation. Nature 454, 428–435.

Michael, O.S., Oluranti, O.I., Oshinjo, A.M., Adetunji, C.O., Adetunji, J.B., Esiobu, N.D., 2022. Microbiota transplantation, health implications, and the way forward. In: Microbiomes and Emerging Applications, first ed.st Edition. Imprint CRC Press, ISBN 9781003180241, p. 19. https://doi.org/10.1201/9781003180241-5. First Published 2022.

Miller, S.J., Zaloga, G.P., Hoggatt, A.M., Labarrere, C., Faulk, W.P., 2005. Short-chain fatty acids modulate gene expression for vascular endothelial cell adhesion molecules. Nutrition 21, 740—748.

Miyamoto, J., Hasegawa, S., Kasubuchi, M., et al., 2016. Nutritional signaling via free fatty acid receptors. Int. J. Mol. Sci. 17, E450.

Miyoshi, J., Chang, E.B., 2017. The gut microbiota and inflammatory bowel diseases. Transl. Res. 179, 38—48.

Moreno-Indias, I., Sánchez-Alcoholado, L., Pérez-Martínez, P., Andrés-Lacueva, C., Cardona, F., Tinahones, F., Queipo-Ortuño, M.I., April 2016. Red wine polyphenols modulate fecal microbiota and reduce markers of the metabolic syndrome in obese patients. Food Funct. 7 (4), 1775—1787. https://doi.org/10.1039/c5fo00886g.

Mueller, N.T., Bakacs, E., Combellick, J., Grigoryan, Z., Dominguez-Bello, M.G., 2015. The infant microbiome development: mom matters. Trends Mol. Med. 21 (2), 109—117. https://doi.org/10.1016/j.molmed.2014.12.002.

Murawaki, Y., Kobayashi, M., Koda, M., Kawasakia, H., 2000. Effects of lactulose on intestinal bacterial flora and fecal organic acids in patients with liver cirrhosis. Hepatol. Res. 17, 56—64.

Murphy, E.A., Velazquez, K.T., Herbert, K.M., 2015. Influence of high-fat diet on gut microbiota: a driving force for chronic disease risk. Curr. Opin. Clin. Nutr. Metab. Care 18 (5), 515—520.

Muyzer, G., 1999. DGGE/TGGE a method for identifying genes from natural ecosystems. Curr. Opin. Microbiol. 2, 317—322.

Nadal, I., Donat, E., Ribes-Koninckx, C., Calabuig, M., Sanz, Y., 2007. Imbalance in the composition of the duodenal microbiota of children with coeliac disease. J. Med. Microbiol. 56, 1669—1674.

Nash, M.J., Frank, D.N., Friedman, J.E., December 13, 2017. Early microbes modify immune system development and metabolic homeostasis-the "restaurant" hypothesis revisited. Front. Endocrinol. 8, 349. https://doi.org/10.3389/fendo.2017.00349.

Neish, A.S., 2009. Microbes in gastrointestinal health and disease. Gastroenterology 136, 65—80.

Newgard, C.B., An, J., Bain, J.R., Muehlbauer, M.J., Stevens, R.D., Lien, L.F., Haqq, A.M., Shah, S.H., Arlotto, M., Slentz, C.A., et al., 2009. A branched-chain amino acid-related metabolic signature that differentiates obese and lean humans and contributes to insulin resistance. Cell Metabol. 9, 311—326.

Nie, Y.F., Hu, J., Yan, X.H., 2015. Cross-talk between bile acids and intestinal microbiota in host metabolism and health. J. Zhejiang Univ. - Sci. B 16 (6), 436—446.

Nilsson, N.E., Kotarsky, K., Owman, C., Olde, B., 2003. Identification of a free fatty acid receptor, FFAR2, expressed on leukocytes and activated by short-chain fatty acids. Biochem. Biophys. Res. Commun. 303, 1047—1052.

Noack, J., Kleessen, B., Proll, J., et al., 1998. Dietary guar gum and pectin stimulate intestinal microbial polyamine synthesis in rats. J. Nutr. 128, 1385—1391.

Nogal, A., Valdes, A.M., Menni, C., 2021. The role of short-chain fatty acids in the interplay between gut microbiota and diet in cardio-metabolic health. Gut Microb. 13 (1), 1—24. https://doi.org/10.1080/19490976.2021.1897212.

Nøhr, M.K., Egerod, K.L., Christiansen, S.H., Gille, A., Offermanns, S., Schwartz, T.W., Møller, M., 2015. Expression of the short chain fatty acid receptor GPR41/FFAR3 in autonomic and somatic sensory ganglia. Neuroscience 290, 126—137.

Noor, S.O., et al., 2010. Ulcerative colitis and irritable bowel patients exhibit distinct abnormalities of the gut microbiota. BMC Gastroenterol. 10, 134.

Nord, C.E., 1990. Studies on the ecological impact of antibiotics. Eur. J. Clin. Microbiol. Infect. Dis. 9, 517—518.

O'Donnell, J.A., Zheng, T., Meric, G., Marques, F.Z., March 2023. The gut microbiome and hypertension. Nat. Rev. Nephrol. 19 (3), 153—167. https://doi.org/10.1038/s41581-022-00654-0.

Oelschlaeger, T.A., 2010. Mechanisms of probiotic actions—a review. Int. J. Med. Microbiol. 300, 57—62.

Olaniyi, K.S., Amusa, O.A., Areola, E.D., Olatunji, L.A., 2020. Suppression of HDAC by sodium acetate rectifies cardiac metabolic disturbance in streptozotocin—nicotinamide-induced diabetic rats. Exp. Biol. Med. 245 (7), 667—676.

Omolekulo, T.E., Michael, O.S., Olatunji, L.A., 2019. Sodium acetate improves disrupted glucoregulation and hepatic triglyceride content in insulin-resistant female rats: involvement of adenosine deaminase and dipeptidyl peptidase-4 activities. Naunyn-Schmiedeberg's Arch. Pharmacol. 392, 103—116.

Paliy, O., Kenche, H., Abernathy, F., Michail, S., June 2009. High-throughput quantitative analysis of the human intestinal microbiota with a phylogenetic microarray. Appl. Environ. Microbiol. 75 (11), 3572–3579. https://doi.org/10.1128/AEM.02764-08.

Palmer, C., Bik, E.M., DiGiulio, D.B., Relman, D.A., Brown, P.O., July 2007. Development of the human infant intestinal microbiota. PLoS Biol. 5 (7), e177. https://doi.org/10.1371/journal.pbio.0050177.

Palmnäs, M.S.A., et al., 2014. Low- dose aspartame consumption differentially affects gut microbiota- host metabolic interactions in the diet- induced obese rat. PLoS One 9, e109841.

Park, J., Kim, M., Kang, S.G., Jannasch, A.H., Cooper, B., Patterson, J., Kim, C.H., 2015. Short-chain fatty acids induce both effector and regulatory T cells by suppression of histone deacetylases and regulation of the mTORS6K pathway. Mucosal Immunol. 8, 80–93.

Patterson, E., Ryan, P.M., Cryan, J.F., et al., 2016. Gut microbiota, obesity and diabetes. Postgrad Med. J. 92, 286–300.

Payne, A.N., Chassard, C., Banz, Y., Lacroix, C., 2012. The composition and metabolic activity of child gut microbiota demonstrate differential adaptation to varied nutrient loads in an in vitro model of colonic fermentation. FEMS Microbiol. Ecol. 80, 608–623, 2012.

Perler, B.K., Friedman, E.S., Wu, G.D., February 10, 2023. The role of the gut microbiota in the relationship between diet and human health. Annu. Rev. Physiol. 85, 449–468. https://doi.org/10.1146/annurev-physiol-031522-092054.

Pluznick, J., 2014. A novel SCFA receptor, the microbiota, and blood pressure regulation. Gut Microbes 5 (2), 202–207. https://doi.org/10.4161/gmic.27492.

Pluznick, J.L., Protzko, R.J., Gevorgyan, H., et al., 2013. Olfactory receptor responding to gut microbiota-derived signals plays a role in renin secretion and blood pressure regulation. Proc. Natl. Acad. Sci. U. S. A. 110, 4410–4415.

Proctor, C., Thiennimitr, P., Chattipakorn, N., Chattipakorn, S.C., February 2017. Diet, gut microbiota and cognition. Metab. Brain Dis. 32 (1), 1–17. https://doi.org/10.1007/s11011-016-9917-8.

Przybyłowicz, K.E., Danielewicz, A., July 30, 2022. Eating habits and disease risk factors. Nutrients 14 (15), 3143. https://doi.org/10.3390/nu14153143.

Psichas, A., Sleeth, M.L., Murphy, K.G., et al., 2015. The short chain fatty acid propionate stimulates GLP-1 and PYY secretion via free fatty acid receptor 2 in rodents. Int. J. Obes. 39, 424–429.

Rea, D., Coppola, G., Palma, G., Barbieri, A., Luciano, A., Del Prete, P., Rossetti, S., Berretta, M., Facchini, G., Perdonà, S., Turco, M.C., 2018. Microbiota effects on cancer: from risks to therapies. Oncotarget 9 (25), 17915.

Rehbinder, E.M., Lødrup Carlsen, K.C., Staff, A.C., et al., 2018. Is amniotic fluid of women with uncomplicated term pregnancies free of bacteria? Am. J. Obstet. Gynecol. 219 (3), 289.e1–289.e12.

Rezac, S., Kok, C.R., Heermann, M., Hutkins, R., August 24, 2018. Fermented foods as a dietary source of live organisms. Front. Microbiol. 9, 1785. https://doi.org/10.3389/fmicb.2018.01785.

Ritzhaupt, A., Wood, I.S., Ellis, A., Hosie, K.B., Shirazi-Beechey, S.P., 1998. Identification of a monocarboxylate transporter isoform type 1 (MCT1) on the luminal membrane of human and pig colon. Biochem. Soc. Trans. 26, S120.

Roca-Saavedra, P., Mendez-Vilabrille, V., Miranda, J.M., Nebot, C., Cardelle-Cobas, A., Franco, C.M., Cepeda, A., February 2018. Food additives, contaminants and other minor components: effects on human gut microbiota-a review. J. Physiol. Biochem. 74 (1), 69–83. https://doi.org/10.1007/s13105-017-0564-2.

Rohrmann, S., Linseisen, J., 2016. Processed meat: the real villain? Proc. Nutr. Soc. 75 (3), 233–241.

Round, J.L., Lee, S.M., Li, J., Tran, G., Jabri, B., Chatila, T.A., Mazmanian, S.K., 2011. The Toll-like receptor 2 pathway establishes colonization by a commensal of the human microbiota. Science 332, 974–977.

Roytio, H., Mokkala, K., Vahlberg, T., Laitinen, K., 2017. Dietary intake of fat and fibre according to reference values relates to higher gut microbiota richness in overweight pregnant women. Br. J. Nutr. 118 (5), 343–352.

Russell, W.R., Gratz, S.W., Duncan, S.H., Holtrop, G., Ince, J., Scobbie, L., Duncan, G., Johnstone, A.M., Lobley, G.E., Wallace, R.J., et al., 2011. High-protein, reduced-carbohydrate weight-loss diets promote metabolite profiles likely to be detrimental to colonic health. Am. J. Clin. Nutr. 93, 1062–1072.

Russell, S.L., Gold, M.J., Hartmann, M., Willing, B.P., Thorson, L., Wlodarska, M., Gill, N., Blanchet, M.R., Mohn, W.W., McNagny, K.M., Finlay, B.B., 2012. Early life antibiotic-driven changes in microbiota enhance susceptibility to allergic asthma. EMBO Rep. 13, 440—447.

Ryan, P.M., Stanton, C., Ross, R.P., et al., 2019. Paediatrician's perspective of infant gut microbiome research: current status and challenges. Arch. Dis. Child. 104 (7), 701—705.

Said, H.S., Suda, W., Nakagome, S., Chinen, H., Oshima, K., Kim, S., Kimura, R., Iraha, A., Ishida, H., Fujita, J., Mano, S., Morita, H., Dohi, T., Oota, H., Hattori, M., 2014. Dysbiosis of salivary microbiota in inflammatory bowel disease and its association with oral immunological biomarkers. DNA Res. 21, 15—25.

Samuel, B.S., Shaito, A., Motoike, T., Rey, F.E., Backhed, F., Manchester, J.K., et al., 2008. Effects of the gut microbiota on host adiposity are modulated by the short-chain fatty-acid binding G protein-coupled receptor, Gpr41. Proc. Natl. Acad. Sci. U.S.A. 105, 16767—16772.

Sanchez, J.I., Marzorati, M., Grootaert, C., Baran, M., Van Craeyveld, V., Courtin, C.M., Broekaert, W.F., Delcour, J.A., Verstraete, W., Van de Wiele, T., 2009. Arabinoxylan-oligosaccharides (AXOS) affect the protein/carbohydrate fermentation balance and microbial population dynamics of the Simulator of Human Intestinal Microbial Ecosystem. Microb. Biotechnol. 2, 101—113.

Santhakumar, A.B., Battino, M., Alvarez-Suarez, J.M., March 2018. Dietary polyphenols: structures, bioavailability and protective effects against atherosclerosis. Food Chem. Toxicol. 113, 49—65. https://doi.org/10.1016/j.fct.2018.01.022.

Sekirov, I., Russell, S.L., Antunes, L.C., Finlay, B.B., 2010. Gut microbiota in health and disease. Physiol. Rev. 90, 859—904.

Selma, M.V., Espín, J.C., Tomás-Barberán, F.A., August 12, 2009. Interaction between phenolics and gut microbiota: role in human health. J. Agric. Food Chem. 57 (15), 6485—6501. https://doi.org/10.1021/jf902107d.

Sheflin, A.M., Whitney, A.K., Weir, T.L., 2014. Cancer-promoting effects of microbial dysbiosis. Curr. Oncol. Rep. 16 (10), 406.

Shen, W., Wolf, P.G., Carbonero, F., Zhong, W., Reid, T., Gaskins, H.R., McIntosh, M.K., August 2014. Intestinal and systemic inflammatory responses are positively associated with sulfidogenic bacteria abundance in high-fat-fed male C57BL/6J mice. J. Nutr. 144 (8), 1181—1187. https://doi.org/10.3945/jn.114.194332.

Sikorska, H., Smoragiewicz, W., 2013. Role of probiotics in the prevention and treatment of ethicillin-resistant Staphylococcus aureus infections. Int. J. Antimicrob. Agents 42, 475—481.

Singh, N., Gurav, A., Sivaprakasam, S., Brady, E., Padia, R., Shi, H., Thangaraju, M., Prasad, P.D., Manicassamy, S., Munn, D.H., et al., 2014. Activation of Gpr109a, receptor for niacin and the commensal metabolite butyrate, suppresses colonic inflammation and carcinogenesis. Immunity 40, 128—139.

Singh, R., Zogg, H., Wei, L., Bartlett, A., Ghoshal, U.C., Rajender, S., Ro, S., January 30, 2021. Gut microbial dysbiosis in the pathogenesis of gastrointestinal dysmotility and metabolic disorders. J. Neurogastroenterol. Motil. 27 (1), 19—34. https://doi.org/10.5056/jnm20149.

Slavin, J., April 22, 2013. Fiber and prebiotics: mechanisms and health benefits. Nutrients 5 (4), 1417—1435. https://doi.org/10.3390/nu5041417.

Smith, E.A., Macfarlane, G.T., 1997. Dissimilatory amino acid metabolism in human colonic metabolism. Anaerobe 3, 327—337.

Smith, M.I., et al., 2013a. Gut microbiomes of Malawian twin pairs discordant for kwashiorkor. Science 339, 548—554.

Smith, P.M., Howitt, M.R., Panikov, N., Michaud, M., Gallini, C.A., Bohlooly-Y, M., Glickman, J.N., Garrett, W.S., 2013b. The microbial metabolites, short chain fatty acids, regulate colonic Treg cell homeostasis. Science 341, 569—573.

Sonnenburg, E.D., Smits, S.A., Tikhonov, M., Higginbottom, S.K., Wingreen, N.S., Sonnenburg, J.L., 2016. Diet-induced extinctions in the gut microbiota compound over generations. Nature 529 (7610), 56—64.

Soret, R., Chevalier, J., De Coppet, P., Poupeau, G., Derkinderen, P., Segain, J.P., Neunlist, M., 2010. Short-chain fatty acids regulate the enteric neurons and control gastrointestinal motility in rats. Gastroenterology 138, 1772–1782.

Stenman, L.K., Waget, A., Garret, C., Klopp, P., Burcelin, R., Lahtinen, S., 2014. Potential probiotic Bifidobacterium animalis lactis 420 prevents weight gain and glucose intolerance in diet-induced obese mice. Benef. Microbes 5, 437–445.

Suez, J., et al., 2014. Artificial sweeteners induce glucose intolerance by altering the gut microbiota. Nature 514, 181–186.

Suez, J., Korem, T., Zilberman- Schapira, G., Segal, E., Elinav, E., 2015. Non- caloric artificial sweeteners and the microbiome: findings and challenges. Gut Microb. 6, 149–155.

Sun, L., Zhang, X., Zhang, Y., Zheng, K., Xiang, Q., Chen, N., Chen, Z., Zhang, N., Zhu, J., He, Q., 2019. Antibiotic-induced disruption of gut microbiota alters local Metabolomes and immune responses. Front. Cell. Infect. Microbiol. 9, 99.

Tan, S., Caparros-Martin, J.A., Matthews, V.B., Koch, H., O'Gara, F., Croft, K.D., Ward, N.C., July 4, 2018. Isoquercetin and inulin synergistically modulate the gut microbiome to prevent development of the metabolic syndrome in mice fed a high fat diet. Sci. Rep. 8 (1), 10100. https://doi.org/10.1038/s41598-018-28521-8.

Tang, W.H., Wang, Z., Levison, B.S., Koeth, R.A., Britt, E.B., Fu, X., Wu, Y., Hazen, S.L., April 25, 2013. Intestinal microbial metabolism of phosphatidylcholine and cardiovascular risk. N. Engl. J. Med. 368 (17), 1575–1584. https://doi.org/10.1056/NEJMoa1109400.

Thaiss, C.A., Itav, S., Rothschild, D., Meijer, M.T., Levy, M., Moresi, C., Dohnalová, L., Braverman, S., Rozin, S., Malitsky, S., Dori-Bachash, M., Kuperman, Y., Biton, I., Gertler, A., Harmelin, A., Shapiro, H., Halpern, Z., Aharoni, A., Segal, E., Elinav, E., December 22, 2016. Persistent microbiome alterations modulate the rate of post-dieting weight regain. Nature 540 (7634), 544–551. https://doi.org/10.1038/nature20796.

Thangaraju, M., Carswell, K.N., Prasad, P.D., Ganapathy, V., 2009. Colon cancer cells maintain low levels of pyruvate to avoid cell death caused by inhibition of HDAC1/HDAC3. Biochem. J. 417, 379–389.

Theis, K.R., Romero, R., Winters, A.D., et al., 2019. Does the human placenta delivered at term have a microbiota? Results of cultivation, quantitative real-time PCR, 16S rRNA gene sequencing, and metagenomics. Am. J. Obstet. Gynecol. 220 (3), 267.e1–267.e39.

Thomas, D.W., Greer, F., 2010. Probiotics and prebiotics in pediatrics. Pediatrics 126, 1217–1231.

Thorburn, A.N., McKenzie, C.I., Shen, S., Stanley, D., Macia, L., Mason, L.J., Roberts, L.K., Wong, C.H.Y., Shim, R., Robert, R., et al., 2015. Evidence that asthma is a developmental origin disease influenced by maternal diet and bacterial metabolites. Nat. Commun. 6, 7320.

Thursby, E., Juge, N., 2017. Introduction to the human gut microbiota. Biochem. J. 474 (11), 1823–1836.

Upadrasta, A., Madempudi, R.S., 2016. Probiotics and blood pressure: current insights. Integr. Blood Press. Control 9, 33–42.

Uzbay, T., 2019. Germ-free animal experiments in the gut microbiota studies. Curr. Opin. Pharmacol. 49, 6–10.

Van Den Abbeele, P., Venema, K., van de Wiele, T., Verstraete, W., Possemiers, S., 2013. Different human gut models reveal the distinct fermentation patterns of arabinoxylan versus inulin. J. Agric. Food Chem. 61, 9819–9827.

Vangay, P., Ward, T., Gerber, J.S., Knights, D., 2015. Antibiotics, pediatric dysbiosis, and disease. Cell Host Microbe. 17 (5), 553–564.

Victor 3rd, D.W., Quigley, E.M., 2016. The microbiome and the liver: the basics. Semin. Liver Dis. 36 (4), 299–305.

Vieira, A.T., Macia, L., Galvão, I., et al., 2015. A Role for Gut Microbiota and the metabolite-sensing receptor GPR43 in a murine model of gout. Arthritis Rheumatol. 67, 1646–1656.

Vijay, A., Valdes, A.M., April 2022. Role of the gut microbiome in chronic diseases: a narrative review. Eur. J. Clin. Nutr. 76 (4), 489–501. https://doi.org/10.1038/s41430-021-00991-6.

Vinolo, M.A., Rodrigues, H.G., Hatanaka, E., Hebeda, C.B., Farsky, S.H., Curi, R., 2009. Short-chain fatty acids stimulate the migration of neutrophils to inflammatory sites. Clin. Sci. (Lond.) 117, 331–338.

Vinolo, M.A., Rodrigues, H.G., Nachbar, R.T., Curi, R., 2011a. Regulation of inflammation by short chain fatty acids. Nutrients 3, 858–876.

Vinolo, M.A., Rodrigues, H.G., Hatanaka, E., Sato, F.T., Sampaio, S.C., Curi, R., 2011b. Suppressive effect of short-chain fatty acids on production of proinflammatory mediators by neutrophils. J. Nutr. Biochem. 22, 849–855.

Wang, B., Yao, M., Lv, L., Ling, Z., Li, L., 2017. The human microbiota in health and disease. Engineering 3 (1), 71–82.

Wolever, T.M., Josse, R.G., Leiter, L.A., Chiasson, J.L., 1997. Time of day and glucose tolerance status affect serum short-chain fatty acid concentrations in humans. Metabolism 46, 805–811.

Wolk, A., 2017. Potential Health Hazards of eating red meat. J. Intern. Med. 281 (2), 106–122.

Yamashita, H., Fujisawa, K., Ito, E., Idei, S., Kawaguchi, N., Kimoto, M., Hiemori, M., Tsuji, H., 2007. Improvement of obesity and glucose tolerance by acetate in Type 2 diabetic Otsuka Long-Evans Tokushima Fatty (OLETF) rats. Biosci. Biotechnol. Biochem. 71, 1236–1243.

Yao, Q., Li, H., Fan, L., Zhang, Y., Zhao, S., Zheng, N., Wang, J., February 8, 2021. Dietary regulation of crosstalk between gut microbiome and immune response in inflammatory bowel disease. Foods 10 (2), 368. https://doi.org/10.3390/foods10020368.

Yoneda, N., Yoneda, S., Niimi, H., Ueno, T., Hayashi, S., Ito, M., Shiozaki, A., Urushiyama, D., Hata, K., Suda, W., Hattori, M., Kigawa, M., Kitajima, I., Saito, S., 2016. Polymicrobial amniotic fluid infection with Mycoplasma/Ureaplasmaand other bacteria induces severe intraamniotic inflammation associated with poor perinatal prognosis in preterm labor. Am. J. Reprod. Immunol. 75, 112–125.

Yoshida, H., Ishii, M., Akagawa, M., September 15, 2019. Propionate suppresses hepatic gluconeogenesis via GPR43/AMPK signaling pathway. Arch. Biochem. Biophys. 672, 108057. https://doi.org/10.1016/j.abb.2019.07.022.

You, H., Tan, Y., Yu, D., Qiu, S., Bai, Y., He, J., Cao, H., Che, Q., Guo, J., Su, Z., May 17, 2022. The therapeutic effect of SCFA-mediated regulation of the intestinal environment on obesity. Front. Nutr. 9, 886902. https://doi.org/10.3389/fnut.2022.886902.

Zaiss, M.M., Rapin, A., Lebon, L., Dubey, L.K., Mosconi, I., Sarter, K., Piersigilli, A., Menin, L., Walker, A.W., Rougemont, J., et al., 2015. The intestinal microbiota contributes to the ability of helminths to modulate allergic inflammation. Immunity 43, 998–1010.

Zambell, K.L., Fitch, M.D., Fleming, S.E., 2003. Acetate and butyrate are themajor substrates for de novo lipogenesis in rat colonic epithelial cells. J. Nutr. 133, 3509–3515.

Zeng, S., Huang, Z., Hou, D., Liu, J., Weng, S., He, J., November 6, 2017. Composition, diversity and function of intestinal microbiota in pacific white shrimp (Litopenaeusvannamei) at different culture stages. PeerJ 5, e3986.

Zhang, C., Zhang, M., Pang, X., Zhao, Y., Wang, L., Zhao, L., 2012. Structural resilience of the gut microbiota in adult mice under high-fat dietary perturbations. ISME J. 6, 1848–1857.

Zhang, L., Hu, Y., Xu, Y., et al., 2019. The correlation between intestinal dysbiosis and the development of ankylosing spondylitis. Microb. Pathog. 132, 188–192.

Zheng, X., Qiu, Y., Zhong, W., Baxter, S., Su, M., Li, Q., Xie, G., et al., 2013. A targeted metabolomic protocol for short-chain fatty acids and branched-chain amino acids. Metabolomics 9, 818–827.

Zheng, J., Xiao, X., Zhang, Q., Mao, L., Yu, M., Xu, J., 2015. The placental microbiome varies in association with low birth weight in full-term neonates. Nutrients 7 (8), 6924–6937.

Zhou, D., Pan, Q., Shen, F., Cao, H.X., Ding, W.J., Chen, Y.W., Fan, J.G., 2017. Total fecal microbiota transplantation alleviates high-fat diet-induced steatohepatitis in mice via beneficial regulation of gut microbiota. Sci. Rep. 7 (1), 1529.

Zhu, W., Gregory, J.C., Org, E., Buffa, J.A., Gupta, N., Wang, Z., Li, L., Fu, X., Wu, Y., Mehrabian, M., Sartor, R.B., McIntyre, T.M., Silverstein, R.L., Tang, W.H.W., DiDonato, J.A., Brown, J.M., Lusis, A.J., Hazen, S.L., March 24, 2016. Gut microbial metabolite TMAO enhances platelet hyperreactivity and thrombosis risk. Cell 165 (1), 111–124. https://doi.org/10.1016/j.cell.2016.02.011.

Zhuang, L., Chen, H., Zhang, S., et al., 2019. Intestinal microbiota in early life and its implications on childhood health. Dev. Reprod. Biol. 17 (1), 13–25.

Index

Note: "Page numbers followed by f indicate figures, t indicate tables, and b indicate boxes."

Printed in the United States
by Baker & Taylor Publisher Services